Progress in Computer Science
No. 3

Edited by
E. Coffman
R. Graham
D. Kuck

Birkhäuser
Boston · Basel · Stuttgart

Applied Probability–Computer Science: The Interface Volume II

Sponsored by Applied Probability Technical Section
College of the Operations Research Society of America
The Institute of Management Sciences
January 5-7, 1981
Florida Atlantic University
Boca Raton, Florida

Ralph L. Disney,
Teunis J. Ott,
editors

1982

Birkhäuser
Boston • Basel • Stuttgart

Editors:

Ralph L. Disney
Department of Industrial Engineering
and Operations Research
Virginia Polytechnic Institute
and State University
Blacksburg, Virginia 24061, USA

Teunis J. Ott
Bell Laboratories
Holmdel, New Jersy 07733, USA

Library of Congress Cataloging in Publication Data
Main entry under title:

Applied probability-computer science.

(Progress in computer science ; no. 2-3)
1. Electronic data processing--Congresses.
2. Electronic digital computers--Programming--Congresses. 3. Probabilities--Congresses.
4. Queuing theory--Congresses. I. Disney, Ralph L., 1928- . II. Ott, Teunis J. III. Applied Probability Technical Section-College of the Operations Research Society of America, the Institute of Management Sciences. IV. Series.
QA75.5.A66 1982 0001.64 82-18506

ISBN 3-7643-3067-8 (v.1)
ISBN 3-7643-3093-7 (v.2)

CIP-Kurztitelaufnahme der Deutschen Bibliothek

Applied probability-computer science, the interface / sponsored by Applied Probability Techn. Sect., College of the Operations Research Soc. of America, The Inst. of Management Sciences. - Boston ; Basel ; Stuttgart : Birkhäuser

1981. January 5 - 7, 1981, Florida Atlantic University, Boca Raton, Florida.
Vol. 2 (1982).
(Progress in computer science ; Vol. 3)
ISBN 3-7643-3093-7

NE: Florida Atlantic University Boca Raton, Fla.
GT

ISBN 3-7643-3093-7

Printed in USA

TABLE OF CONTENTS

NETWORKS OF QUEUES, II
Jean Walrand, Chairman

QUEUEING MODELS IN PERFORMANCE ANALYSIS, II
Benjamin Melamed, Chairman

TABLE OF CONTENTS BY AUTHORS

PROBABILISTIC ANALYSIS OF DATABASES

Kenneth Sevcik, Chairman

D. E. Willard

S. C. Christodoulakis

A LOG LOG N SEARCH ALGORITHM FOR NONUNIFORM DISTRIBUTIONS

Dan E. Willard

Extended Abstract

Searching an ordered file is a very common operation in data processing. Given N records, stored in contiguous locations in memory and ordered by numeric keys $Y_1 < Y_2 \ldots < Y_N$, one often wishes to find a particular record whose key equals y. Several recent papers [2,5,6,9] have shown how an algorithm called interpolation search performs this operation in log log N expected runtime when the keys in the ordered file are generated by the uniform probability distribution. In this paper, we study how retrieval can be efficiently performed for nonuniform probability distributions.

1. Statement of Main Results

Two new results are presented in this paper. The first is that the log log N asymptotic retrieval time of interpolation search does not remain in force for most nonuniform probability distributions. Our second result is more surprising, and it is that this log log N expected runtime can be reestablished for nearly all nonuniform probability distributions by using a modified version of interpolation search.

More specifically, let u denote a probability density function over the real line. An ordered file of cardinality N will be said to be _generated by_ u iff this file is constructed by taking N records whose keys are independently determined by u and storing these records in ascending order. The probability density u will be said to be regular iff there exists some vector (b_1,b_2,b_3,b_4) such that u and its first derivative satisfy the following conditions:

$$u(y) \geq b_1 > 0 \text{ whenever } b_3 < y < b_4 \tag{1}$$

$$|u'(y)| \leq b_2 \text{ whenever } b_3 < y < b_4 \tag{2}$$

$$u(y) = 0 \text{ whenever } y \leq b_3 \text{ or } y \geq b_4. \tag{3}$$

The main algorithm of this paper will attain log log N expected runtime on files generated by arbitrary regular probability densities.

Our results should not be confused with remarks made in earlier papers [2,6] to the effect that results on uniform distributions can easily be extended to nonuniform distributions if the distribution function $D(y) = \int_{-\infty}^{y} u(x)\,dx$ is employed to map an initial nonuniform distribution onto a uniform distribution. The disadvantage of this method is that it relies on detailed information about u (or D) that is typically unavailable or expensive. This paper considers the more difficult problem where such information is inaccessible and shows that log log N asymptotic runtime can be achieved without it.

2. Description of Algorithm

All algorithms for searching ordered files can be regarded as iterative procedures whose i-th iteration searches file $F_i = (Y_{L_i} <\ldots< Y_{R_i})$ by generating a cut index C_i and comparing the cut value Y_{C_i} with the search key y. If $Y_{C_i} = y$ then the search terminates successfully; otherwise, the next iteration will examine either the $(Y_{L_i} <\ldots< Y_{C_i})$ or $(Y_{C_i} <\ldots< Y_{R_i})$ subfile according to whether or not $Y_{C_i} < y$. The only difference between binary search, interpolation search, and our proposed procedure is the specific rule for computing the cut index.

Under binary search [4] the cut index is the middle position in subfile $F_i = (Y_{L_i} < \ldots < Y_{R_i})$. It can therefore be defined as:

$$C_i^{BIN} = \left\lceil (L_i + R_i)/2 \right\rceil . \tag{4}$$

Under interpolation search [6] a comparable request for record y will produce a cut index that intuitively represents the expected position of y assuming that the $R_i - L_i - 1$ untested interior keys of subfile $F_i = (Y_{L_i} < \ldots < Y_{R_i})$ are uniformly distributed between the previously tested boundary keys Y_{L_i} and Y_{R_o}. In order to define this concept formally, let N_i and ℓ_i denote the following two quantities:

$$N_i = R_i - L_i - 1 \tag{5}$$

$$\ell_i = Y_{R_i} - Y_{L_i} . \tag{6}$$

In this notation, the cut index, C_i^{INT}, of interpolation search is

$$C_i^{INT} = \left\lceil L_i + \frac{y - Y_{L_i}}{\ell_i} N_i \right\rceil . \tag{7}$$

Our algorithm for searching ordered files will be called the RETRIEVE(α, θ, ϕ) procedure. It is defined as follows: Let Δ_i, C_i^+ and C_i^- denote the following three quantities where the parameters α, θ and ϕ, satisfying $0 < \alpha \leq 1$ and θ, $\phi > 0$, are fixed constants used to fine-tune the runtime coefficient:

$$\Delta_i = \left\lceil \theta \, \ell_i^{\alpha} N_i + \phi \sqrt{N_i} \right\rceil \tag{8}$$

$$C_i^+ = \left\lceil C_i^{INT} + \Delta_i \right\rceil , \tag{9}$$

$$C_i^- = \left\lfloor C_i^{INT} - \Delta_i \right\rfloor . \tag{10}$$

During its i-th iteration, RETRIEVE will set the cut index C_i equal to

I) the smaller of C_i^+ and $R_i - 1$ when $i = 1 \pmod 3$,

II) the larger of C_i^- and $L_i + 1$ when $i = 2 \pmod 3$,

and

III) C_i^{BIN} when $i = 0 \pmod 3$.

It will then use this cut index to reduce the search space in a manner similar to binary search and interpolation search.

In the full-length version of this paper [8], we show that if $0 < \alpha < 1$, $\theta > 0$, and $\phi \geq 2$, then RETRIEVE has an expected runtime log log N with a coefficient that depends on α, θ, ϕ, and the vector (b_1, b_2, b_3, b_4) bounding u. This extended abstract will outline the key ideas behind this proof and define another algorithm, called FIND, which has a better coefficient than RETRIEVE but whose proof is more complicated.

3. Motivation for Developing the Procedure RETRIEVE

Background information from the literature on numerical analysis is presented in this section. Citing these results, we explain that interpolation search must be inefficient on nonuniform distributions and explain the intuition behind the added efficiency of RETRIEVE.

The analog of interpolation search in numerical analysis is called the method of regula falsi. Given a function g and a value y, this method consists of an iterative procedure that conducts repeated interpolations to find an approximate solution for $g(x) = y$. The method of result falsi is known to require $\Omega[\log(1/\varepsilon)]$ runtime[1] to make approximations with an accuracy of ε [7].

It is easy to show this result implies that interpolation search has a logarithmic lower bound on its <u>expected</u> runtime for searching nonuniformly generated files. But binary search has a better runtime which is logarithmic in the worst-case. Interpolation search should therefore be avoided on nonuniformly generated files.

Our goal in this paper is to explain how the algorithm RETRIEVE has an asymptotically better performance than both binary and interpolation search on nonuniformly generated files. Part of the intuition behind the log log N runtime of RETRIEVE can be understood by noting that since u is continuous, it is very similar to the uniform distribution on very short intervals (where it is essentially constant). Also, observe that the efficiency of interpolation search increases as the probability distribution becomes more nearly uniform. The algorithm RETRIEVE was designed to take advantage of these properties by having the cuts produced by its rules I and II resemble those of interpolation search increasingly as ℓ_i gets smaller. The principal theme of this paper is that such a method of gradual transformation into interpolation search leads to a remarkable improvement in runtime.

Although it was not initially conceived in this manner, some partial analogs of RETRIEVE can be found in the literature on the "modified" method of regula falsi [7]. These algorithms are based on the observation that the unmodified version of regula falsi produces a relatively inefficient calculation of the root of a function - with an unfortunate bias toward searching mostly on one side of the designated root. Numerical analysts [7] have found that altering their search points so that there is a more symmetric convergence upon the root will dramatically improve the efficiency of regula falsi. The positive and negative increments Δ_i of the indices C_i^+ and C_i^- of RETRIEVE lead to a related type of gain in efficiency. The mathematical machinery of this paper can thus be viewed as the synthesis of the probability techniques which have been applied to interpolation search with a relaxation method whose partial analog can be found in the literature on regula falsi.

4. Runtime Analysis of RETRIEVE

In this section, we outline the proof of RETRIEVE's $O(\log \log N)$ runtime. More details can be found in the unabridged version of this paper [8].

It will be useful to characterize the subfile $F_i = (Y_{L_i} < \ldots < Y_{R_i})$ that is searched by the i-th iteration of this algorithm by the vector $(L_i, R_i, Y_{L_i}, Y_{R_i})$. This vector, called the <u>state at the i-th iteration</u>, will be denoted by S_i. Intuitively, S_i encompasses all the information known about F_i at the beginning of the i-th iteration of RETRIEVE. The previously defined quantities N_i and ℓ_i are functions of S_i.

The term <u>j-periodic iteration</u> will be used in this paper to refer to any iteration i such that $i = 1 \bmod j$. The term <u>j-cycle</u> will refer to a sequence of j iterations which begins with some j-periodic iteration i and ends with iteration $i + j - 1$. This terminology is convenient because the algorithm RETRIEVE has an inherently periodic nature, each of the cutting rules I through III being executed once during every 3-cycle. The following proposition is one of the two main theorems of this paper.

<u>Theorem 1</u>. A sufficient condition for an iterative searching algorithm A to search an ordered file F in expected runtime $O(\log \log N)$ is that there exist an integer $j \geq 2$ and constants $p > 0$, $\lambda < 1$, $0 < \alpha < 1$, $K \geq 1$, r and K' satisfying $0 < (2r)^{\alpha/2} < 1/K'$ and enabling the following two conditions to hold for every j-periodic iteration i:

I) If $\ell_i < r$ then there exists a probability greater than p that state S_{i+j} will satisfy

$$N_{i+j} \leq K \text{ MAX}\left(\sqrt{N_i};\ (\ell_i{}^{\alpha})N_i\right) \tag{11}$$

$$\ell_{i+j} \leq K' \text{ MAX}\left(\ell_i/\sqrt{N_{i+1}};\ \ell_i{}^{1+\alpha}\right), \tag{12}$$

and

II) If $\ell_i \geq r$ then there exists a probability greater than p that it satisfy

$$\ell_{i+j} \leq \lambda \ell_i \ . \tag{13}$$

The heart of our analysis consists of first proving Theorem 1 and then showing that algorithm RETRIEVE meets the theorem's requirements for log log N runtime. These two topics are discussed in great detail in the full-length version of this paper. We highlight the main parts of this analysis in the next several paragraphs.

Sketch of Proof of Theorem 1. It is useful to introduce one further definition. The j-cycle beginning with iteration i will be said to be strongly reducing iff state S_{i+j} satisfies one of the following three inequalities:

1) $N_{i+j} < K\sqrt{N_i}$

2) $\ell_{i+j} < K'\ell_i^{1+\alpha/2}$ and $r > \ell_i \geq N_i^{-1/(2\alpha)}$

3) $\ell_{i+j} < \lambda\ell_i$ and $\ell_i \geq r$.

Strongly reducing j-cycles can be proved to satisfy the following two conditions:

I) Let ℓ_1 denote the initial ℓ-value of algorithm A. For each ℓ_1 and λ, α, K, K' and r satisfying the hypothesis of Theorem 2, there exists a constant C such that no more than C · log log N strongly reducing j-cycles may occur before algorithm A terminates. (This observation holds because strongly reducing j-cycles produce such sharp reductions in either the value of N or ℓ that no more than O(log log N) events of types 1, 2, or 3 can occur.)

II) Each j-cycle during this search will have at least a probability p of being strongly reducing.[2]

The combination of assertions I and II can be shown to imply that the expected number of j-cycles before algorithm A terminates is no greater than (C/p) log log N. Since each j-cycle includes j iterations of algorithm A, the expected number of its iterations must be less than or equal to (C j/p) log log N. Hence, its expected runtime is O(log log N).

The second part of our analysis consists of showing that specifically the algorithm RETRIEVE satisfies the hypotheses of Theorem 1, and therefore it has a runtime in O(log log N). In our discussion, we let

$$b_5 = 3\ b_2/b_1 \ . \tag{14}$$

Given a cut index C_i and a state S_i whose components are the usual attributes L_i, R_i, Y_{L_i}, Y_{R_i}, ℓ_i, and N_i, the following new notation will be used:

$$\delta_i = 2\cdot\ell_i/\sqrt{N_i} + b_5\ell_i^{\ 2} \ , \tag{15}$$

$$\bar{y}_i = Y_{L_i} + \ell_i(C_i-L_i)/N_i \ . \tag{16}$$

Note that $\bar{y}_i$ represents the expected value of the key Y_{C_i} when the records belonging to the subfile $F_i = (Y_{L_i} < \ldots < Y_{R_i})$ are generated by the uniform probability distribution.

We will say the cut value Y_{C_i} of the i-th iteration is <u>good</u> iff it is sufficiently close to expected value $\bar{y}_i$ to satisfy:

$$\left| Y_{C_i} - \bar{y}_i \right| < \delta_i \ . \tag{17}$$

The probability of this event is indicated below:

<u>Theorem 2</u>. Let F denote a file that is generated by a probability density u whose regular bound is (b_1,b_2,b_3,b_4). The cut of any iteration i has a probability exceeding 3/4 of satisfying equation (17) (and thus being "good").

The proof of Theorem 2 is given in the unabridged version of this paper [8]; its intuitive justification is quite simple. Note that there are two reasons for the actual value of Y_{C_i} to differ from the quantity $\bar{y}_i$, which designates its expected value under the uniform distribution. These are that u is not uniform and that Y_{C_i} will differ from its expected value simply because of random variations. It is easy to apply theorems from numerical analysis and probability to show that the first difference will be of order ℓ_i^2 and the second will have a mean and median size of order $\ell_i/\sqrt{N_i}$. The central point is that the sum of these two quantities has the same asymptotic magnitude as δ_i; therefore the expected difference between Y_{C_i} and $\bar{y}_i$ has this magnitude. And consequently Theorem 2 must be satisfied by some number at least near δ_i in magnitude. The full-length version of this article discusses this subject in more detail and formally proves that precisely δ_i satisfies Theorem 2.

Now, we will explain how to apply Theorem 2 to prove that the algorithm RETRIEVE satisfies the hypothesis of Theorem 1 and therefore operates in time $O(\log \log N)$. In our discussion, we take $i = 1 \bmod 3$, $j = 3$ and $p = 9/16$. The unabridged version of this article shows that if $0 < \alpha < 1$, $\theta \geq 0$, $\phi \geq 2$, $\ell_i < r$, and the i-th and (i+1)-st iterations of RETRIEVE produce good cuts then the cuts Y_{C_i} and $Y_{C_{i+1}}$ will bracket y respectively from the right and left in a manner assuring that the state S_{i+j} will satisfy equations (11) and (12), for suitably chosen constants r, K, and K' (whose value depends on α, θ, ϕ, and u's bounding vector). Now, since Theorem 2 implies that the i-th and (i+1)-st iterations will produce good cuts with a probability of at least $p = 9/16$, this result shows that RETRIEVE must satisfy requirement I of Theorem 1. Since the iteration i+2 of RETRIEVE will produce a cut index of type C^{BIN}, it is relatively easy to prove RETRIEVE also satisfies requirement II. Using this reasoning, the fulllength version of our article shows that RETRIEVE meets all the requirements of Theorem 1 when it is applied to a regular probability density with the parameters $0 < \alpha < 1$, $\theta > 0$, and $\phi \geq 2$. For these parameters, it will therefore have an expected time $O(\log \log N)$.

5. A Brief Description of the Algorithm FIND

We have also proven an expected time $O(\log \log N)$ for numerous other algorithms, which generally have a better coefficient than RETRIEVE although longer proofs. For instance, let Δ_i have the same definition as in equation (8), and C_i^{INF} and C_i^{SUP} designate the cut indices:

$$C_i^{INF} = L_i + \Delta_i \tag{18}$$

$$C_i^{SUP} = R_i - \Delta_i \tag{19}$$

Consider an iterative search algorithm FIND(α,θ,ϕ) whose cut index C_i is defined equal to

I) the median value of set $\left\{C_i^{INF},\ C_i^{INT},\ C_i^{SUP}\right\}$ if $C_i^{INF} \leq C_i^{SUP}$

and

II) C_i^{BIN} otherwise.

This algorithm achieves a complexity $O(\log \log N)$ with a better coefficient than RETRIEVE. The remainder of this article will outline the similarities and distinctions between these two algorithms.

The proof of a complexity $O(\log \log N)$ for the procedure FIND is similar to RETRIEVE insofar as it also rests on showing that this procedure satisfies the requirements of Theorem 1. The two algorithms also have another characteristic in common. During every 3-cycle where $\ell_i < r$ and the algorithm FIND produces good cuts, one of these cuts will lie slightly to the left of y and another to its right in a manner assuring that the state S_{i+j} satisfies equations (11) and (12). The analysis of FIND differs from RETRIEVE chiefly because it is impossible to predict, a priori, which of these three iterations lie to the left and right based on the assumption that they produce good cuts. That is, FIND differs from RETRIEVE because its cutting rules do not depend on the rigid constraint i mod 3. Chiefly because its cutting rules are more flexible, the algorithm FIND has a better coefficient but a more complicated proof of its time O(log log N) than RETRIEVE. Other possible algorithms which achieve expected time O(log log N) are also discussed in the unabridged version of our paper [8].

6. Acknowledgment

I would like to thank Professor S. Bing Yao for suggesting that I study interpolation search. I thank Y. Edmund Lien and Eric Wolman for their suggestions on presentation. I also thank William Gewirtz and John Ying for their comments on a previous draft.

7. References

[1] Feller, W., An Introduction to Probability Theory and Its Applications, Vol. 1, third ed., Wiley, New York, 1968.

[2] Gonnet, G., Rogers, L., and George, J., "An Algorithmic and Complexity Analysis of Interpolation Search," Acta Inform., 13, 1980, pp. 39-52.

[3] Kendall, M. G. and Stuart A., Advanced Theory of Statistics, Vol. 1, fourth ed., Macmillan, New York, 1977.

[4] Knuth, D. E. The Art of Computer Programming, Vol. 3: Sorting and Searching, Addison-Wesley, Reading, MA, 1973.

[5] Pearl, Y. and Reingold, E. M. "Understanding the Complexity of Interpolation Search," Inform. Process. Lett., 6, 1977, pp. 219-222.

[6] Pearl, Y., Itai, A., and Avni, H. "Interpolation Search - A Log Log N Search," Comm. ACM, 21, 1978, pp. 550-554.

[7] Ralston, A. and Rabinowitz, P., A First Course in Numerical Analysis, second ed., McGraw Hill, New York, 1978, pp. 338-344.

[8] Willard, D. E., "Searching Nonuniformly Generated Files in Log Log N Runtime", the unabridged version of this paper.

[9] Yao, A. C. and Yao, F. F., "The Complexity of Searching an Ordered Random Table," Proc. Seventeenth Annual Symp. Foundations of Comp. Sci., 1976, pp. 173-177.

8. Endnotes

[1] A function F_1 is said to have a lower bound (f_2) iff all the values of F_1 are at least proportional to F_2.

[2] Note that the numbers ℓ_i and N_i always satisfy one of the following three conditions:

A) $\ell_i < N_i^{-1/(2\alpha)}$ and $\ell_i < r$

B) $r > \ell_i \geq N_i^{-1/(2\alpha)}$;

C) $\ell_i \geq r$.

The proof of assertion II rests on applying the hypotheses of Theorem 1 to show that in cases A, B and C, there is a probability exceeding p of a j-cycle satisfying conditions 1, 2, and 3, respectively.

Bell Laboratories, Holmdel, New Jersey 07733

A MULTIVARIATE STATISTICAL MODEL FOR DATA BASE PERFORMANCE EVALUATION

S. C. Christodoulakis

Abstract

A variety of data base performance evaluation studies require an estimation of the number of records qualifying in a query or a set of queries (record selectivities). This paper presents a multivariate statistical model for the estimation of record selectivities. The model takes into account non-uniformity and correlations of the attribute values in the data base, as well as non-uniformity and correlations of the attribute values in queries.

1. Introduction

The performance of a data base system depends on a large number of factors. They include the data model and the data language, the storage structure and accessing algorithms, the system software and hardware characteristics, as well as the data base contents, the placement of data on devices and the queries issued by the user population [Sevcik 1981]. The impact of these factors on system performance is currently not very well understood.

Three important data base performance issues that have been examined extensively in the literature are: (1) the selection of suitable strategies for accessing the data in relational data base environments ([Selinger et al. 1979], [Yao 1979], [Goodman et al. 1979]), (2) the selection of a set of indices in an inverted data base environment ([Schkolnick 1975], [Hammer and Chan 1976], [Anderson and Berra 1977]), and (3) the development of performance predictors for physical data base design ([Gambino and Gerritsen 1977], [Teorey and Das 1976], [Teorey and Oberlander 1978]). Analytic models that have been proposed for the analysis of these problems require an estimate of the number of records qualifying in a query (usually referred to as record selectivity), and the average number of records qualifying in a set of queries (average record selectivity). The performance measures are directly or indirectly expressed in terms of these quantities. In order to estimate these quantities, the data base contents, as well as the queries issued by the user population have to be modelled.

Most of the previous models for estimating record selectivities are based on the uniformity and independence of attribute values in a file assumption ([Selinger et al. 1979], [Goodman et al. 1979], [Yao and De Jong 1978]). A more detailed model is suggested by [Kerschberg et al. 1979]. Recent models for estimating average record selectivities are based on uniformity and independence of attribute values in a file, as well as on uniformity and independence of attribute values in query classes [Anderson and Berra 1977]. An alternative approach for estimating average record selectivities is suggested by [Hammer and Chan 1976].

In this paper we will present a model for the estimation of record selectivities. In section 2, we describe a model of the data base contents, and we present estimations of record selectivities based on this model. In section 3, we describe a model of queries in the data base and we present a method to estimate average selectivities. Finally, in section 4, we present our conclusions and suggestions for further research.

2. A Multivariate Statistical Model for the Estimation of Record Selectivities

In this section we present a multivariate statistical model for the estimation of record selectivities in flat files. We regard the records of a flat file as points of an n-dimensional space, where n is the number of attributes of the file. Each attribute corresponds to an axis of the n-dimensional space. Attribute values of quantitative attributes (like SALARY, AGE, HEIGHT) correspond directly to points of the axis. Attribute values of categoric attributes (like RESPONSIBILITY-LEVEL, SEX, DEGREE), are mapped into discrete values in the corresponding axis. Grouping of several values into a single value may be desirable in order to avoid excess mapping if the number of categories is very large. The attribute values of a record of the file determine a point in this space. Dense sections of the space correspond to attribute value combinations that involve a large number of records. We hereafter refer to this space as the *attribute space*.

We wish to model the distribution of records in attribute space with a continuous multi-dimensional distribution. This distribution will then be used to derive estimates of record selectivities. It is desirable to use only a small number of parameters to describe the distribution of points in the attribute space in order to avoid the excess IO produced by the transfer of a large set of values at run time. For the same reason (run time efficiency), we would like a density function that is cheap to evaluate. Other desirable criteria are adaptivity and intuitive significance of parameters. Adaptivity of the parameters allows us to periodically update the parameter values to reflect the values of the new records inserted in the data base without scanning all the data base again. Intuitive significance of parameters allows the parameters to be approximated in the absence of actual data

(data base design phase), as well as to draw useful inferences for the data base design based on the values of these parameters.

Traditionally parametric and non-parametric techniques have been used for describing the distribution of points in an n-dimensional space ([Andrews 1972], [Duta and Hart 1973], [Tou and Gonzalez 1974]). Parametric techniques assume that the density has a known distribution and use the available samples to estimate the parameters of the distribution. Non-parametric techniques do not make an a priori assumption about the density distribution that the data follow, but they use the available samples to determine the shape of the distribution. Parametric techniques have the advantage of being mathematically tractable and being cheaper to evaluate than the non-parametric ones, but if the data do not follow the assumed distributions large errors may result. On the other hand, non-parametric techniques may be more accurate for a small number of variables at the expense of keeping a large number of terms to approximate the distribution; the coefficients also do not have intuitive meaning. Finally, the large number of attributes and the size of existing data bases make iterative techniques impractical.

We have used an intermediate approach. We describe the distribution of records in the attribute space using a member of a family of diverse distributions. An important parameter of the family of distributions is the convariance matrix of the variables. The covariance matrix is a symmetric, positive definite matrix, and is defined as $C = E(\bar{x} - \bar{m})(\bar{x} - \bar{m})$, where $\bar{m}$ is the mean vector of the distribution, prime denotes transposition, and E denotes expectation. For example, in the two-dimensional case $(n = 2)$, the covariance matrix is:

$$\begin{bmatrix} \sigma_1^2 & \rho_{12}\sigma_1\sigma_2 \\ \rho_{12}\sigma_1\sigma_2 & \sigma_2^2 \end{bmatrix}$$

where σ_1 and σ_2 are the standard deviations of the two variables, and ρ_{12} is their correlation. Thus, this model takes into account the correlation of attribute values.

The family of distributions includes the multivariate normal distribution and the multivariate extensions of the Pearson Type 2 and Type 7 distributions ([Tou and Gonzalez 1974], [Cooper 1964], [Elderton 1953]).

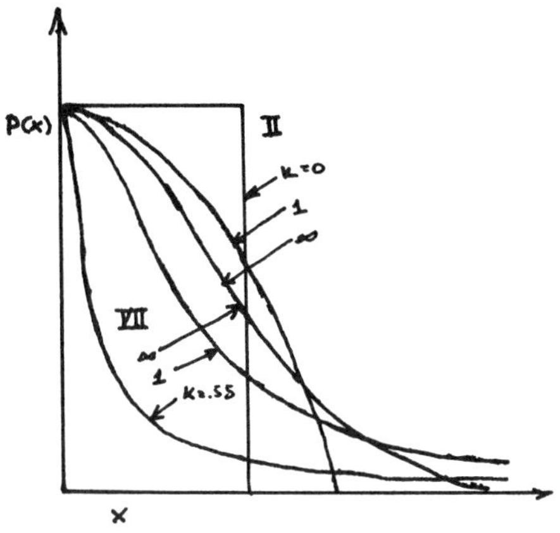

Figure 1. A Family of Univariate Distributions

In one dimension the above family includes a wide range of probability density functions. The various probability functions involved are shown in figure 1, normalized to have the same peak value (there is a symmetric left side for each distribution shown in the figure). The univariate Pearson Type 2 involves symmetric distributions ranging from the uniform distribution for $k = 0$ to parabolic for $k = 1$, and approaching the normal for large k. This distribution is non zero only over a finite region, and thus, well suited for describing the distribution of attribute values in a data base environment. In one dimension the Type 7 Pearson distributions include the normal for high values of k, the t distribution for half integer values of k and an appropriate scale parameter, and the Cauchy distribution for $k = 1$.

The multivariate normal distribution is important and is central to the family of our distributions. The probability density function of the multivariate normal distribution is given by:

$$p(x) = \frac{1}{(2\pi)^{n/2}|C|^{1/2}} \exp[-(1/2)(\bar{x} - \bar{m})'C^{-1}(\bar{x} - \bar{m})]$$

where C is the covariance matrix of the variables and $\bar{m}$ is the mean vector. The points of constant density are hyperellipsoids for which $r^2 = (\bar{x} - \bar{m})'C^{-1}(\bar{x} - \bar{m})$ is constant. r^2 is usually called the Mahalanobis distance from $\bar{x}$ to $\bar{m}$. Figure 2 shows curves of constant density for the normal distribution in a two-dimensional space.

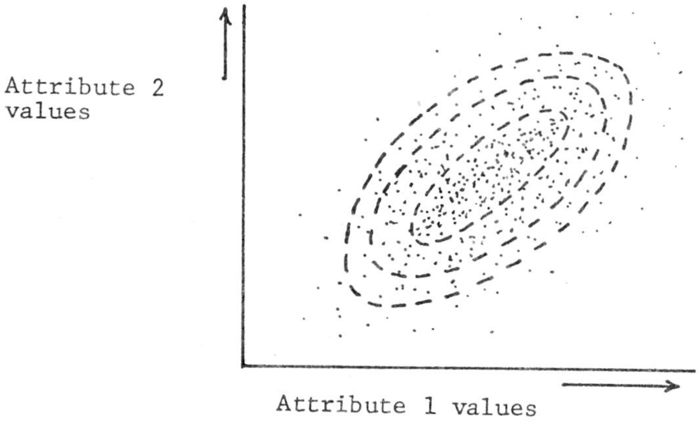

Figure 2. Normal Density Curves

The univariate Pearson Type 2 distribution is

$$p(x) = \frac{\omega}{B(.5,k+1)} [1 - \omega^2(x - m)^2]^k$$

with $|x - m| \leq \frac{1}{\omega}$

where B is the Beta function, and ω is a scaling constant. A multivariate extension of this distribution is

$$p(\bar{x}) = \begin{cases} h(\bar{x}) & \text{in region } T \\ 0 & \text{elsewhere} \end{cases}$$

where T is the interior of the hyperellipsoid $(\bar{x}-\bar{m})'W(\bar{x}-\bar{m}) = 1$, and

$$h(\bar{x}) = \frac{\Gamma(k+n)/2+1)}{\pi^{n/2}\Gamma(k+1)} |W|^{1/2}[1-(\bar{x}-\bar{m})'W(\bar{x}-\bar{m})]^k$$

where Γ is the Gamma function. The scaling matrix W is given by

$$W = \frac{1}{2k+n+2} C^{-1}$$

and $k \geqq 0$ is a parameter.

The univariate Pearson Type 7 distribution is

$$p(x) = \frac{\omega}{B(1/2,k-1/2)} [1+\omega^2(x-m)^2]^{-k}.$$

A multivariate extension of this distribution is

$$p(\bar{x}) = \frac{\Gamma(k)}{\pi^{n/2}\Gamma(k-n/2)} |W|^{1/2}[1+(\bar{x}-\bar{m})'W(\bar{x}-\bar{m})]^{-1}$$

with $2k > n$.

For $2k > n+q$, the q'th moment exists [Cooper 1964]. In particular, if $2k > n+2$ the covariance matrix exists and $W = \frac{1}{2k-n-2} C^{-1}$.

In addition to the variety of distributions which it includes, this family of distributions has some more desirable properties for our purpose. Since queries on the data base refer to only a subset of all the attributes, it is important that the parameters of the distribution followed in the subspace specified by the attributes appearing in the query be determined efficiently. The family of distributions described has this property. The marginal distribution of the multivariate normal is also normal, and the marginals of Type 2 and Type 7 Pearson distributions remain of the same Type. In going from a space of dimensionality n to a space of dimensionality $m < n$, the portion of the original n dimensional covariance matrix corresponding to the m-subspace is preserved. The same is true for the inverse scaling matrix, i.e., W^{-1}. The parameter k' of the distribution becomes in the m-subspace k, where

$k = k' + (n - m)/2$ for Type 2, and

$k = k' - (n - m)/2$ for Type 7

A member of this family of distributions can be selected for approximating the probability density in a subspace by considering the distributions followed in each dimension of this subspace and using an inversion of the above formula to estimate the parameter k. A second important property of this family of distributions is that the estimation of parameters, and the selection of a member of the family can be done in one file pass. During this pass, the mean values and the fourth moments for each attribute, as well as the covariance matrix, are estimated. Details of the selection of a member of the family of distributions to approximate the density in a subspace of the attribute space, as well as the estimation of the parameters of this distribution appear in [Cooper 1964], and [Christodoulakis 1981]. A third advantage is that the parameters of the distributions are adaptive. Finally, the parameters of the distributions have an intuitive significance since they refer to quantities which are well understood and well described in the literature, like means, standard deviations, and correlations.

Next, we describe the estimation of record selectivities. Given a conjunctive query specifying any number of attributes of the file, we would like to estimate the expected number of records qualifying for the query. Formally, a conjunctive query Q has the form

$$Q = (q, \bar{x}, \Delta\bar{x})$$

where q is the m-dimensional attribute subspace specified by the attributes participating in the query, and $\bar{x} = (x_{i_1}, x_{i_2}, \ldots, x_{i_m})'$ and $\Delta\bar{x} = (\Delta x_{i_1}, \Delta x_{i_2}, \ldots, \Delta x_{i_m})'$ are vectors in q. The interpretation is that the conjunctive query Q is asking for all the records of the file which have attribute values $\bar{y} = (y_{i_1}, y_{i_2}, \ldots, y_{i_m})'$ in the subset q of their attributes, such that $x_{i_j} \leq y_{i_j} < x_{i_j} + \Delta x_{i_j}$ for $j = 1, \ldots, m$. This formulation allows asking for ranges of attribute values as well as for equality of attribute values for categoric attributes (in this case Δx_i will be the difference of two consecutive discrete values in the axis of the attribute). The expected number of records qualifying for the query Q is therefore given by

$$E(Q) = N \int_{x}^{x+\Delta x} p(\bar{y}) dy$$
$$= N \int_{x_{i_1}}^{x_{i_1}+\Delta x_{i_1}} (\ldots (\int_{x_{i_m}}^{x_{i_m}+\Delta x_{i_m}} p(y_{i_1}, \ldots, y_{i_m}) dy_{i_m}) \cdots) dy_{i_1}$$

where N is the number of records in the file. For small values of $\Delta x_{i_1}, \ldots, \Delta x_{i_m}$ the above quantity can be approximated by

$$E(Q) = Np(\bar{x}) \Delta x_{i_1} \Delta x_{i_2} \cdots \Delta x_{i_m}$$

where $p(\bar{x})$ is the value of the probability density function at the point $\bar{x}$ of the space. This formula can be used for the estimation of record selectivities in conjunctive queries. The estimation of record selectivities of other types of queries can be based on this result [Christodoulakis 1981].

As an example, consider the application of the model for the estimation of record selectivities in an engineering data base. A recent report [Engineering Report 1979], provides several statistics such as years of experience, responsibility level, and salary information for this population of 14500 engineers. All the attributes have highly unimodal distribution of values rather than uniform, as well as highly correlated attributes. If the information about the professional engineers of Ontario were organized in a data base (one record per engineer), then the uniformity and independence assumptions could lead to serious errors in the estimation of the number of records that qualify in a query. With our model the number of engineers in responsibility level "A" is estimated to be 1091 compared to the actual 1038, a relative error of .02 in comparison to the relative error of .39 under the uniformity assumption. (We use in this paper a normalized relative error defined as $|a-e|/(a+e+cons)$, where a is the actual occurences, e is the estimated occurences, and cons is a constant set to avoid division by zero, as well as to minimize the effect of large relative errors for small number of actual occurences. This relative error takes values between zero and one hundred percent). The estimated number of engineers in responsibility level "A" and more than 15 years of experience is zero, compared to the actual 6. The relative error is .05 instead of .98 under the uniformity and independence assumption, and .97 given by the model of Kerschberg et al. [Kerschberg et al. 1979]. In this example, our model improved considerably the estimation of record selectivities compared to the other models.

In Table 1 we tabulate data from an inventory file of 1080 records with 11 attributes. The table shows the distributions followed by the attribute values of the attributes. (Type is the Type of the distribution and K is the parameter k determining the specific member of the family selected). We have assigned numeric values to all the categoric attributes so that the distribution of attribute values in each axis are approximately symmetric unimodal. The majority of the distributions shown have high peaks as indicated by the parameter k of the distribution, which indicates again that the uniformity assumption is unrealistic in real data base environments. The table also shows the average normalized relative error in the estimation of selectivities over all equality queries on a given atrribute (Aveu), using the uniformity assumption. (For non-discrete value attributes the range of attribute values has been divided into subranges corresponding to queries). The average normalized relative error using our model is also shown (Avem). The last column in the table (Prwin), shows the proportion of the equality queries on an attribute, in which our model achieves a lower error in the estimation of selectivities. A similar set of performance measures, (but more specific to query optimizers), has been proposed in a recent performance evaluation of the System R optimizer ([Astrahan et al. 1980]). A comparison of the two approaches shows that our model considerably improves the estimation of selectivities for single attribute queries.

Table 1

Att#	Type	K	Aveu	Avem	Prwin
1	2	5.0	.73	.03	1.00
2	7	6.0	.70	.04	.95
3	7	4.8	.58	.06	.88
4	2	0.0	.52	.15	.93
5	7	3.8	.60	.15	.92
6	7	3.7	.65	.13	.90
7	7	3.2	.55	.20	.81
8	7	3.3	.55	.16	.85
9	2	28.3	.63	.06	.95
10	2	4.1	.43	.03	1.00
11	7	9.5	.58	.10	.90
Av.			.63	.11	.91

Table 2 compares the accuracy of the our model with the uniformity and independence assumption model in conjunctive queries on a subset of the possible combinations of two attributes. A member of the family of bivariate distributions was used to approximate the density in a two-dimensional space. When the marginals were of a different Type, the bivariate normal was used to approximate the density distribution. In the file used, a significant propertion of the combination of attribute values of two attributes did not involve any records of the file at all, which makes models using only mean values inappropriate. Our results suggest that a significant improvement can be achieved using our model in the estimation of selectivities in multiattribute queries.

Table 2

Att#	Aveu	Avem	Prwin
1,2	.25	.04	.94
2,3	.37	.08	.92
3,4	.38	.07	.91
4,5	.32	.07	.91
5,6	.43	.11	.92
6,7	.33	.00	.87
7,8	.19	.07	.87
8,9	.20	.07	.87
9,10	.55	.12	.85
10,11	.53	.17	.82
Aver	.35	.09	.89

3. Estimation of Average Selectivities

In this section we present a model for the estimation of average record selectivities. We separate the queries into classes as in [Anderson and Berra 1977] and [Hammer and Chan 1976] according to the attributes participating in a conjunctive or a disjunctive query. We view the queries of a class as points in an m-dimensional sub-space of the attribute space, where m is the number of attributes specified in this class of queries. We call this sub-space the query class space. We use the same family of probability density functions to describe the distribution of this class of queries. We use $q(\bar{x})$ to denote the probability density in the query class space.

For simplicity (but without loss of generality) we assume here that the attribute values are distinct and that all the queries are equality queries. The number of records qualifying in a query $(q,\bar{x},\Delta\bar{x})$ is given by $Np(\bar{x})\Delta x_1 \Delta x_2 \cdots \Delta x_m$ as in the previous section. The expected number of records qualifying in a query of a class c therefore is

$$E(N_c) = \int q(\bar{x})(Np(\bar{x})\Delta x_1 \cdots \Delta x_m)dx$$
$$= N\Delta x_1 \cdots\cdot \Delta x_m \int q(\bar{x})p(\bar{x})dx$$

The expected number of records qualifying in any subset of the attributes of the class (included in a canditate indexing set say), can be estimated by considering only the relevant attributes in the above formula. If all the queries in the data base were single attribute queries, this approach of estimating average record selectivities will produce similar results with the method of Hammer and Chan. However, in multiattribute retrieval envoronments, the two approaches are considerably different.

The integral in the formula can be approximated for the different distributions considered [Christodoulakis 1981]. The solution in the case that $p(\bar{x})$ and $q(\bar{x})$ are normal distributions is given by ([Christodoulakis 1981])

$$E(Nc) = N \frac{\Delta x_1 \cdots \Delta x_m e^{-M/2}}{(2\pi)^{m/2}|\Sigma|^{1/2}|S|^{1/2}|C|^{1/2}}$$

where S is the covariance matrix in the attribute subspace, Σ is the covariance matrix in the query class space, C is a positive definite matrix, and M is a constant.

As an example consider single attribute equality queries on the attribute RESPONSIBILITY-LEVEL of the engineering data base. The average record selectivity depends on which are the more frequent queries. If most queries are asking for RESPONSIBILITY-LEVEL = "D", the average selectivity will be near to the actual record occurences of RESPONSIBILITY-LEVEL = "D" (4345), while if most queries are asking for RESPONSIBILITY-LEVEL = "A", the average selectivity will be near to the record occurences of RESPONSIBILITY-LEVEL = "A" (1355). If the queries were uniformly spread over all the possible values, the average selectivity would be equal to the mean value of record occurences per responsibility level (2416). These three estimates are considerably different. Figure 3 shows the average record selectivity as estimated by our model, as function of the standard deviation of the RESPONSIBILITY-LEVEL values asked in queries (we assumed that the means coincided).

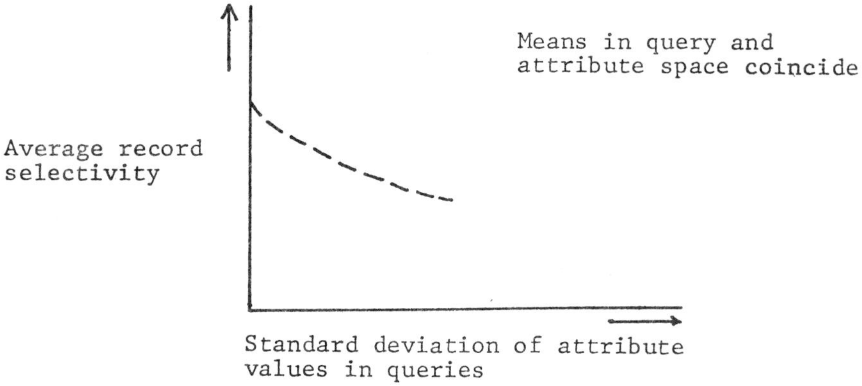

Figure 3. Average Selectivity in Single Attribute Queries

In multiattribute queries, the average record selectivity depends on the correlations of attribute values in the data base, as well as on the correlation of the attribute values asked in the queries on the data base. Figure 4 shows the average record selectivity as a function of the correlations of the attribute values in the file and in the queries, as estimated by our model (in this experiment the means did not coincide).

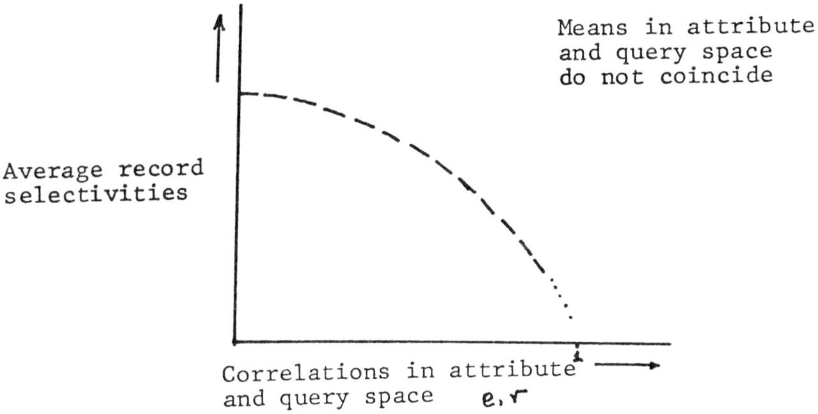

Figure 4. Average Record Selectivity in Conjunctive Queries

4. Summary and Conclusions

We have presented the problem of estimation of record selectivities. We have pointed out that data bases contain attributes with high peaks in the distribution of their attribute values, correlated attributes, and that certain combinations of attribute values do not appear in the data base. We have shown that under these circumstances the assumptions used in previous models for predicting selectivities may lead to significant errors.

We have presented a statistical model to estimate record selectivities. An important parameter of the distributions that we have used is the covariance matrix of the attributes. Thus we take into account correlations of data values. The set of parameters is adaptive, intuitively meaningful, and can be estimated in one file pass. We have experimentally shown that this model gives better approximations than the previous models. The uniformity and independence assumption model can be seen as a special case of our model. In fact, it can be combined with our model if desirable (for example if an attribute with a very large number of discrete values exists). Our model is more useful in large data base environments with frequent multi-attribute queries, as well as multi-file environments [Christodoulakis 1981].

We have also described the use of average record selectivities in data base performance evaluation, and we have pointed out that the various methods previously proposed for estimating average record selectivities may lead to errors. We have presented a different approach for estimating average record selectivities. Our approach takes into account non uniformity and correlations of attribute values in both, the file, and the queries.

We have not considered another important data base performance evaluation problem, namely the estimation of the number of blocks containing the qualifying records in a query. Existing models estimate the number of blocks using the random selection assumption [Yao 1978]. We will leave this issue for future considerations.

5. Acknowledgements

I am grateful to Professor K. Sevcik for motivation, support, and help in the preparation of this paper.

6. References

[1] M. M. Astrahan, M. W. Blasgen, D. D. Chamberlin, K. P. Eswaran, J. N. Gray, P. P. Griffiths, W. F. King, R. A. Lorie, P. R. McJones, J. W. Mehl, G. R. Putzolu, I. L. Traiger, B. W. Wade, V. Watson (1976), "System R: Relational Approach to Database Management," ACM TODS 1, 2, pp. 97-137.

[2] M. M. Astrahan, W. Kim, M. Scholnick (1980), "Evaluation of the System R Access Path Selection Mechanism," to be presented in IFIP (Australia).

[3] T. W. Anderson (1958), An Introduction to Multivariate Statistical Analysis, John Wiley, New York.

[4] H. D. Anderson and P. B. Berra (1977), "Minimum Cost Selection of Secondary Indexes for Formated Files," ACM TODS 2, 1, pp. 68-90.

[5] H. Andrews (1972), Introduction to Mathematical Techniques in Pattern Recognition, Wiley-Intersicnece.

[6] M. W. Glasgen and K. P. Eswaran (1977), "Storage and Access, in Relational Data Bases," IBM Systems J., no. 4, pp. 363-377.

[7] M. W. Glasgen, M. M. Astrahan, D. D. Chamberlin, J. N. Gray, W. F. King, B. G. Lindsay, R. A. Lorie, J. W. Mehl, T. G. Price, G. R. Putzolu, M. Schkolnick, P. G. Selinger, D. R. Slutz, H. R. Strong, I. L. Traiger, R. W. Wade, R. A. Yost (1979), "System R: An Architectural Update," IBM report RJ2581(33481). IBM research laboratory, San Jose, California.

[8] A. F. Cardenas (1975), "Analysis and Performance of Inverted Data Base Structures," Comm. ACM, 18, pp. 253-263.

[9] S. Christodoulakis, "Estimating Selectivities in Data Bases," Ph.D. Thesis University of Toronto, CSRG #136, 1981.

[10] P. W. Cooper (1963), "Statistical classification with quadratic forms," Biometrika, 50.

[11] P. W. Cooper (1964), "Hyperplanes, Hyperspheres, and Hyperquadrics as Decision Boundaries," Computer and Informa tion Sciences, Tou and Wilcox editors, Spartan books, inc.

[12] R. Demolomb (1980), "Estimation of the Number of Tuples Satisfying a Query Expressed in Predicate Calculus Language," Proceedings VLDB, pp. 55, 58.

[13] R. O. Duta and D. E. Hart (1973), Pattern Classification and Scene Analysis, Wiley-Interscience.

[14] W. P. Elderton (1953), Frequency Curves and Correlation, Harren, New York.

[15] R. Epstein, M. Stonebraker, E. Wong (1978), "Distributed Query Processing in a Relational Data Base System," ACM SIGMOD, pp. 169-180.

[16] T. J. Cambino and R. Cerritsen (1977), "A Data Base Design Decision Support System," Proceedings VLDB, pp. 534-544.

[17] N. Goodman, P. A. Bernstein, E. Wong, C. L. Reeve, J. B. Rothnic (1979), "Query Processing in SDD-1: A System for Distributed Databases," technical report, Computer Corporation of America.

[18] M. Hammer and A. Chan (1976), "Index Selection in a Self Adaptive Data Base Management," ACM SIGMOD, pp. 1-8.

[19] R. Henver and S. B. Yao (1979), "Query Processing in Distributed Database Systems," IEEE Trans. Software Engrg. SE5.

[20] L. Kerschberg, P. D. Ting, S. B. Yao (1979), "Optimal Distributed Query Processing," submitted for publication.

[21] W. F. King (1974), "On the Selection of Indices for a File," IBM research report RJ 1341(20850), IBM Research Laboratory, San Jose, California.

[22] J. W. Liu (1976), "Algorithms for Parsing Search Queries in Systems with Inverted File Organizations," ACM TODS 1, 4, pp. 299-316.

[23] V. Y. Lum, H. Ling (1971), "An Optimization Problem on the Selection of Secondary Keys," Proceedings ACM National Conference, pp. 349-356.

[24] F. Nakamura, I. Yoshida, H. Kondo (1975), "A Simulation Model for Data Base System Performance Evaluation," Proceedings NCC, AFIPS, pp. 459-465.

[25] E. A. Ozkarahan, S. A. Schuster, K. C. Sevcik (1977), "Performance Evaluation of a Relational Associative Processor," ACM TODS 2, 2, pp. 175-196.

[26] M. T. Pezzaro (1980), "Analytic Evaluation of Physical Database Designs," submitted for publication.

[27] A. Putkoven (1978), "On the Selection of the Access Path in Inverted Database Organization," Information Systems, 4, pp. 219-225.

[28] M. Schkolnick (1975), "The Optimal Selection of Secondary Indices for Files," Information Systems, 1, pp. 141-146.

[29] M. Schkolnick (1978), "A Survey of Physical Database Design Methodology and Techniques," Proceedings VLDB, pp. 474-487.

[30] P. G. Selinger, M. M. Astrahan, D. D. Chamberlin, R. A. Lorie, T. G. Price (1979), "Access Path Selection in a Relational Database Management System," IBM research report RJ2429.

[31] K. C. Sevcik (1981), "Data Base System Performance Prediction Using An Analytical Model," Proceedings VLDB, 1981.

[32] M. Stonebraker (1974), "The Choice of Partial Inversions and Combined Lndices," Internat. J. Comput. Inform. Sci., 3, pp. 167-188.

[33] T. J. Teorey and K. S. Das (1976), "Application of an Analytical Model to Evaluate Storage Structures," Proceedings ACM SIGMOD, pp. 9-19.

[34] T. J. Toorey and L. B. Oberlander (1978), "Network Database Evaluation using Analytical Modeling," Proceedings 1978 NCC, AFIPS, pp. 833-842.

[35] S. Todd and J. Vorhofstad (1977), "An Optimizer for a Relational Database System-Description and Evaluation," technical report, IBM United Kingdom Scientific Centre, Peterlee, Durhan.

[36] S. Tou and R. Gonzalez (1974), Pattern Recognition Principles, Addison Wesley Publ. comp.

[37] D. C. Tsichritzis and F. H. Lochovsky (1977), Data Base Management Systems, Academic Press.

[38] S. B. Yao (1977), "An Attribute Based Model for Database Access Path Cost Analysis," ACM TODS 2, 1, pp. 45-67.

[39] S. B. Yao (1977a), "Approximating Block Accesses in Data Base Organizations," Comm. ACM, 20, 4, pp. 260-261.

[40] S. B. Yao (1979), "Optimization of Query Evaluation Algorithms," ACM TODS, 4, 2.

[41] S. B. Yao and D. De Jong (1978), "Evaluation of Data Base Access Paths," Proceedings ACM SIGMOD, pp. 66-77.

[42] E. Wong and K. Youssefi (1976), "Decomposition- A Strategy for Query Processing," ACM TODS 1, 3, pp. 307-327.

[43] P. C. Yue and C. K. Wong (1975), "Storage Cost Considerations in Secondary Index Selection," Internat. J. Comput. Inform. Systems, 4, pp. 307-315.

Computer Systems Research Group, University of Toronto, CANADA.

Discussant's Report on
"A Multivariate Statistical Model for Data Base
Performance Evaluation,"
by S. C. Christodoulakis

The main point of this paper is that in order to estimate record selectivities, it is necessary to consider the possibility that attribute values may be non-uniform and cross-correlated. Hence, the traditional assumptions of uniformity and independence are non-realistic and have to be revised. The paper suggests the use of a family of multivariate Pearson type-2 and type-7 distributions and the multinormal distribution to represent the joint distribution of attributes. The point is well taken and skillfully presented, so my comments should not be taken as objections but rather as some further thoughts on the issue.

To make my first point clear, consider the following situation. Consider a relation which contains the attributes NAME, SEX, MARITAL-STATUS, #-OF-CHILDREN and some other attributes. Assume the following statistics for a subsample of 600 persons:

SEX	MARITAL-STATUS	#-OF-CHILDREN	Number of Persons Qualifying
Male	Married	0	100
"	"	>0	100
"	Single	0	100
"	"	>0	0
Female	Married	0	100
"	"	>0	100
"	Single	0	80
"	"	>0	20

By recording only correlations between pairs of variables, it is impossible to capture the fact that single males don't have children (in this specific sample). Hence, a representation of these three attributes by a three-dimensional normal distribution (for example) is not realistic. Does this call for the use of distributions with a higher dimensionality?

This question is analogous to the question whether a multivariate model should replace the traditional one-dimensional method: there is a tradeoff involved, and there is no clear-cut answer. The answer depends on the relationship between the (expected) benefits of using a more sophisticated model (i.e., better predictions and hence better performance) vs. the expected costs of using the more sophisticated model (higher complexity, more data collection--especially in the initial design phase, and of course the cost of maintaining and updating the parameters of the distributions). In order to be able to better understand this tradeoff, it is necessary to be able to analyze the expected costs and benefits. One step in this direction is the behavior shown in Figure 3 (relative error as a function of the correlation coefficient), which was constructed from sample data. If this behavior could be generalized, this would consist of a further step in this direction.

Another comment relates to the treatment of categoric (i.e., nominal) attributes. The treatment suggested is to define (or redefine, or map) the categories so that the distribution of attribute values on each axis is approximately unimodal symmetric. This treatment has the same disadvantages which the author attributes to the model of Kerschberg, Ting and Yao (1979) (where attribute values were sorted on frequency of occurrence). The decision whether a unimodal-symmetric density is more appropriate than a monotone density clearly depends on the circumstances and may not be answered in general. A similar problem also arises with respect to the representation of attributes whose values are character strings (e.g., names) in the attribute space, since they have to be mapped into the real axis.

Author's Response to Discussant's Report

The comments of the discussant are nice and motivate further research.

The need for a more detailed model comes from the fact that data bases by their nature contain non-uniform and dependent attributes. The impact of the error on the estimation of selectivities on the system cost can be large, and it depends on the frequency and the nature of queries (for example the selectivities estimated for restrictions followed by joins can be very large if independence is assumed). Moreover, attribute dependencies are useful in order to examine the effect of non-random placement of qualifying records among the blocks of the file [Christodoulakis 1980].

Since keeping exact information on any possible combination of attribute values is unrealistic, we need to approximate the actual distribution. The family of distributions used for describing the distributions in the attribute space has several advandages described below: 1) It includes a wide range of probability density functions, and it takes into account non-uniformity and correlations among attribute values. Correlations of attribute values are typical of large data bases describing populations. In such data bases large errors in the estimation of selectivities may result. 2) The number of variables (parameters) required for describing the distribution is not combinatorial. The covariance matrix has a size of $n(n+1)/2$, where n is the number of attributes. In contrast, if nonparametric techniques are used, in addition to the larger number of parameters required to describe the distribution in the multivariate space, integration is required in order to find the distribution in subspaces specified by queries. Finally, we are only interested in approximating the number of qualifying records such that we avoid large errors (for example, in the example in the discussant's comments the marital status versus children relationship can capture the fact that only a small number of single persons have children. A little more will be captured by the fact that more women than men have children). 3) The parameters of the distributions used have an intuitive significance since they refer to quantities which are well understood and well described in the literature, like means, standard deviations, and correlations. Due to the mathematical tractability and the intuitive significance of the parameters, the impact of data characteristics on data base design can be studied analytically, and the results can be interpreted. Finally, when the data base exists, the parameters can be estimated in one file pass (iterative methods are unrealistic for large data bases).

Discussant: Dr. Haim Mendelson, Graduate School of Management, University of Rochester, Rochester, New York 14627.

THE USE OF SAMPLE PATHS IN PERFORMANCE ANALYSIS

Ward Whitt, Chairman

S. Stidham
H. C. Tijms & M. H. Van Hoorn
H. Daduna & R. Schassberger

SAMPLE-PATH ANALYSIS OF QUEUES

Shaler Stidham, Jr.

Abstract

In this paper we provide a survey of sample-path methods in queueing theory, particularly in connection with "distribution-free" analysis such as (i) relations between customer averages and time averages, such as $L = \lambda W$ (Little's formula); (ii) relations between the stationary distribution of a process and an imbedded process; and (iii) the phenomenon of insensitivity.

We compare these results to those obtained in the computer-science literature by operational analysis, in which behavior of processes over finite time intervals is emphasized. We study probabilistic assumptions under which relations between ensemble averages can be inferred from such finite-time behavior. (Without some probabilistic assumptions, operational analysis is limited to describing past behavior, rather than predicting future behavior.) The theory of random (marked) point processes provides a natural set of weak probabilistic assumptions. We discuss the relationship between the sample-path approach and the point-process approach.

1. Introduction

The analysis of sample-path behavior has always played an important role in the theory of stochastic processes. The relation between sample averages and ensemble averages (expectations) has been at the heart of the development of modern probability theory, as exemplified by the ergodic theorem. In applied probability careful study of sample-path properties had led to deeper understanding in such areas as Markov processes, weak convergence theory, bounds and approximations, and optimal design and control of queueing systems (Stidham (1970), Brumelle (1971), Weber (1980), Sobel (1981)). Recently, interest among computer scientists in sample-path behavior of queueing networks has been stimulated by the development and exploitation of "operational analysis" by Denning and Buzen (1978) and others.

The purpose of this paper is to provide a survey of a class of applications of sample-path analysis to queueing systems. In this context sample-path analysis will mean: the rigorous study of the properties of individual realizations of stochastic processes, including time-dependent and asymptotic frequencies and averages and the connection between them and corresponding probabilities and expectations, via laws of large numbers (ergodic theorems).

We shall focus our attention on the use of sample-path analysis to derive distribution-free results. Such results include:

(i) relations between time averages and customer averages, e.g., "Little's formula" ($L = \lambda W$);

(ii) relations between the distributions of a continuous-time process and an imbedded discrete-time process, e.g., the virtual and actual waiting-time processes;

(iii) the insensitivity phenomenon: the property that the equilibrium state distribution in certain stochastic systems (including many queueing networks) depends only on the means of certain random variables (such as service times).

A key to proving distribution-free properties of queueing systems by sample-path arguments is the relation $H = \lambda G$ (Stidham (1970), Brumelle (1971), Stidham (1979), Heyman and Stidham (1980)), a generalization of $L = \lambda W$. After reviewing the sample-path approach to $L = \lambda W$ (Section 2), we present a self-contained sample-path analysis of $H = \lambda G$, including some results that have not appeared previously in the open literature (Section 3). In Section 4, the sample-path approach to $H = \lambda G$ is compared to probabilistic methods, with particular emphasis on models based on the theory of random marked point processes (Matthes (1963 a,b), König and Matthes (1963), König, Matthes and Nawrotzki (1967), Franken (1976), Miyazawa (1979), Franken, König, Arndt, and Schmidt (1981), Rolski (1981)).

Section 5 contains several examples of distribution-free results that can be obtained via $H = \lambda G$. We compare these results and the methods used to obtain them to the corresponding methods and results from operational analysis in Section 6.

Although the relations and formulas derived by operational analysis are demonstrably valid, the approach has become controversial and is mistrusted by many theoreticians and practitioners because of the unorthodox definition of some of the quantities measured and the potential for misunderstanding and misuse as a predictive, rather than purely descriptive, tool. The missing link, if operational analysis is to be useful for predicting future behavior of stochastic systems, is a mechanism for connecting observed behavior over finite time intervals with behavior over infinite time intervals and with probabilistic properties of an existing or proposed system.

The missing link is provided by a rigorous analysis of asymptotic behavior on sample paths, together with ergodic theory, exemplified by what we have called <u>sample-path analysis</u>. The use of this approach to derive distribution-free results for queueing systems in fact predates the development of operational analysis by several years, as examplified by Stidham (1972b), which was by no means the first example of the rigorous use of a sample-path argument in queueing theory.

2. Background: Little's Formula on Sample Paths

The queueing formula, $L = \lambda W$, says that the average number of customers in the system (L) equals the average number arriving per unit time (λ) multiplied by the average time spent by a customer in the system (W). The key to the interpretation of the formula, and to the appropriate method of proof, lies in the meaning given to the words "average" and "system". In classical queueing theory, "average" is usually taken to be synonomous with "steady-state mean". The "system" may be the queue plus the service mechanism, or the queue by itself, or the service mechanism by itself.

Little (1961) was apparently the first to attempt a rigorous proof of $L = \lambda W$ for a general input-output system. He interpreted "steady-state" to mean that the stochastic processes involved (number in system, interarrival time, waiting time) are strictly stationary and metrically transitive. Hence the ergodic theorem applies, so that with probability one (e.g., for almost all realizations) the limiting average interarrival time and waiting time equal their respective means. This makes it possible to exploit deterministic relationships among the various processes, which are valid for each individual realization (sample path) of the evolution of the queueing system.

The importance of sample-path relationships was not brought into sharp focus, however, in Little's proof of $L = \lambda W$, (cf. also Brumelle (1971)), since the proof mixed sample-path arguments with arguments that depend on the probabilistic structure (strict stationarity) of the processes. The same is true of the later proof by Jewell (cf. also

Stidham (1970), (1972)) which depends on the system probabilistically regenerating itself at the start of each busy period. On the other hand, sample-path arguments of a heuristic nature have been almost obligatory in textbooks (cf. Gross and Harris (1974), Kleinrock (1975)) and have also appeared in journal articles (cf. Eilon (1969), Maxwell (1970)).

By a "sample-path" proof we mean one in which L, λ, and W are respectively interpreted as limiting time-average number of customers in the system, limiting average number of arrivals per unit time, and limiting customer-average time in system, and in which $L = \lambda W$ is shown to hold as a deterministic relation on each sample path, provided the sample path meets certain regularity conditions. In general, by its very nature a sample-path argument requires no probabilistic assumptions. Indeed, probabilistic assumptions are irrelevant, since by focussing attention on a particular sample path we are in effect assuming that the behavior of the system over time is completely known to us. It is as if we were standing at "time infinity", looking back at the complete history of the system, a history which is now deterministic precisely because it has already happened. In this sense, sample-path analysis is not really a part of probabilistic modeling, but an (often useful) accessory to it.

To make use of the sample-path version of $L = \lambda W$ in the context of the steady state in a probabilistic model, one must verify that the probabilistic assumptions of the model guarantee that, with probability one (i.e., for almost every sample path), (i) the limiting averages, L, λ, and W, exist, (ii) they coincide with the respective mean steady-state values, and (iii) the sample-path regularity conditions (if any) are satisfied. (See Section 4.)

Apparently the first published rigorous proof of the sample-path version of $L = \lambda W$ appeared in Newell (1971). Although it requires a regularity condition that turns out to be superfluous, the proof is simple and intuitive appealing. It uses the fact that, if the system is empty at time t, the total amount of waiting in the interval [0,t] can be computed in two ways: by integrating the number of customers in the system at time s over $s \in [0,t]$, or by adding the waiting times of all customers who have arrived in [0,t]. This immediately yields $L = \lambda W$, with the averages taken over the finite interval [0,t]. The validity of $L = \lambda W$ over $[0,\infty]$ follows, provided the limiting averages exist and the system returns to the empty state infinitely often.

If the system is not empty at time t, then the second method of computing the total waiting in [0,t] yields a larger value than the first. Most heuristic sample-path proofs of $L = \lambda W$ contain an (unproved) assertion that the difference between the two values is $o(t)$ as $t \to \infty$. (Eilon (1969) ignores the difference, in effect, by redefining W. This makes $L = \lambda W$ trivially true, but leaves the user with the equally difficult problem of showing that his definition and the usual one are equivalent. The same approach has been taken more recently in the setting of "operational analysis". (See Section 6 of this paper.)).

Newell's assumption that the system returns to the empty state infinitely often is satisfied by many queueing models[1], but it is an unnecessary restriction. This was shown by Stidham (1972b), (1974), who proved the sample-path version of $L = \lambda W$, under "minimal" assumptions: that the limiting averages exist and are finite.

The proofs in Stidham (1972b), (1974) were motivated by economic considerations (queueing design and control models) and perhaps also can be best understood by means of an economic interpretation. Suppose that a cost of one dollar is incurred for each time unit spent by each customer in the system. Let $I_n(t)$ be the rate at which customer n

$(n=1,2,\dots)$ incurs cost at time t $(t \geqq 0)$. Then

$$I_n(t) = \begin{cases} 1, & \text{if } t_n \leqq t < t_n + w_n \\ 0, & \text{otherwise,} \end{cases}$$

where t_n is the time of arrival of customer n $(0 \leqq t_n \leqq t_{n+1} < \infty,\ n \geqq 1)$ and

$$w_n = \int_0^\infty I_n(t)\, dt = \text{waiting time in system of customer } n \ (n \geqq 1),$$

$$\ell(t) = \sum_{n=1}^{\infty} I_n(t) = \text{number of customers in system at time } t \ (t \geqq 0),$$

$$a(t) = \sup\{n : t_n \leqq t\} = \text{number of arrivals in } [0,t] \ (t \geqq 0).$$

Define $U(t) := \sum_{n:t_n \leqq t} w_n$, $V(t) := \sum_{n:t_n + w_n \leqq t} w_n$, $t \geqq 0$. Thus $U(t)$ $(V(t))$ is the total cost that would be incurred in $[0,t]$ if all the costs associated with each customer were charged at the time of his arrival (departure). From this observation it follows that

$$U(t) \geqq \int_0^t \ell(s)\, ds \geqq V(t), \quad t \geqq 0. \tag{2.1}$$

(See Stidham (1974) for a more formal proof.)

Lemma 2.1. Suppose

$$w_n/t_n \to 0, \text{ as } n \to \infty . \tag{2.2}$$

Then the existence of any one of the limits $\lim_{t\to\infty} U(t)/t$, $\lim_{t\to\infty} \int_0^t \ell(s)ds/t$, $\lim_{t\to\infty} V(t)/t$, implies the existence of the other two and

$$\lim_{t\to\infty} U(t)/t = \lim_{t\to\infty} \int_0^t \ell(s)ds/t = \lim_{t\to\infty} V(t)/t.$$

Proof. From (2.1), it suffices to show that (2.2) implies that the limits of $U(t)/t$ and $V(t)/t$ must coincide, if either exists. Let $\varepsilon > 0$ be given. It follows from (2.2) that there exists an N such that $w_n < t_n\, \varepsilon$ for $n > N$. Then

$$U(t) = V(t) = \sum_{n:t_n+w_n \leqq t} w_n$$

$$\geqq \sum_{n:t_n(1+\varepsilon)\leqq t} w_n - \sum_{n\leqq N} w_n ,$$

so that

$$\lim_{t\to\infty} U(t)/t \geqq \lim_{t\to\infty} V(t)/t \geqq (1+\varepsilon)^{-1} \lim_{t\to\infty} U(t)/t.$$

The desired result is now immediate, since ε was arbitrary.

We shall also need the following lemma (cf. Lemma 3 of Stidham (1972b)).

Lemma 2.2. For any $\bar{\lambda}$, $0 \leqq \bar{\lambda} \leqq \infty$, the following are equivalent:

(i) $a(t)/t \to \bar{\lambda}$, as $t \to \infty$,

(ii) $t_n/n \to 1/\bar{\lambda}$, as $n \to \infty$.

Proof. See the proof of Lemma 3 of Stidham (1972b), which mimics the proof of the a.s. version of the elementary renewal theorem. There is a minor flaw however, in the proof that (i) implies (ii). The flaw occurs in the assertion that $a(t_n) = n$, which need not be true if there are batch arrivals (in which case we can only say that $a(t_n) \geqq n$). This difficulty can be overcome as follows:

Suppose that (i) holds. Suppose also that $t_n \to \infty$ as $n \to \infty$. (If not, then $\bar{\lambda} = \infty$ and $t_n/n \to 0 = 1/\bar{\lambda}$, so that (ii) holds trivially.) Let $\varepsilon > 0$ be given. Then

$$\lim_{n\to\infty} a(t_n)/t_n = \lim_{n\to\infty} a(t_n-\varepsilon)/t_n = \bar{\lambda}. \tag{2.3}$$

But, from the definition of a(t),

$$a(t_n-\varepsilon) \leqq n \leqq a(t_n). \tag{2.4}$$

Dividing each term in (2.4) by t_n and using (2.3) we conclude that $n/t_n \to \bar{\lambda}$, and $n \to \infty$, so that (ii) holds.

Condition (2.2) is central to the sample-path proof of $L = \lambda W$, as will be revealed in Theorem 2.3 below. Define the following limiting averages, when they exist:

$$\bar{\ell} := \lim_{t\to\infty} \int_0^t \ell(s)\,ds/t$$

$$\bar{\lambda} := \lim_{t\to\infty} a(t)/t$$

$$\bar{w} := \lim_{n\to\infty} \sum_{k=1}^{n} w_k/n.$$

Theorem 2.3. Suppose $\bar{\lambda} < \infty$, $\bar{w} < \infty$. Then $\bar{\ell}$ exists and $\bar{\ell} = \bar{\lambda}\,\bar{w}$.

Proof. It follows from $\bar{w} < \infty$ that $\lim_{n\to\infty} w_n/n = 0$. $\bar{\lambda} < \infty$ implies that $\lim_{n\to\infty} t_n/n = 1/\bar{\lambda} > 0$ (cf. Lemma 2.2). Combining these two results yields (2.2), so that Lemma 2.1 applies and $\bar{\ell} = \lim_{t\to\infty} U(t)/t$. But $\lim_{t\to\infty} U(t)/t = (\lim_{t\to\infty} \sum_{k=1}^{a(t)} w_k/a(t)) \cdot (\lim_{t\to\infty} a(t)/t) = \bar{\lambda}\,\bar{w}$, so that $\bar{\ell} = \bar{\lambda}\,\bar{w}$.

3. The Relation $H = \lambda G$

In the economic interpretation of $L = \lambda W$ we postulated that each customer incurs cost at the rate of one dollar per unit time in the system, so that the cost rate of customer n at time t is given by the function $I_n(t)$, which is simply the indicator of whether the customer is present. It is natural to think of more general cost-rate functions, for example, in queueing design and control models in which the cost incurred by a customer is a non-linear function of his time in the system. But the study of more general cost rates also leads to unexpected payoffs, in the form of sample-path versions of some well-known and some less familiar conservation relations, relations between continuous-time processes and imbedded discrete-parameter processes, and the insensitivity phenomenon. In this section we shall introduce a sample-path version of $H = \lambda G$: the relation between time averages and customer averages that results from allowing more general cost rates.[2]

The basic data for our model are the pairs $(t_n, f_n(\cdot))$ $(n = 1,2,\ldots)$ where $0 \leqq t_n \leqq t_{n+1} < \infty$, $t_n \to \infty$ as $n \to \infty$, and $f_n: \mathbb{R} \to \mathbb{R}$ is Lebesque integrable, for each $n \geqq 1$.

We interpret t_n as the time of arrival of customer n and $f_n(t)$ as the cost rate incurred by customer n at time t. Define

$$h(t): = \sum_{n=1}^{\infty} f_n(t), \quad t \in \mathbb{R} \tag{3.1.1}$$

$$g_n: = \int_0^{\infty} f_n(t)dt, \quad n \in \mathbb{N} \tag{3.1.2}$$

and assume that h is Lebesque - integrable on finite t - intervals. As before, define

$$a(t): = \sup \{n : t_n \leqq t\}$$

Finally, define the following limiting averages, when they exist:

$$\overline{h}: = \lim_{t\to\infty} \int_0^t h(s)ds/t$$

$$\overline{g}: = \lim_{n\to\infty} \sum_{k=1}^{n} g_k/n$$

$$\overline{\lambda}: = \lim_{t\to\infty} a(t)/t$$

We can write

$$\int_0^t h(s)ds = \int_0^t \sum_{k=1}^{\infty} f_k(s)ds$$

$$= \int_0^t \sum_{k=1}^{a(t)} f_k(s)ds + \int_0^t \sum_{k=a(t)+1}^{\infty} f_k(s)ds$$

$$= \int_0^{\infty} \sum_{k=1}^{a(t)} f_k(s)ds + \int_0^t \sum_{k=a(t)+1}^{\infty} f_k(s)ds - \int_t^{\infty} \sum_{k=1}^{a(t)} f_k(s)ds$$

so that

$$\int_0^t h(s)ds = \sum_{k=1}^{a(t)} g_k + r(t) \tag{3.2}$$

where

$$r(t) := \int_0^t \sum_{k=a(t)+1}^{\infty} f_k(s)ds - \sum_{k=1}^{a(t)} \int_t^{\infty} f_k(s)ds.$$

From (3.2) it follows that

$$\int_0^t h(s)ds/t = (a(t)/t) \cdot (\sum_{k=1}^{a(t)} g_k/a(t)) + r(t)/t.$$

Since $t_n < \infty$ for all $n \geqq 1$ implies $a(t) \to \infty$ as $t \to \infty$, we have

Theorem 3.1. Suppose $\overline{\lambda}$ and $\overline{g}$ exist. Then $\overline{h}$ exists and $\overline{h} = \overline{\lambda}\,\overline{g}$ iff $r(t)/t \to 0$ as $t \to \infty$.

Remarks.

(i) The requirement in Theorem 3.1 that $\overline{\lambda}$ and $\overline{g}$ exist permits $\overline{\lambda} = \infty$ and $\overline{g} = \pm\infty$.

(ii) The condition that $t_n < \infty$ (and hence $a(t) \to \infty$) is implied by $\overline{\lambda} > 0$. The condition that $t_n \to \infty$ is implied by $\overline{\lambda} < \infty$.

Theorem 3.1 reveals that $r(t)/t \to 0$ is the weakest possible sufficient condition for $\overline{h} = \overline{\lambda}\,\overline{g}$, when the limits exist. In applications, however, it is often convenient to work with stronger sufficient conditions that are easier to verify. To this end define

$$r_n := \int_0^{t_n} \sum_{k=n}^{\infty} f_k(s)ds - \sum_{k=1}^{n-1} \int_{t_n}^{\infty} f_k(s)ds.$$

and note that

$$r(t) = r_{a(t)+1} - \int_t^{t_{a(t)+1}} \sum_{k=1}^{\infty} f_k(s)ds. \tag{3.3}$$

Corollary 3.2. Suppose $\overline{\lambda} < \infty$ and $\overline{g}$ exist. If

$$r_n/n \to 0, \tag{3.4.1}$$

$$\int_{t_n}^{t_{n+1}} \left| \sum_{k=1}^{\infty} f_k(s) \right| ds/n \to 0, \tag{3.4.2}$$

as $n \to \infty$, then $\overline{h}$ exists and $\overline{h} = \overline{\lambda}\,\overline{g}$.

Proof. Divide both sides of the equality in (3.3) by t and then multiply the right-hand side by $a(t)/a(t)$. Since $a(t) \to \infty$ and

$a(t)/t \to \bar{\lambda} < \infty$ as $t \to \infty$, (3.4.1) and (3.4.2) imply $r(t)/t \to 0$. Now apply Theorem 3.1.

Corollary 3.3. Suppose $\bar{\lambda} < \infty$ and $\bar{g}$ exist. If

$$\int_0^{t_n} \Big| \sum_{k=n}^{\infty} f_k(s) \Big| ds/n \to 0 \tag{3.5.1}$$

$$\int_{t_n}^{\infty} \Big| \sum_{k=1}^{n} f_k(s) \Big| ds/n \to 0 \tag{3.5.2}$$

as $n \to \infty$, then $\bar{h}$ exists and $\bar{h} = \bar{\lambda}\,\bar{g}$.

Proof. Obviously, (3.5.1) - (3.5.2) imply (3.4.1). It suffices to prove that they also imply (3.4.2). Now

$$\int_{t_n}^{t_{n+1}} \Big| \sum_{k=1}^{\infty} f_k(s) \Big| ds \leqq \int_{t_n}^{\infty} \Big| \sum_{k=1}^{n} f_k(s) \Big| ds + \int_0^{t_{n+1}} \Big| \sum_{k=n+1}^{\infty} f_k(s) \Big| ds. \tag{3.6}$$

Each of the two terms on the right-hand side of (3.6) is $o(n)$, by (3.5.2) and (3.5.1) respectively.

If $f_k \geqq 0$ for all k, then the above results can be sharpened. In particular, we have the following corollary, which contains Theorem 1 of Brumelle (1971).

Corollary 3.4. Suppose $f_k \geqq 0$ for all k, and suppose $\bar{\lambda} < \infty$ and $\bar{g}$ exist. Then $\bar{h}$ exists and $\bar{h} = \bar{\lambda}\,\bar{g}$ iff $r_n/n \to 0$ as $n \to \infty$.

Proof. When $f_k \geqq 0$, it can easily be shown that $r_n/n \to 0$ iff $r(t)/t \to 0$.

So, in the case where $f_k \geqq 0$ for all k, establishing $\bar{h} = \bar{\lambda}\,\bar{g}$ reduces to verifying the key statement $r_n/n \to 0$. But even this may be quite difficult in applications. More easily verified, yet widely applicable, sufficient conditions would be useful to have. Stidham (1972b) and Heyman and Stidham (1980) intoduced the following condition, which is a natural generalization of the condition that $w_n/n \to 0$ for the special case of $\bar{\ell} = \bar{\lambda}\,\bar{w}$:

There exists a sequence $\{z_n, n = 1,2,...\}$ of non-negative numbers such that $z_n/n \to 0$, as $n \to \infty$, and $f_n(t) = 0$ for $t \notin [t_n, t_n + z_n]$. (3.7)

Theorem 3.5. Suppose $f_k \geqq 0$ for all k, that $\overline{\lambda} < \infty$ and $\overline{g}$ exist, and that condition (3.7) holds. Then $\overline{h}$ exists and $\overline{h} = \overline{\lambda}\,\overline{g}$.

Proof. An easy proof, mimicking the proof of $\overline{\ell} = \overline{\lambda}\,\overline{w}$ based on condition (2.5), may be found in Heyman and Stidham (1980). Alternatively, one can verify that (3.7) implies (3.5.1) - (3.5.2) and then use Corollary 3.3.

The virtue of Theorem 3.5, as we shall see in the discussion of applications in Section 5, is that it provides a sufficient condition for $\overline{h} = \overline{\lambda}\,\overline{g}$ that (i) is satisfied in typical applications, (ii) is easily checked, and (iii) often leads to probabilistic versions with weaker regularity conditions than those used in a direct approach based on (3.5.1) - (3.5.2).

4. Relation to Probabilistic Models; the RMPP Approach

Now consider a stochastic input-output system defined on some probability space (Ω, F, P_o). Specifically, for each $\omega \in \Omega$, let $t_n(\omega)$ be the time of arrival of customer n, and $f_n(t,\omega)$ the cost rate incurred by customer n at time t. Assume that the bivariate sequence $\{t_n(\omega), f_n(\cdot,\omega))\}$ satisfies the conditions of Section 3, with probability one (P_o). Similarly define $h(t,\omega)$, $g_n(\omega)$, and $r_n(\omega)$.

The results of Section 3 can now be applied to each sample path $\omega \in \Omega$ of this stochastic system. In particular we have as a consequence of Corollary 3.4:

Theorem 4.1. Suppose that, with probability one (P_o), $f_k(\cdot,\omega) \geqq 0$ for all k, and

$$t_n(\omega)/n \to 1/\overline{\lambda}(\omega) > o \qquad (4.1.1)$$

$$\sum_{k=1}^{n} g_k(\omega)/n \to \bar{g}\ (\omega) \qquad (4.1.2)$$

$$r_n(\omega)/n \to o \qquad (4.1.3)$$

as $n \to \infty$. Then, with probability one (P_o),

$$\int_o^t h(s,\omega)\ ds/t \to \bar{h}(\omega) = \bar{\lambda}(\omega)\bar{g}(\omega).$$

Moreover, if $\bar{\lambda}(\omega) = \lambda$ with probability one (P_o), and $H = E_{P_o}\ [\bar{h}(\omega)]$ and $G = E_{P_o}\ [\bar{g}(\omega)]$ exist, then

$$H = \lambda G.$$

Corollary 4.2. The conclusions of Theorem 4.1 hold if (4.1.3) is replaced by:

There exists a sequence $\{z_n(\omega),\ n=1,2,\ldots\}$ such that $f_n(t,\omega) = 0$ for $t \notin [t_n(\omega),\ t_n(\omega) + z_n(\omega)]$ and $z_n(\omega)/n \to 0$, as $n \to \infty$, with probability one (P_o). (4.1.4)

Natural sufficient conditions for the existence of the limits in (4.1.1) and (4.1.2) are strict stationarity or the existence of regeneration points (see Stidham (1972b), Heyman and Stidham (1980)).

A quite different approach to $L = \lambda W$ and $H = \lambda G$, using purely probabilistic arguments and avoiding completely the analysis of sample paths, has been developed by Franken, Rolski, Miyazawa, and others, based on the theory of stationary random marked point processes (RMPP). We sketch the RMPP approach below, referring the reader interested in details to the monographs of Franken, König, Arndt, and Schmidt (1979) and Rolski (1981) (for the RMPP approach to $L = \lambda W$) and to Miyazawa (1979) and Stidham (1979) (for the RMPP approach to $H = \lambda G$).

A random marked point process (RMPP) is a bivariate sequence $\emptyset = \{(t_n,\ k_n),\ n = 0,\ \pm 1,\ \pm 2,\ldots\}$, where $\{t_n\}$ is an increasing sequence of random time points, labeled so that $t_o \leq 0 < t_1$, and $\{k_n\}$ is a sequence of random elements from a measurable space.[3] Each t_n records the time at which a certain event occurs and k_n records some information about the event. For example, t_n could be the arrival time of customer n, and k_n could be a characteristic such as the class or service time of customer n.

The theory of RMPP's has been applied to queueing models to derive all three types of distribution-free results mentioned in the introduction. The approach depends crucially on the notion of a stationary distribution P and its associated Palm distribution P_o. Roughly speaking, the stationary distribution P describes invariance under arbitrary time shifts, whereas P_o is defined only on sequences $\{(t_n, k_n)\}$ with $t_o = 0$ and describes invariance under shifts by an amount t_1. P_o may be interpreted as giving the conditional state distribution of the RMPP, given that an event occurs at time 0. Thus, the Palm distribution provides a natural mechanism for studying the behavior of a discrete-parameter process imbedded in a continuous-time process.

For several well-known queueing systems in which the bivariate sequence $\{(t_n, S_n)\}$ of arrival times and service times is a RMPP, Franken (1976) and others have proved a stationary version of $L = \lambda W$:

$$E_P[\ell(0)] = \lambda_P E_{P_o}[w_o]. \tag{4.2}$$

Here $\ell(0)$ is the number in the system at time $t = 0$, w_o is the sojourn time of customer 0, and λ_P is the intensity of the RMPP under the stationary distribution P (in this case, the mean arrival rate of customers). The proof of (4.2) depends on Campbell's theorem for RMPP's and the inversion formula for Palm probabilities; cf. Franken et al. (1979).

The same approach can be used (cf. Miyazawa (1979), Stidham (1979)) to prove a stationary version of $H = \lambda G$. Let $\{(t_n, f_n(\cdot)),$ $n = 0, \pm 1, \pm 2, \ldots\}$ be a bivariate random sequence satisfying (3.1.1) and (3.1.2). Define the functions k_n: $\mathbb{R} \to \mathbb{R}$, for $n = 0, \pm 1, \pm 2, \ldots$, by

$$k_n(\cdot) := f_n\ (t_n + \cdot), \tag{4.3}$$

so that $k_n\ (t - t_n)$ is the "cost rate" for customer n at time t. The following theorem is proved in Stidham (1979):

Theorem 4.3. Suppose $\emptyset = \{(t_n, k_n), n = 0, \pm 1, \pm 2, \ldots\}$ is a RMPP with stationary probability measure P and corresponding Palm probability measure P_o, where k_n is defined by (4.3). Let $h(t)$, $t \in \mathbb{R}$, and g_n, $n = 0, \pm 1, \pm 2, \ldots$, be defined by (3.1.3) and (3.1.4). Then

$$E_P\ [h(0)] = \lambda_P\ E_{P_o}\ [g_o]. \tag{4.4}$$

Now, with $T_n := t_{n+1} - t_n$, it follows from the definition of the measure P_o that the bivariate random sequence $\{(T_n, k_n), n = 0, \pm 1, \pm 2, \ldots\}$ is strictly stationary with respect to P_o and, in addition $\lambda_P = (E_{P_o}\ [T_o])^{-1}$ (see Stidham (1979), Franken et al. (1979)). Hence, both $\{T_n\}$ and $\{g_n\}$ are strictly stationary (P_o) and the ergodic theorem implies that $t_n/n \to E_{P_o}\ [T_o]$ and $\sum_{k=1}^{n} g_k/n \to E_{P_o}\ [g_o]$ as $n \to \infty$, with probability one (P_o), provided both sequences are also metrically transitive.

For the case $f_n \geqq 0$, the following theorem proved in Stidham (1979) shows that strict stationarity of $\{(T_n, k_n)\}$ is by itself sufficient to yield the sample-path version of $H = \lambda G$.

Theorem 4.1. Suppose $f_n \geqq 0$ and $\{(T_n, k_n), n = 0, \pm 1, \pm 2, \ldots\}$ is strictly stationary (P_o), where k_n is defined by (4.3). Then each of the conditions (3.5.1) - (3.5.4) holds with probability one (P_o).

This result shows that the sample-path approach to proving $\overline{h} = \overline{\lambda}\ \overline{g}$ is more powerful than the RMPP approach, in the sense that its conditions are weaker. Note also that Theorem 4.4 does not require metric transitivity of $\{(T_n, k_n)\}$.

5. Distribution-Free Results via $H = \lambda G$

In this section we illustrate the power of sample-path analysis by using the sample-path version of $H = \lambda G$ to derive several distribution-free results for queueing systems.

1. Our first example is from Brumelle (1971). (See also Wolff (1969), Stidham (1972a), Franken et al. (1979)). Consider a G/G/c queueing system with input $\{(T_n, S_n)\}$. Here $T_n = t_n - t_{n-1}$ and S_n is the service requirement (work to be done) for customer n. Let d_n denote the delay (time in queue) of customer n. Define $f_n(t)$ as follows:

$$f_n(t) := \begin{cases} S_n, & \text{if } t_n \leq t \leq t_n + d_n \\ S_n - (t - t_n - d_n), & \text{if } t_n + d_n \leq t < t_n + d_n + S_n \\ 0, & \text{otherwise.} \end{cases} \tag{5.1}$$

Thus, $f_n(t)$ is the work of customer n that is in the system at time t. An amount of work, S_n, enters the system at time t_n, remains constant until customer n enters service (at time $t_n + d_n$), then decreases at unit rate while customer n is in service.

With this definition of $f_n(t)$, $h(t) = \sum_{n=1}^{\infty} f_n(t)$ is the total amount of work in the system at time t (the virtual waiting time if c = 1 and the discipline is FIFO) and $g_n = \int_0^\infty f_n(t)dt = S_n d_n + 1/2\, S_n^2$, the total "cost" associated with customer n, assuming cost is incurred at a rate equal to the amount ofthe customer's work that is in the system.

Theorem 5.1. For the G/G/c system with f_n defined by (5.1), suppose that

$$t_n/n \to 1/\overline{\lambda} > 0 \tag{5.2.1}$$

$$\sum_{k=1}^{n} S_k/n \to \overline{S} < \infty \tag{5.2.2}$$

$$\sum_{k=1}^{n} d_k/n \to \bar{d} < \infty \tag{5.2.3}$$

as $n \to \infty$. Then

$$\bar{g} = \lim_{n\to\infty} \sum_{k=1}^{n} [S_k\, d_k + 1/2\; S_k^{\,2}]/n$$

exists, $\bar{h} = \lim_{t\to\infty} \int_0^t h(s)\, ds/t$ exists, and $\bar{h} = \bar{\lambda}\;\bar{g}$.

Proof. With $z_k := d_k + S_k$, (5.2.2) and (5.2.3) imply (3.7), so that the desired result follows from Theorem 3.5.

Theorem 5.1 describes behavior along a particular sample path. A probabilistic counterpart is given by the following corollary (cf. Theorem 6 of Brumelle (1971)).

Corollary 5.2. Suppose a G/G/c system has strictly stationary, metrically transitive input $\{(T_n, S_n), n = 0, \pm 1, \pm 2, \ldots\}$ such that $\infty > E\,[T_0] = 1/\lambda > 0$ and $E\,[S_0] < c \cdot E\,[T_0]$. Then a strictly stationary version of $\{d_n\}$ exists and, if $E\,[d_0] < \infty$, it follows that $\bar{h} = \lambda\bar{g}$ with probability one and $H = \lambda G$, where $H = E\,[\bar{h}]$ and $G = E\,[\bar{g}]$.

Proof. The existence of a stationary version of $\{d_n\}$ follows from Loynes (1962). That conditions (5.2.1) - (5.2.3) hold with probability one follows from the assumptions concerning $E\,[T_0]$, $E\,[S_0]$, and $E\,[d_0]$.

It should be noted that our requirement that $E\,[d_0] < \infty$ is considerably weaker than Brumelle's conditions, which involve finiteness of second moments. This is an illustration of the advantage of working with the sufficient condition (3.7), rather than trying to verify $r_n/n \to 0$ directly as Brumelle does.

Corollary 5.3. Consider a G/G/c system in which S_0 and d_0 are independent, in addition to the assumptions of Corollary 5.2. Then, with probability one,

$$H = \lambda\; [E\,[S_0]\; E\,[d_0] + 1/2\; E\,[S_0^{\,2}]]. \tag{5.3}$$

The relation (5.3) has been put to many uses. In GI/GI/1 FIFO queues it gives a relation between the mean steady-state virtual and actual waiting times in the queue. In M/G/1 queues, in which virtual and actual waiting times have the same distribution (so that $E[d_o] = H$), it yields a simple proof of the Pollaczek-Khintchine formula: $E[d_o] = \lambda E[S_o^2]/ 2(1-\lambda E[S_o])]$. This same approach has been used to obtain formulas for the mean delays in fixed-priority queues (Wolff (1969)), alternating-priority queues (Stidham (1972a)), more complicated systems with priorities and feedback (Simon (1981)), and in mean-value analysis of queueing networks (Reiser (1979)).

2. To motivate our second example, consider an open network of queues, with nodes labeled $m = 1,2,\ldots,M$. Let $\bar{\lambda}$ be the arrival rate to the network. Let $\bar{\ell}_m$ denote the long-run average number of customers at node m. Often (e.g., in a Jackson network) $\bar{\ell}_m$ can be readily computed. A particular arrival to the system may make several distinct visits to node m before leaving the system. We would like to compute $\bar{w}_m$, the long-run average total time spent by a customer in node m (summed over all his visits to the node). If the relation $\bar{\ell}_m = \bar{\lambda}\,\bar{w}_m$ holds, then we can use it to derive $\bar{w}_m$.

The problem is, although $\ell_m = \bar{\lambda}\,\bar{w}_m$ looks very much like an example of $L = \lambda W$, the "system" in question (node m) does not satisfy the conditions needed in our sample-path approach to $L = \lambda W$ (Section 2), when $\bar{\lambda}$ and $\bar{w}_m$ are defined as we have defined them. Specifically, condition (2.3) implicitly requires that the sojourn of each customer in "the system" be uninterrupted. Multiple distinct visits are not permitted. This is not a superfluous condition, as the counterexamples in Stidham (1978) and Heyman and Stidham (1980) indicate.

It might be argued that the prohibition on multiple visits is not really a restriction on the applicability of $L = \lambda W$, since in the case of multiple visits by a customer, that customer should be treated as a new "arrival" each time he visits. If this convention is followed, then condition (2.3) once again applies and $L = \lambda W$ holds.

While restoring the faith of the true believer in $L = \lambda W$, this convention is of no use to us in solving the problem at hand, since it requires us also to redefine "sojourn time" as the time spent in the system per visit. Hence we cannot use such a version of $L = \lambda W$ to calculate $\overline{w}_m$ - the total time spent at node m.

Theorem 3.5 provides a way out of this dilemma, since we need only find a bound z_n on the time from customer n's first entrance into node m until his last departure, such that $z_n/n \to 0$. The easiest way to do this is to use the sojourn time of customer n in the entire network. As long as the long-run average over all customers of this sojourn time is finite, condition (3.7) will be satisfied and hence we will know that $\overline{\ell}_m = \overline{\lambda}\,\overline{w}_m$.

3. Our third example illustrates how $H = \lambda G$ can be used to get relations between the distribution of a continuous-time process and that of an imbedded discrete-time process. This example is a generalization of one given in Heyman and Stidham (1980). For a more detailed analysis, see Stidham (1981).

Consider an input-output system with the property that discrete units (customers) arrive and depart one at a time. The system is said to be in state k when there are k units in the system, $k = 0,1,\ldots$. As usual denote the customer arrival time sequence by $\{t_n\}$ and assume it satisfies the conditions of Section 3. Define the following quantities, when they exist ($k = 0,1,\ldots$):

$$\overline{\lambda}: = [\lim_{n\to\infty} t_n/n]^{-1}$$

$$p_k: = \lim_{t\to\infty} (\text{amount of time that system is in state k during } [0,t])/t$$

π_k: $= \lim_{n\to\infty}$ (number of arrivals that find system in state k among the first n arrivals)/n

r_k: $= \lim_{t\to\infty}$ (number of departures from system while in state k during [0,t]/(time spent in state k during [0,t]).

For obvious reasons, r_k is called the departure rate in state k.

We can use $\overline{h} = \overline{\lambda}\,\overline{g}$ to obtain a relation between $\{p_k\}$ and $\{\pi_k\}$ as follows:

Fix a state $k \geqq 1$ and define $t_m(k-1)$: = time of the m-th arrival (after t = 0) to find the system in state k - 1, $m \geqq 1$, $k \geqq 1$. Define the functions $f_m(t)$ ($m \geqq 1$, $t \geqq 0$) by

$$f_m(t) := \begin{cases} 1, & \text{if } t_m(k-1) \leqq t < t_{m+1}(k-1) \text{ and system is in state k} \\ & \quad \text{at time t} \\ 0, & \text{otherwise} \end{cases}$$

Then $h(t) = \sum_{m=1}^{\infty} f_m(t) = \underset{\sim}{1}$ (system is in state k at time t) for $t \geqq 0$, and $\overline{h} = p_k$, if the limit exists. Also, $g_m = \int_0^\infty f_m(t)dt$ = length of time system spends in state k between $t_m(k-1)$ and $t_{m+1}(k-1)$, $m \geqq 1$. It can easily be shown that

$$\overline{g} = \lim_{m\to\infty} \sum_{j=1}^{m} g_j/n = r_k^{-1} \tag{5.4}$$

and

$$\lim_{m\to\infty} t_m(k-1)/m = (\overline{\lambda}\,\pi_{k-1})^{-1} \tag{5.5}$$

if the limits exist.

Theorem 5.4. Suppose $0 < \overline{\lambda} < \infty$ and that π_{k-1} and p_k exist. Then

$$r_k\, p_k = \overline{\lambda}\,\pi_{k-1}. \tag{5.6}$$

Proof. Use (5.4), (5.5), and Theorem 3.5, with $z_n = t_{n+1}(k-1) - t_n(k-1)$.

The key to the usefulness of (5.6) in applications lies in the evaluation of r_k.

a) Suppose the departure process is a pure death process, with state-dependent death rate μ_k, $k = 0,1,\ldots$. Then $r_k = \mu_k$ and (5.6) becomes

$$\bar{\lambda}\ \pi_{k-1} = \mu_k\ p_k$$

In the special case of a GI/M/c queue with mean interarrival time λ^{-1} and mean service rate μ at each server, this further simplifies to

$$\lambda\ \pi_{k-1} = \min\ (k,c)\ \mu\ p_k,$$

as in Heyman and Stidham (1980).

b) When the arrival process is Poisson with mean rate λ, we can use the property that "Poisson arrivals see time averages" (see Wolff (1981) and references therein) to conclude that $\pi_{k-1} = p_{k-1}$ and hence rewrite (5.6) as

$$\lambda\ p_{k-1} = r_k\ p_k.$$

This same "birth-death" type equation has been derived by "operational analysis" (see Denning and Buzen (1978) and Section 6 of this paper). The operational-analysis argument for (5.7) is based on "flow-balance" considerations rather than $H = \lambda G$. The equation (5.7) can be used to solve recursively for p_k in terms of p_0:

$$p_k = \lambda^k\ (\sum_{j=1}^{k} r_j)^{-1}\ p_0.$$

The difficulty with this approach is that it is not usually possible to calculate the departure rates $\{r_k\}$, independently of $\{p_k\}$. (The significant exception, of course, is the special case of an M/M/1 queue, where $r_k \equiv \mu$. See also Marie (1980) and Marie and Stewart (1980) for applications to systems with Coxian service times.) In fact, (5.7) may be viewed as an alternative definition of r_k. The option of using the original sample-path definition of r_k is not really satisfactory, since it depends on the structural characteristics (e.g., the arrival rate) for the observed system, and hence has no predictive value when these characteristics change. These issues are discussed further in the next section.

c) Consider a single-server queue with arrival rate $\overline{\lambda}$, mean service time μ^{-1}, and LCFS-PR queue discipline. In this case, it can be shown (Stidham (1981)) that $r_k = \mu$ and hence (5.6) becomes

$$\overline{\lambda}\ \mu_{k-1} = \mu\ p_k \tag{5.8}$$

In the case of Poisson arrivals, (5.8) reduces to

$$\lambda\ p_{k-1} = \mu\ p_k .$$

Thus, we have a sample-path proof of the insensitivity and product form of the equilibrium probabilities in the M/G/1 (LCFS-PR) system. Efforts to extend this method to prove insensitivity for other disciplines (such as processor-sharing) have so far been unsuccessful.

An alternate sample-path approach to insensitivity is given in Stidham (1976) for finite-source (Engset) and infinite-server systems. Traditional approaches to insensitivity rely on reversibility (Kelly (1976), (1979)) or local balance. The latter approach was pioneered by König and Matthes (1963) and König, Matthes and Nawrotzki (1967), using the theory of RMPP, for a very general class of queueing models. Schassberger (1978a), (1978b) developed an essentially equivalent class of models, called <u>generalized semi-Markov schemes</u> (GSMS), and gave alternative proofs of insensitivity, based on phase-type approximations of arbitrary distributions.

Schassberger also showed that a general Jackson network (Jackson (1963), Lemoine (1977) is a special case of a GSMS. Jansen and König (1980) also showed how the RMPP-GSMS approach specializes to queueing networks, yielding insensitivity results derived earlier by Baskett et al. (1975), Chandy et al. (1979), and others under stronger assumption of independent random variables.

6. Comparison with Operational Analysis

Denning and Buzen (1978) (see also the references cited therein for earlier work) proposed a technique which they call operational analysis, for the study of queues. Operational analysis involves recording the operating characteristics of a stochastic system over a finite time interval [0,T] in terms of such sample measures as the arrival rate (number of arrivals in [0,T] divided by T), the departure rate (number of service completions in [0,T] divided by T), server utilization (busy time divided by T), and average service time (busy time divided by number of service completions).

Based on these measures Denning and Buzen derive a finite-time-interval version of $L = \lambda W$. They also extend the analysis to networks of queues, deriving a sample-path version of the traffic equation.

It should be noted that, although the version of $L = \lambda W$ derived by Denning and Buzen holds under all conditions, their definitions of such quantities as average service time and average sojourn time only coincide with the standard definitions of customer averages if the system begins and ends the interval [0,T] in the empty state. As in Eilon (1969), the non-standard definitions of certain averages in Denning and Buzen (1978) make the proofs of such relations as $L = \lambda W$ trivial. The problem, as pointed out in Stidham (1972b), then becomes not one of proving the relation but rather one of showing that in the limit, as $T \to \infty$, the non-standard definition agrees with the standard one.

Denning and Buzen (1978) also show that the relative frequencies of the various states over the interval [0,T] satisfy global balance equations of the same form as those satisfied by the equilibrium probabilities, with the conditional flow rates from one state to another defined as the sample-average number of transitions per unit time spent in the first state during [0,T]. The global balance equations hold exactly under the condition of job-flow balance; i.e., when the system begins and ends the interval [0,T] in the same state, e.g., the empty state.

Under three additional assumptions - one step behavior, device homogeneity, and routing homogeneity - Denning and Buzen verify that the global-balance equations have a product-form solution. This is a remarkable result, in that it does not require any explicit probabilistic assumptions or restrictions on the queue discipline (cf. some of the results in Section 5).

In Buzen and Denning (1980) operational analysis is used to derive relations between the queue-length distributions as seen by an arriving customer and by an outside observer, both for the case of individual queues and for closed networks with product-form equilibrium distributions.

Thus operational analysis has been used to derive distribution-free results of all three types cited above.

Like operational analysis, sample-path analysis begins by observing the behavior of a system over a finite time interval. In the sample-path approach, however, the analysis is explicitly extended to asymptotic behavior over an infinite time horizon, and from there to stochastic (ensemble) behavior for systems satisfying specific probabilistic assumptions.

It remains to develop an exhaustive study of the relationship between sample-path analysis and operational analysis, and in particular to elucidate further the implications for stochastic modeling of such assumptions from operational analysis as device homogeneity. Some careful analysis along these lines has already been carried out by, among others, Bryant (1981) and Hofri (1979), who point out the dependence of the operationally defined departure rate on other system parameters, such as the arrival rate. (See also the discussion of Example 3 in the previous section.) This dependence calls into question the usefulness of operational analysis as a tool for predicting future behavior of an existing or proposed system. Moreover, the homogeneity assumption (on-line behavior - off-line behavior) required to circumvent this deficiency has been shown by Bryant to hold asymptotically with probability one only if the arrival process is Poisson. The studies in Bryant (1981) and Hofri (1979) are carried out in terms of simple models such as M/G/1. Using the tools of sample-path analysis, we hope to pursue these topics in the future in a more general setting, again without the necessity of restriction to a particular class of stochastic models.

Closely related to operational analysis is mean-value analysis (cf. Lavenberg and Reiser (1979), Reiser (1979), (1981), Reiser and Lavenberg (1980), Schweitzer (1981) and papers cited therein), a technique, based on derivations from the product-form equilibrium probabilities, for recursively computing mean sojourn times and response times in Jackson networks. Buzen and Denning (1980) have given operational-analysis derivations of some of the same properties. Here again, we expect that a careful use of sample-path analysis will make clear the boundaries of applicability of such techniques, as well as provide rigorous derivations that are easier to understand than those based on RMPP theory, and do not depend on detailed stochastic assumptions.

We hope to make further investigations along these lines in a future paper.

7. References

[1] Baskett, F., Chandy, J. M., Muntz, R. R., and Palacios, J. (1975). Open, Closed, and Mixed Networks with Different Classes of Customer. J. Assoc. Comput. Mach., 22, pp. 248-260.

[2] Brumelle, S. (1971). On the Relation Between Customer and Time Averages in Queues. J. Appl. Probab., 8, pp. 508-520.

[3] Bryant, R. (1981). On Homogeneity and On-Line = Off-Line Behavior in M/G/1 Queueing Systems. IEEE Trans. Software Engrg. SE-7.

[4] Buzen, J. and Denning, P. J. (1980). Measuring and Calculating Queue Length Distributions. IEEE Trans. Comput., C-18, pp. 33-44.

[5] Chandy, K. M., Howard, J. H., and Towsley, D. F. (1977). Product Form and Local Balance in Queueing Networks. J. Assoc. Comput. Mach., 24, pp. 250-263.

[6] Denning, P. J., and Buzen, J. P. (1978). The Operational Analysis of Queueing Network Models. Comput. Surveys, 10, pp. 225-261.

[7] Eilon, S. (1969). A Simpler Proof of $L = \lambda W$. Oper. Res., 17, pp. 915-917.

[8] Franken, P. (1976). A New Approach to Investigation of Stationary Queueing Systems. Preprint No. 6/76, Sektion Mathemetik der Humboldt Universität, Berlin.

[9] Franken, P., König, D., Arndt, U., and Schmidt, V. (1981). Queues and Point Processes. Akademie-Verlag, Berlin. J. Wiley and Sons, New York.

[10] Gross, D., and Harris, C. (1974). Fundamentals of Queueing Theory. J. Wiley and Sons, New York.

[11] Heyman, D. P., and Stidham, S., Jr. (1980). The Relation Between Customer and Time Averages in Queues. Oper. Res., 28, pp. 983-994.

[12] Hofri, M. (1979). The Output Rate of Single-Server Devices under Inhomogeneous Poisson Load. Technical Report No. 150, Faculty of Computer Sciences, Technion-IIT, Haifa, Israel.

[13] Jackson, J. R. (1963). Jobshop-like Queueing Systems. Management Sci., 10, pp. 131-142.

[14] Jansen, U., and König, D. (1980). Insensitivity and Steady-State Probabilities in Product Form for Queueing Networks. Elektron. Informationsverab. Kybernet., EIK 16, pp. 385-397.

[15] Jewell, W. (1967). A Simple Proof of: $L = \lambda W$. Oper. Res., 15, pp. 1109-1116.

[16] Kelly, F. P. (1976). Networks of Queues. Adv. in Appl. Probab., 8, pp. 416-432.

[17] Kelly, F. P. (1979). Reversibility and Stochastic Networks. J. Wiley and Sons, New York.

[18] Kleinrock, L. (1975). Queueing Systems, Vol. 1: Theory. J. Wiley and Sons, New York.

[19] König, D., and Matthes, K. (1963). Verallgemeinerungen der Erlangschen Formeln. Math. Nachr., 26, pp. 45-56.

[20] König, D., Matthes, K., and Nawrotzki, K. (1967), "Verallgemeinerungen der Erlangschen und Engsetschen Formeln," (Eine Methode in der Bedienungstheorie), Akademie-Verlag, Berlin.

[21] Lavenberg, S. and Reiser, M. (1979). Stationary State Probabilities at Arrival Instants for Closed Queueing Networks with Multiple Types of Customers. Report RC7592, IBM Research Center, Yorktown Heights, New York.

[22] Lemoine, A. (1977). Networks of Queues -- A Survey of Equilibrium Analysis. Management Sci., 24, pp. 464-481.

[23] Little, J. (1961). A Proof of the Queuing Formula: $L = \lambda W$. Oper. Res., 9, pp. 383-387.

[24] Loynes, R. (1962). The Stability of a Queue with Non-Independent Interarrival and Service Times. Proc. Cambridge Phil. Soc., 58, pp. 497-520.

[25] Marie, R. (1980). Calculating Equilibrium Probabilities in $\lambda(n)/C_k/1/N$ Queues. In Proc. Performance '80, G. S. Graham (Ed.), Toronto, Canada.

[26] Marie, R. and Stewart, W. (1980). Maintenance-Like Queueing Systems with General Service-Time Distributions. Department of Computer Science, North Carolina State University.

[27] Matthes, K. (1963a). Stationare Zufällige Puntfolgen. I. Jber. Deutsch. Math.-Verein, 66, pp. 66-79.

[28] Matthes, K. (1963b). Unbeschränkt Teilbare Verteilungsgesetze Stationärer Zufälliger Puntfolgen. Wiss. Z. Tech. Hochsch. Electro. Ilmenau, 9, pp. 235-238.

[29] Maxwell, W. (1970). On the Generality of the Equation L = λW. Oper. Res., 18, pp. 172-174.

[30] Miyazawa, M. (1979). A Formal Approach to Queueing Processes in the Steady State and their Applications. J. Appl. Probab., 16, pp. 332-346.

[31] Newell, G. (1971). Applications of Queueing Theory. Chapman and Hall, London.

[32] Reiser, M. (1979). A Queueing Network Analysis of Computer Communication Networks with Window Flow Control. IEEE Trans. Comm., COM-27, pp. 1199-1209.

[33] Reiser, M. (1981). The Significance of the Decomposition and the Arrival Theorems for the Evaluation of Closed Queeuing Networks. TIMS Conference on Applied Probability-Computer Science: The Interface, Florida Atlantic University, Boca Raton, January 1981.

[34] Reiser, M. and Lavenberg, S. (1980). Mean-Value Analysis of Closed Multi-Chain Queueing Networks. J. Assoc. Comput. Mach., 27, pp. 313-322.

[35] Rolski, T. (1981). Stationary Random Processes associated with Point Processes. Lecture Notes in Statistics, 5. Springer-Verlag, New York, Heidelberg, Berlin.

[36] Schassberger, R. (1978). Insensitivity of Steady-State Distributions of Generalized Semi-Markov Processes with Speeds. Adv. in Appl. Probab., 10, pp. 836-851.

[37] Schassberger, R. (1978). The Insensitivity of Stationary Probabilities in Networks of Queues. Adv. in Appl. Probab., 10, pp. 906-912.

[38] Schweitzer, P. (1981). Bottleneck Determination in Networks of Queues. TIMS Conference, Ref. (33).

[39] Simon, B. (1981). Priority Queues with Feedbacks. Conference on Applied Probability Models for Complex Stochastic Systems, North Carolina State University, January 1981.

[40] Sobel, M. (1981). The Optimality of Full-Service Policies. To appear in Oper. Res.

[41] Stidham, S., Jr. (1970). On the Optimality of the Single-Server Queueing System. Oper. Res., 18, pp. 708-732.

[42] Stidham, S., Jr. (1972a). Regenerative Processes in the Theory of Queues, with Applications to the Alternating-Priority Queue. Adv. in Appl. Probab., 4, pp. 542-577.

[43] Stidham, S., Jr. (1972b). L = λW: A Discounted Analogue and a New Proof. Oper. Res., 20, pp. 708-732.

[44] Stidham, S., Jr. (1974). A Last Word on L = λW. Oper. Res., 22, pp. 417-421.

[45] Stidham, S., Jr. (1976). A Simple Analysis of the G/G/K/K/K Queueing System. NCSU-IE Technical Report No. 76-10, Department of Industrial Engineering, North Carolina State University.

[46] Stidham, S., Jr. (1978). Queueing Systems where $L \neq \lambda W$. NCSU-IE Technical Report No. 78-6, North Carolina State University.

[47] Stidham, S., Jr. (1979). On the Relation Between Time Averages and Customer Averages in Stationary Random Marked Point Processes. NCSU-IE Technical Report No. 79-1, Department of Industrial Engineering, North Carolina State University, To appear in Math. Oper. Res.

[48] Stidham, S., Jr. (1981). Paper in preparation.

[49] Weber, R. (1980). On the Marginal Benefit of Adding Servers to G.GI/in Queues. Management Sci., 26, pp. 946-950.

[50] Whitt, W. (1971). Embedded Renewal Processes in the GI/G/s Queue.

[51] Wolff, R. (1969). Work-Conserving Priorities. J. Appl. Probab., 7, pp. 327-337.

[52] Wolff, R. (1981). Poisson Arrivals See Time Averages. Department of Industrial Engineering and Operations Research, University of California, Berkeley. To appear in Oper. Res.

8. Endnotes

[1] But not all. See, e.g., Whitt (1971).

[2] The material in this section is based primarily on Brumelle (1971), Heyman and Stidham (1980), and Stidham (1979), and helpful discussions with Dr. Richard F. Serfozo, Bell Laboratories.

[3] Henceforth, we shall omit the argument from the notation for random variables, except to make reference to a particular realization.

Research partially supported by National Science Foundation Grant ENG 78-24420.

Department of Industrial Engineering and Graduate Program in Operations Research, North Carolina State University, P. O. Box 5511, Raleigh, NC 27650.

COMPUTATIONAL METHODS FOR SINGLE-SERVER AND MULTI-SERVER QUEUES WITH MARKOVIAN INPUT AND GENERAL SERVICE TIMES

H. C. Tijms

M. H. Van Hoorn

Abstract

We first consider a wide class of single-server queues with state dependent Markovian input including the finite capacity M/G/1 queue and the machine repair problem. We specify efficient and stable algorithms to compute the steady-state probabilities and the moments of the waiting time. Next we discuss the multi-server queue with Poisson input and general service times. We present for the steady-state probabilities good quality approximations to be computed from a stable recursive algorithm. As a by-product we obtain simple approximations for the delay probability and the moments of the waiting time. For the output process we derive tractable and good approximations for the moments of the interdeparture time. Also we discuss extensions to the finite capacity M/G/c queue and the machine repair problem with multiple repairmen having general repair times.

1. Introduction

Until a few years ago the work in queueing analysis primarily concerned analytical results and little work was done on computationally tractable results and approximations. However, in recent years substantial contributions have been made to the field of algorithmic analysis of queues, cf. [3]-[4], [14]-[18] and [21]-[22] amongst others.

This paper presents a stable recursive method to compute the exact values of the steady-state probabilities in a wide class of single server queues with Markovian input and to compute approximate values of the steady-state probabilities in multi-server queues with Markovian input and general service times.

In section 2 we give the algorithmic analysis for a class of single-server queues with state-dependent Markovian input, cf. also [32]. This class covers a number of important single-server queueing models with random and quasi-random input including the finite capacity M/G/1 queue (cf. [18] and [19]) and the machine servicing problem (cf. [15]). We specify stable and efficient algorithms to compute the steady-state probabilities and the waiting time moments in these useful queueing models for the performance analysis of computer systems.

In section 3 we first consider the M/G/c queue. For this multi-server queue exact methods to compute the steady-state probabilities are only available for deterministic service times (cf. [6] and [18]) and for phase-type service times (cf. [9]-[10], [12]-[13], [21] and [29]). However, in particular for phase-type service times the exact methods are only computationally feasible to a limited extent by the dimensionality of the equilibrium state equations for the continuous-time Markov chain representation. In general we may not expect computationally tractable methods can be developed so we have to resort to approximations. For the mean queue size approximations are given in [1], [23] and [30] amongst others. Approximations for the state probabilities are discussed in [8], [11] and [22].

We present for the steady-state probabilities different and improved approximations. These approximations obtained by direct probabilistic arguments are computed by a stable recursive algorithm.

The good quality approximations yield as byproduct simple approximations for the delay probability and the moments of the actual waiting time. For the output process for which so far only complicated analytical results were known (cf. [24]) we derive as new result simple and good approximations for the moments of the interdeparture time. Further, for the case of deterministic services times we give somewhat different approximations that in particular result in a very accurate approximation for the delay probability improving the widely used Erlang delay probability approximation. Finally, we obtain for the first time tractable approximations for the finite capacity M/G/c queue and for the machine repair problem with multiple repairmen having general repair times.

2. Computational Methods for a General Class of Single Server Queues

Consider a single server queueing system where customers singly arrive according to a state-dependent Markovian input process with rate λ_j when j customers are in the system, i.e., the interarrival times are exponentially distributed with mean $1/\lambda_j$ when j customers are present. Each customer joins the system and the service times of the customers are independent random variables having a common probability distribution function F with $F(0) = 0$. To satisfy conditions for statistical equilibrium it is assumed that $\limsup_{n\to\infty} \lambda_n ES < 1$ where S denotes the service time of a customer. The server is never idle when customers are present.

This single server queueing model with state-dependent arrival rate covers a number of important systems including the finite capacity M/G/1 queue having only place for N customers (take $\lambda_j = \lambda$ for $0 \leq j < N$ and $\lambda_j = 0$ for $j \geq N$) and the machine servicing problem with a single repairman and N identical machines having exponential running times with mean $1/\eta$ (take $\lambda_j = (N-j)\eta$ for $0 \leq j < N$ and $\lambda_j = 0$ for $j \geq N$).

We first introduce some notation. Unless stated otherwise it is assumed that the system is empty at epoch 0. Define the following random variables

T = the next epoch at which the system becomes empty,

T_n = amount of time during which n customers are in the system in the busy cycle $(0,T]$, $n \geq 0$,

N = number of customers served in the busy cycle $(0,T]$,

N_n = number of service completion epochs at which the customer served leaves n other customers behind in the system in the busy cycle $(0,T]$, $n \geq 0$.

Further, define the state probabilities

$p_n = \lim_{t\to\infty} \Pr$ {at time t there are n customers in the system}, $n \geq 0$,

$\pi_n = \lim_{k\to\infty} \Pr$ {the kth customer sees upon arrival n other customers in the system}, $n \geq 0$.

The above limits exist and the limiting distributions are probability distributions, cf. [27]. Note that in general $\pi_n \neq p_n$ except for Poisson input (i.e., $\lambda_j = \lambda$ for all $j \geq 0$). By the theory of regenerative processes (cf. [25] and [27]) and the up- and down crossing result that the long-run fraction of customers seeing upon arrival n other customers in the system equals the long-run fraction of customers leaving upon departure n other customers behind, we have

$$p_n = ET_n/ET \text{ and } \pi_n = EN_n/EN \text{ for } n \geq 0. \tag{2.1}$$

Further, by noting that the queueing system is equivalent to the queueing system in which customers arrive according to a Poisson process with rate $\lambda^* = \max_j \lambda_j$ and an arriving customer does not join the system with probability $1-\lambda_n/\lambda^*$ when n other customers are present and by using the property that Poisson arrivals see time averages (cf. [27]), it readily follows with conditional probabilities that (cf. also [32]),

$$\pi_n = \lambda_n p_n \Big/ \sum_{j=0}^{\infty} \lambda_j p_j \quad \text{for } n \geq 0. \tag{2.2}$$

Also, noting that EN/ET equals the long-run average number of customers served per unit time and hence equals the long-run average number of customers joining the system per unit time, it follows from suitable versions of Little's formula that (cf. [32])

$$EN/ET = \sum_{j=0}^{\infty} \lambda_j p_j, \tag{2.3}$$

$$\Big(\sum_{j=0}^{\infty} \lambda_j p_j\Big) ES = \text{long-run average number of busy servers} = 1 - p_0. \tag{2.4}$$

Note that (2.1)-(2.3) and the first relation in (2.4) also apply to the multi-server case.

To derive a recurrence relation between the probabilities p_n and π_n, we define the following quantity. For $n \geq k \geq 0$, let

A_{nk} = expected amount of time during which n customers are in the system until the next service completion epoch given that at epoch 0 a service is completed with k customers left behind in the system. (2.5)

Then, by partioning the busy cycle (0,T] by means of the service completion epochs and using Wald's theorem (cf. [25]), it follows that

$$ET_n = \sum_{k=0}^{n} A_{nk}\, EN_k \quad \text{for } n \geq 0. \tag{2.6}$$

Noting that $p_0 ET = 1/\lambda_0$ by (2.1) and using the relations (2.1)-(2.3) and (2.6) we find the following recursive scheme

$$p_0 ET = 1/\lambda_0, \quad p_n ET = \sum_{k=0}^{n} \lambda_k A_{nk} p_k ET \quad \text{for } n \geq 1. \tag{2.7}$$

By the numerically stable algorithm (2.7) we can recursively compute the numbers $p_0 ET$, $p_1 ET, \ldots$ once we have evaluated the quantities A_{nk} to be discussed below. Next, using (2.2)-(2.4), the state probabilities p_n

and π_n for $n \geq 0$ can be obtained in any desired accuracy by normalization. The above approach based on regenerative analysis is a fertile approach which can be applied to many variants of the M/G/1 queue, cf. [7] and [33].

Alternatively a recursive relation for the state probabilities π_n can be obtained as follows. Note that for $n \geq 1$ the long-run fraction of services having the property that at the completion exactly n customers are left behind equals the long-run fraction of services having the property that at its beginning at most n customers are present and during its execution the number of customers present exceeds the level n (cf. also [5] for a similar up- and downcrossings argument for the virtual waiting time process). Hence, using the second part of (2.1),

$$\pi_n = \pi_0 b_{n1} + \sum_{k=1}^{n} \pi_k b_{n-k+1,k} \text{ for } n \geq 1, \tag{2.8}$$

where b_{nk} is the probability that at least n customers arrive during a service for which at the beginning k customers are present. The approaches (2.7) and (2.8) are of an equal simplicity but the regenerative approach (2.7) seems better suited for such variants as single-server queues with group arrivals, cf. [33].

We next discuss two wide cases in which eficient methods to compute the numbers A_{nk}, $n \geq k \geq 1$ can be given. Note that $A_{n0} = A_{n1}$ for $n \geq 1$.

CASE 1. A PHASE-TYPE SERVICE TIME. Suppose that the service time distribution function F is a finite mixture of Erlang distribution functions, i.e.,

$$F(t) = \sum_{\ell=1}^{r} q_\ell E_{m_\ell,\mu_\ell}(t) \text{ with } E_{m,\mu}(t) = 1 - \sum_{j=0}^{m-1} e^{-\mu t} \frac{(\mu t)^j}{j!} . \tag{2.9}$$

Hence we can imagine that with probability q_ℓ the service of a customer requires the completion of m_ℓ independent phases having each an exponential distribution with mean $1/\mu_\ell$. It is known that each probability distribution function concentrated on $(0,\infty)$ can be approximated with any prescribed accuracy by a phase-type distribution function as (2.9), cf. also [3]. Define $A_{nk}^{(\ell)}(i)$ is the expected amount of time during which n customers are present in a (remaining) service time that starts with k customers present and consists of i independent phases having each an exponential distribution with mean $1/\mu_\ell$. Considering such a remaining service time and using the memoryless property of the exponential distribution and the property that with probability $\mu_\ell/(\mu_\ell+\lambda_k)$ the current phase is completed before a customer arrives, we have for any $n \geq 1$, $1 \leq \ell \leq r$ and $1 \leq i \leq m_\ell$,

$$A_{nk}^{(\ell)}(i) = \begin{cases} (\mu_\ell+\lambda_n)^{-1} + \mu_\ell(\mu_\ell+\lambda_n)^{-1} A_{nn}^{(\ell)}(i-1), & k = n \\ (\mu_\ell+\lambda_k)^{-1}\{\lambda_k A_{n,k+1}^{(\ell)}(i) + \mu_\ell A_{nk}^{(\ell)}(i-1)\}, & 1 \leq k \leq n-1 \end{cases} \tag{2.10}$$

where $A_{nk}^{(\ell)}(0) = 0$. Hence for any fixed n and ℓ we can recursively compute by a stable algorithm the numbers $A_{nk}^{(\ell)}(i)$ for $i = 1,\ldots,m_\ell$ and $k = n,\ldots,1$. Next we find for any $n \geq 1$

$$A_{nk} = \sum_{\ell=1}^{r} q_\ell A_{nk}^{(\ell)}(m_\ell), \quad 1 \leq k \leq n.$$

Define L_q as the steady-state queue size (excluding any customer in service) and let W_q be the steady-state queueing time of an arbitrary customer (excluding his service time). The moments of L_q follow from the state probabilities p_n. By Little's formula and (2.3)-(2.4), we generally have $EW_q = ESEL_q/(1-p_0)$. For a phase-type service time as (2.9), we can give an efficient algorithm to compute the higher moments of W_q. Therefore, define for $n \geq 1$ and $1 \leq \ell \leq r$,

$$p_{ni}^{(\ell)} = \lim_{t\to\infty} \Pr\{\text{at time t there are n customers present and the residual service time consists of i independent exponential phases having each mean } 1/\mu_\ell\},$$

$$1 \leq i \leq m_\ell,$$

$$\pi_{ni}^{(\ell)} = \lim_{k\to\infty} \Pr\{\text{at the arrival epoch of the k-th customer there are n other customers present and the residual service time consists of i independent exponential phases having each mean } 1/\mu_\ell\},\ 1 \le i \le m_\ell.$$

By the same arguments as used to prove (2.2), we have

$$\pi_{ni}^{(\ell)} = \lambda_n p_{ni}^{(\ell)} \Big/ \sum_{j=0}^{\infty} \lambda_j p_j = \lambda_n ES p_{ni}^{(\ell)}/(1-p_0) \text{ for all } n,\ell,i,$$

The moments of W_q follow from the probabilities $\pi_{ni}^{(\ell)}$, e.g.,

$$EW_q^2 = \sum_{n,\ell,i} \pi_{ni}^{(\ell)} \left[\frac{i(i+1)}{\mu_\ell^2} + (n-1)\left\{\frac{2i}{\mu_\ell} ES + ES^2 + (n-2)(ES)^2\right\}\right],$$

where we assume service in order of arrival. To compute the $p_{ni}^{(\ell)}$, put for abbreviation

$$p_{0i}^{(\ell)} = q_\ell \delta(i,m_\ell) p_0 \quad \text{where} \quad \delta(i,j) = \begin{cases} 1 \text{ if } j = i \\ 0 \text{ if } j \neq i \end{cases}$$

and define $X_1(t)$ = the number of customers present at time t and let $X_2(t) = (\ell,i)$ if at time t a service is in progress and has still i uncompleted phases having each mean $1/\mu_\ell$ and $X_2(t) = (0,0)$ otherwise. For the continuous-time Markov chain $(X_1(t),X_2(t))$ we have for any $n \ge 1$ and $1 \le \ell \le r$,

$$(\lambda_n + \mu_\ell) p_{ni}^{(\ell)} = \lambda_{n-1} p_{n-1,i}^{(\ell)} + \mu_\ell p_{n,i+1}^{(\ell)} + q_\ell \delta(i,m_\ell) \lambda_n p_n,\quad 1 \le i \le m_\ell, \tag{2.11}$$

where $p_{ni}^{(\ell)} = 0$ for $i = m_\ell + 1$. This relation is obtained by equating the rate at which the system leaves the microstate (n,ℓ,i) to the rate at which the system enters this state and inserting into this equation the local balance relation $\lambda_n p_n = \Sigma_h \mu_h p_{n+1,1}^{(h)}$. This latter relation in its turn follows by aggregating the microstates $0,\ldots,n$ into a macrostate and equating the rate at which the system leaves this macrostate to the rate at which the system enters this macrostate. For fixed $1 \le \ell \le r$

the state probabilities $p_{ni}^{(\ell)}$ can be computed by the stable recursive scheme (2.11) for $i = m_\ell,\dots,1$ and $n = 1,2,\dots$ It is important to note that by first computing the probabilities p_n from (2.7) we can compute the $p_{ni}^{(\ell)}$ from a stable and efficient ercursive scheme instead of by solving the difficult system of linear equilibrium equations.

Remark 2.1. For the special case where F is an Erlang distribution function $E_{r,\mu}$, the state probabilities p_n and the moments of W_q can be computed by another algorithm that is simpler than the above one. Therefore note that the number of uncompleted phases in the system uniquely determines the number of customers present. Defining f_j as the steady-state probability that at an arbitrary epoch there are j uncompleted phases in the system, we have

$$p_0 = f_0 \text{ and } p_n = \sum_{j=(n-1)r+1}^{nr} f_j \text{ for } n \geq 1 \tag{2.11}$$

Put for abbreviation $\gamma_j = \lambda_{[j/r]}$ if j/r is an integer and $\gamma_j = \lambda_{[j/r]+1}$ otherwise, where [x] is the largest integer less than or equal to x. Then

$$\mu f_n = \sum_{k=n-r}^{n-1} \gamma_k f_k \text{ for } n \geq 1 \tag{2.12}$$

where $f_j = 0$ for $j < 0$. This recursive relation follows by aggregating the microstates $n, n+1,\dots$ of the continuous-time Markov chain describing the number of uncompleted phases into a macrostate and equating the rate at which the system leaves this macrostate to the rate at which the system enters this macrostate. Finally, noting that $\gamma_k f_k / \Sigma_j \lambda_j p_j = \gamma_k f_k ES/(1-p_0)$ is the steady-state probability that an arriving customer finds k uncompleted phases present, the moments of W_q follows from the state probabilities f_j.

CASE 2. THE FINITE CAPACITY M/G/1 QUEUE AND THE MACHINE REPAIR. For a general service time distribution function we can easily evaluate by numerical integration the numbers A_{nk} for both the finite capacity M/G/1 queue and the machine servicing problem. We only discuss here the latter model. Consider N identical machines with a single repairman where the running times of the machines are independent and exponentially distributed with mean $1/\eta$ and the repair time has the general distribution function F. This problem is a special case of the single-server model having state-dependent input with $\lambda_j = (N-j)\eta$ for $j < N$ and $\lambda_j = 0$ for $j \geq N$ by taking the number of inoperative machines as state for the system. We have for $1 \leq k \leq n \leq N$,

$$A_{nk} = \int_0^\infty (1-F(t))\binom{N-k}{n-k}(1-e^{-\eta t})^{n-k} e^{-\eta t(N-n)}\,dt, \tag{2.13}$$

as follows by noting that $A_{nk} = \int_0^\infty E\chi_t\,dt$ where $\chi_t = 1$ if at time t the first repair is still in progress and n machines are broken down given that at epoch 0 a repair starts with k machines broken down, and $\chi_t = 0$ otherwise.

We conclude this section by presenting some numerical results for the above machine repair problem. For the repair time S we consider the following cases with a phase-type distribution (2.9),

(i) $r = 1$, $m_1 = k$ $(c_S^2 = 1/k)$ for $k = 1,2,3$ and 10,

(ii) $r = 2$, $m_1 = m_2 = 1$, $q_1 = (1/2)[1 + \{(c_S^2 - 1)/(c_S^2 + 1)\}^{1/2}]$, $q_1/\mu_1 = q_2/\mu_2$ for $c_S^2 = 2$, 5, 25 and 100,

(iii) $r = 2$, $m_1 = 1$, $m_2 = 3$, $q_1 = (4-\sqrt{7})/6$, $\mu_1 = \mu_2$ $(c_S^2 = 1/2)$,

(iv) $r = 2$, $m_1 = 1$, $m_2 = 3$, $q_1 = (19-\sqrt{145})/36$, $q_1/\mu_1 = 3q_2/\mu_2$ $(c_S^2 = 2)$,

where $c_S = \{ES^2/(ES)^2-1\}^{1/2}$ denotes the coefficient of variation of S. The latter two cases are denoted by * in table 2.1. In all cases we take $ES = 1$. For various values of $\rho = \eta ES$ and N we give in table 2.1 the mean ER and the coefficient of variation c_R of the response time $R = W_q + S$. Note that, by Little's formula $\lambda'(ER + 1/\eta) = N$ with throughput $\lambda' = (1-p_0)/ES$, the server utilization $1-p_0$ determines ER.

Table 2.1. The Machine Repair Problem with a Single Repairman

$1/\rho$	c_S^2	N=5 ER	N=5 c_R	N=10 ER	N=10 c_R	N=20 ER	N=20 c_R	N=40 ER	N=40 c_R
5	1/10	1.70	0.47	5.03	0.34	15.0	0.13	35.0	0.07
	1/3	1.79	0.54	5.07	0.41	15.0	0.19	35.0	0.11
	1/2	1.84	0.59	5.09	0.45	15.0	0.22	35.0	0.13
	1/2*	1.85	0.58	5.10	0.45	15.0	0.22	35.0	0.13
	1	1.99	0.71	5.19	0.55	15.0	0.29	35.0	0.18
	2	2.17	0.96	5.31	0.73	15.0	0.40	35.0	0.25
	2*	2.13	1.02	5.24	0.75	15.0	0.40	35.0	0.25
	5	2.49	1.46	5.57	1.09	15.0	0.63	35.0	0.39
	25	2.97	3.14	6.06	2.37	15.0	1.43	35.0	0.88
	100	3.16	6.21	6.27	4.72	15.0	2.89	35.0	1.78
10	1/10	1.28	0.39	2.25	0.55	10.0	0.25	30.0	0.10
	1/3	1.33	0.48	2.39	0.61	10.0	0.31	30.0	0.13
	1/2	1.37	0.53	2.48	0.66	10.0	0.34	30.0	0.16
	1/2*	1.37	0.52	2.49	0.65	10.0	0.34	30.0	0.16
	1	1.47	0.68	2.73	0.76	10.0	0.43	30.0	0.21
	2	1.62	0.97	3.06	0.96	10.1	0.57	30.0	0.29
	2*	1.60	1.02	3.00	1.01	10.1	0.57	30.0	0.29
	5	1.93	1.54	3.68	1.37	10.3	0.85	30.0	0.45
	25	2.60	3.34	4.90	2.73	11.1	1.84	30.0	1.01
	100	2.94	6.51	5.52	5.25	11.6	3.67	30.0	2.06
20	1/10	1.13	0.29	1.38	0.46	3.02	0.61	20.0	0.18
	1/3	1.15	0.37	1.45	0.55	3.24	0.66	20.0	0.22
	1/2	1.17	0.42	1.50	0.60	3.39	0.69	20.0	0.25
	1/2*	1.17	0.41	1.50	0.59	3.40	0.69	20.0	0.25
	1	1.22	0.55	1.64	0.75	3.78	0.77	20.0	0.31
	2	1.31	0.85	1.88	1.04	4.33	0.93	20.0	0.41
	2*	1.30	0.90	1.86	1.09	4.25	0.97	20.0	0.42
	5	1.53	1.49	2.41	1.57	5.41	1.25	20.1	0.63
	25	2.21	3.47	3.89	3.04	7.98	2.28	20.8	1.37
	100	2.73	6.72	4.89	5.58	9.78	4.18	21.8	2.75
40	1/10	1.06	0.21	1.15	0.32	1.45	0.50	4.11	0.65
	1/3	1.07	0.27	1.18	0.40	1.54	0.59	4.45	0.69
	1/2	1.08	0.31	1.20	0.46	1.60	0.65	4.67	0.72
	1/2*	1.08	0.30	1.20	0.45	1.60	0.64	4.68	0.71
	1	1.10	0.42	1.27	0.60	1.78	0.79	5.26	0.78
	2	1.15	0.68	1.39	0.92	2.10	1.08	6.12	0.90
	2*	1.15	0.72	1.39	0.98	2.08	1.14	6.04	0.93
	5	1.28	1.31	1.71	1.57	2.88	1.56	7.90	1.14
	25	1.83	3.50	2.95	3.34	5.56	2.72	12.7	1.89
	100	2.47	6.89	4.35	5.91	8.31	4.56	17.1	3.26

3. Approximations for the M/G/c Queue and the Machine Repair Problem

For clarity of presentation we first consider the infinite capacity M/G/c queue with $c > 1$ servers where customers arrive in accordance with a Poisson process with rate λ and the service time S of a customer has a general probability distribution function F with $F(0) = 0$. It is assumed that $\rho = \lambda ES/c < 1$. An arriving customer joins the queue if he finds all c servers occupied or else he is served immediately by one of the free servers. A server will never remain idle if customers are waiting in the queue.

We first introduce some notation. Define the random variables T, T_n, N and N_n and the steady-state probabilities p_n and π_n as in section 2. Then (cf. (2.1)-(2.3)),

$$p_n = ET_n/ET, \quad \pi_n = EN_n/EN \text{ for } n \geq 0, \tag{3.1}$$

$$p_n = \pi_n \text{ for } n \geq 0, \quad EN/ET = \lambda. \tag{3.2}$$

Further, define the delay probability P_W, the mean queue size EL_q and the constant Ω by

$$P_W = \sum_{n=c}^{\infty} p_n, \quad EL_q = \sum_{n=c}^{\infty} (n-c)p_n, \quad \Omega = \left\{ \sum_{k=0}^{c-1} \frac{(\lambda ES)^k}{k!} + \frac{(\lambda ES)^c}{c!(1-\rho)} \right\}^{-1}. \tag{3.3}$$

We write $p_n = p_n(\exp)$, $P_W = P_W(\exp)$ and $EL_q = EL_q(\exp)$ when the service time is exponentially distributed and we have the explicit results

$$p_n(\exp) = \frac{(\lambda ES)^n}{n!}\Omega \text{ for } 0 \leq n \leq c-1, \quad p_n(\exp) = \frac{(\lambda ES)^n}{c!c^{n-c}}\Omega \text{ for } n \geq c, \tag{3.4}$$

$$P_W(\exp) = \frac{(\ ES)^c}{c!(1-\rho)}\Omega, \quad EL_q(\exp) = \frac{(\lambda ES)^c \rho}{c!(1-\rho)^2}\Omega. \tag{3.5}$$

The quantity $P_W(\exp)$ is called the Erlang delay probability and is known to be a good approximation for P_W when the service time has a general distribution, cf. [18].

In general no explicit expression for p_n can be given. However, we can try to set up a recursive scheme as (2.7) to compute the state probabilities. In doing so, we encounter the difficulty that for a service completion epoch at which $j \geq 1$ customers are left behind the distribution function of the time until the next service completion epoch depends on the information of how long the remaining services are already in progress. To overcome this difficulty, we make an approximation by aggregating this required information and using distribution functions depending only on the number of customers left behind. The specification of these distribution functions will determine our approximations. More precisely, we make the following approximation assumption.

APPROXIMATION ASSUMPTION. For any $1 \leq j \leq c-1$, the random variables defined as the smallest of the remaining service times of the j services in progress at those service completion epochs at which j customers are left behind in the system are independent and have common probability distribution function F_j^*. For the service completion epochs at which $j \geq c$ customers are left behind in the system, the times until the next service completion epoch are independent random variables with common probability distribution function F^* where F^* is the same for all $j \geq c$.

Define the quantities A_{nk} for $n \geq k \geq 0$ as in (2.5) with for $k \geq 1$ the stipulation that the approximation assumption applies. Under this assumption, we have

$$ET_n \simeq \sum_{j=0}^{n} EN_j A_{nj} \quad \text{for } n = 0,1,\ldots, \tag{3.6}$$

and so, by (3.1)-(3.2),

$$p_n ET \simeq \sum_{j=0}^{n} \lambda p_j ETA_{nj} \quad \text{for } n = 0,1,\ldots. \tag{3.7}$$

Together (3.7) and the relation $p_0 ET = 1/\lambda$ suggest to define $\{q_n,\ n \geq 0\}$ by

$$q_0 = 1/\lambda \text{ and } q_n = \sum_{j=0}^{n} \lambda q_j A_{nj} \quad \text{for } n = 1,2,\ldots, \tag{3.8}$$

From this stable recursive scheme we can successively compute the numbers $q_0, q_1, \ldots$ once we have evaluated the quantities A_{nk}. Next the state probabilities p_n, $n \geq 0$ can be approximated by

$$p_n(\text{appr}) = q_n / \sum_{j=0}^{\infty} q_j \quad \text{for } n = 0,1,\ldots \,. \tag{3.9}$$

To evaluate the quantities A_{nk} we make in the approximation assumption the following specification

$$1-F_j^*(t) = (1-F_e(t))^j, \quad 1 \leq j < c \text{ and } F^*(t) = F(ct) \tag{3.10}$$

where F_e is the equilibrium distribution of F and is given by

$$F_e(t) = (1/ES)\int_0^t (1-F(x))dx, \quad t \geq 0. \tag{3.11}$$

To motivate this specification, note that if not all c servers are busy the M/G/c queue can be treated as an M/G/∞ queue for which we have the renewal-theoretic result that at an arbitrary epoch the remaining service times of services in progress (if any) are independent random variables with common probability distribution function F_e, cf. p. 161 in [28]. If all c servers are busy we can treat the M/G/c queue with service time S as an M/G/1 queue with service time S/c, cf. also [22]. Note that the approximation assumption is satisfied for the M/M/c queue. In evaluating the quantities A_{nk} we next encounter the computational difficulty that, except for deterministic service times, the closed-form expression for A_{nk} with $k < c$ involves an (n-k+1)-dimensional integral because of the phenomenon that a newly started service may be completed before services in progress. Fortunately, by the specified form of F_j^* for $1 \leq j \leq c-1$, we can establish by induction a very simple expression for q_j for $j \leq c-1$ through which we succeeded in eliminating the multi-dimensional integrals so that the ultimate recursive scheme involves only one-dimensional integrals. The following results have been proved in [31] (see the appendix for a simplified and more generally applicable proof).

Theorem 3.1. Under the approximation assumption with specification (3.10),

$$p_n(\text{appr}) = \frac{(\lambda ES)^n}{n!} p_0(\text{appr}), \quad 0 \leq n \leq c-1, \tag{3.12}$$

$$p_n(\text{appr}) = \lambda p_{c-1}(\text{appr})\, \alpha_{n-c} + \lambda \sum_{j=c}^{n} p_j(\text{appr})\, \beta_{n-j}, \quad n \geq c, \tag{3.13}$$

with $p_0(\text{appr}) = \Omega$ and

$$\alpha_k = \int_0^\infty (1-F_e(t))^{c-1} (1-F(t)) e^{-\lambda t} \frac{(\lambda t)^k}{k!}\, dt, \quad k \geq 0, \tag{3.14}$$

$$\beta_k = \int_0^\infty (1-F(ct)) e^{-\lambda t} \frac{(\lambda t)^k}{k!}\, dt, \quad k \geq 0. \tag{3.15}$$

Hence we can compute the approximations for the state probabilities by a stable recursive scheme where in general any recursion step requires the evaluation of two one-dimensional integrals which are easy to handle by numerical integration. Note that $p_j(\text{appr}) = p_j(\text{exp})$ for $0 \leq j \leq c-1$ and hence

$$P_W(\text{appr}) = P_W(\text{exp}), \tag{3.16}$$

so that as approximation for the delay probability we find the widely used Erlang delay probability which is in general a good approximation. We note that the approximations given in [11] for the state probabilities p_j are also equal to $p_j(\text{exp})$ for $j \leq c-1$ but differ from our approximations for $j > c$. In [11] the approximations are given in a form inconvenient for computational purposes, however it was shown in [31] that these approximations can also be computed by a stable recursive scheme which is obtained from (3.12)-(3.13) by replacing $(1-F_e(t))^{c-1}(1-F(t))$ by $1-F(ct)$ in the integral in (3.14). We note that the latter approximations yield for the mean queue size the same approximation as found in [23].

Denote by L_q the queue size at an arbitrary epoch in the steady-state (excluding the customers in service, if any). Using generating functions we obtain after some algebra from (3.12)-(3.13),

$$EL_q(\text{appr}) = 1/2(1+c_S^2)EL_q(\exp)\{1+(1-\rho)(2c\gamma_1 ES/ES^2-1)\}. \tag{3.17}$$

$$EL_q^2(\text{appr}) = \lambda P_W(\text{appr})[\{\gamma_1 + \frac{\lambda ES^2}{2c^2(1-\rho)}\}\{1 + \frac{\lambda^2 ES^2}{c^2(1-\rho)}\} + \tag{3.18}$$

$$+ \lambda\gamma_2 + \frac{\lambda^2 ES^3}{3c^3(1-\rho)}],$$

where $c_S = \{ES^2/(ES)^2 - 1\}^{1/2}$ denotes the coefficient of variation of the service time S and γ_k is defined by

$$\gamma_k = k \int_0^\infty t^{k-1}(1-F_e(t))^c dt, \ k \geq 1. \tag{3.19}$$

Note that when F is a NBUE-distribution function we have $ES/(c+1) \leq \gamma_1 \leq ES/c$ by $1-F_e(t) \leq 1-F(t) \leq 1$. Similarly, the higher moments of L_q may be derived. From the moments of the queue size L_q we get the moments of the steady-state queueing time W_q of an arbitrary customer (excluding his service time). Under the assumption of service in order of arrival we have (see [20]),

$$EL_q(L_q-1)\ldots(L_q-k+1) = \lambda^k EW_q^k \text{ for } k \geq 1. \tag{3.20}$$

We note that the distribution function of W_q may be approximated by matching of moments, e.g., following [17] we may approximate the waiting time distribution $1-\Pr\{W_q > t|W_q > 0\}$ for the delayed customers by a Weibull distribution function $1-\exp((at)^b)$ by matching the first two moments. In [34] we have derived approximations for the waiting time distribution in the form of a defective renewal equation which integral equation can be effectively solved.

Next we derive for the output process approximations for the moments of the interdeparture time T_D between two consecutive service completion epochs in the steady-state.

<u>Theorem 3.2.</u> Under (3.10),

$$ET_D^m(\text{appr}) = \frac{m!}{\lambda^m}\,[1-P_W(\text{appr})\{\rho - \frac{\lambda^m ES^m}{m!c^m} - (1-\rho)\sum_{i=1}^{m-1}\frac{\lambda^i}{i!}\,\gamma_i\}],\ m \geq 1, \quad (3.21)$$

where γ_i, $i \geq 1$ is given by (3.19). In particular, $ET_D(\text{appr}) = 1/\lambda$ is exact and $ET_D^2(\text{appr}) = (2/\lambda^2)\{1 - \rho P_W(\text{appr}) + (1-\rho)EL_q(\text{appr})\}$.

Proof. Under the condition that the system is empty at epoch 0, define $M_k(t)$ as the probability that the service completions of the first k customers all occur beyond time t, $1 \leq k \leq c$. Also, define $Q(t)$ as the steady-state probability that the time between two consecutive service completion epochs exceeds t. The steady-state probability that at a service completion epoch there are left i customers behind equals $\pi_i = p_i$ (cf. (3.1)-(3.2)) and so, under the approximation assumption with (3.10),

$$Q(t) = \sum_{i=0}^{c-1} p_i(\text{appr})(1-F_e(t))^i M_{c-i}(t) + (1-F(ct))\sum_{i=c}^{\infty} p_i(\text{appr}), \quad (3.22)$$

$$t > 0.$$

By considering what may happen in $(0,\Delta t)$ for Δt small, it follows that for $1 \leq k \leq c-1$,

$$M_k(t+\Delta t) = (1-\lambda\Delta t)M_k(t) + \lambda\Delta t(1-F(t))M_{k-1}(t) + 0(\Delta t),\ t > 0$$

and so, for $1 \leq k \leq c-1$,

$$\frac{dM_k(t)}{dt} = \lambda(1-F(t))M_{k-1}(t) - \lambda M_k(t),\ t > 0, \quad (3.23)$$

where $M_0(t) = 1$ for all t. Put for abbreviation

$$Q_1(t) = \sum_{i=0}^{c-1} p_i(\text{appr})(1-F_e(t))^i M_{c-i}(t),\ t > 0. \quad (3.24)$$

Using (3.23) and the relations

$$\frac{d}{dt}(1-F_e(t))^{i+1} = -\frac{(i+1)}{ES}(1-F(t))(1-F_e(t))^i,\ i \geq 0, \quad (3.25)$$

$$p_{i+1}(\text{appr}) = \frac{\lambda ES}{i+1}\,p_i(\text{appr}),\ 0 \leq i \leq c-2, \quad (3.26)$$

we find after some algebra

$$\frac{dQ_1(t)}{dt} = -\lambda Q_1(t) - \frac{\lambda ES}{c}\, p_{c-1}(\text{appr})\, \frac{d}{dt}\, (1-F_e(t))^c, \quad t > 0.$$

From this first-order differential equation, we get

$$Q_1(t) = \sum_{i=0}^{c-1} p_i(\text{appr})e^{-\lambda t} - \rho p_{c-1}(\text{appr})\int_0^t e^{-\lambda(t-u)}\, d(1-F_e(u))^c, \tag{3.27}$$

$$t > 0.$$

Using the relation (cf. (3.12) and (3.4)-(3.5))

$$\rho p_{c-1}(\text{appr}) = (1-\rho)P_W(\text{appr}) \tag{3.28}$$

and the relation $EX^k = k\int_0^\infty x^{k-1}\Pr\{X > x\}dx$, $k \geq 1$ for any nonnegative random variable X, we obtain (3.21) from (3.22), (3.24) and (3.27)-(3.28) after some algebra. □

Letting $c_D = \sigma(T_D)/ET_D$, we have by Theorem 3.2 that $c_D^2(\text{appr}) = 1-2\rho P_W(\text{appr}) + 2(1-\rho)EL_q(\text{appr})$. In [26] it was empirically established that

$$L_q(\exp) \simeq (1-\rho)^{-1}\rho^{\sqrt{2(c+1)}} \quad \text{for } \rho \text{ close to } 1, \tag{3.29}$$

and so by (3.17) and $P_W(\exp) = (1-\rho)\rho^{-1}EL_q(\exp)$,

$$c_D^2(\text{appr}) \simeq 1-\rho^{\sqrt{2(c+1)}} + c_S^2\rho^{\sqrt{2(c+1)}} \quad \text{for } \rho \text{ close to } 1. \tag{3.30}$$

The above approximations are good quality approximations. This is supported by the findings that $P_W(\text{appr})$ is given by the good quality Erlang delay probability approximation and that $EL_q(\text{appr})$ is competitive to the extremely accurate approximations for the mean queue size specially developed in [1] and [30], cf. [31] for extensive numerical comparisons. Further, by [2] and [16], the following light- and heavy-traffic approximations agree with exact results,

$$\lim_{\rho\to 0} \frac{EL_q(\text{appr})}{EL_q(\exp)} = \frac{c\gamma_1}{ES} \text{ and } \lim_{\rho\to 1} \frac{EL_q(\text{appr})}{EL_q(\exp)} = \frac{ES^2}{2(ES)^2}\ .$$

Also, it can be shown that $\lim_{n\to\infty} p_{n+1}(\text{appr})/p_n(\text{appr})$ is exact, cf. [34].

3.1 The Finite Capacity M/G/c Queue.

Consider the finite capacity M/G/c queueing system having only place for $N \geq c$ customers, i.e., any customer who finds upon arrival N other customers present does not enter and has no effect on the system. No restriction is imposed on $\rho = \lambda ES/c$. The above analysis and results require only obvious modifications. Define now π_n as the steady-state probability that an entering customer finds upon arrival n other customers present, $0 \leq n \leq N-1$. We now have (cf. (2.1)-(2.3) with $\lambda_j = \lambda$ for $j \leq N-1$ and $\lambda_j = 0$ for $j \geq N$),

$$p_n = \frac{ET_n}{ET}, \quad \pi_n = \frac{EN_n}{EN}, \quad \pi_n = \frac{p_n}{1-p_N} \quad \text{for } n \geq 0 \text{ and } \frac{EN}{ET} = \lambda(1-p_N). \tag{3.31}$$

Next a minor modification of the proof of Theorem 3.1 shows that under the approximation assumption with specification (3.10) the approximation $p_n(\text{appr})$ for the finite capacity M/G/c queue is given by (3.12) for $0 \leq n \leq c-1$ and by (3.13) for $c \leq n \leq N-1$ whereas

$$p_N(\text{appr}) = \lambda p_{c-1}(\text{appr}) \sum_{k=N}^{\infty} \alpha_{k-c} + \lambda \sum_{j=c}^{N-1} p_j(\text{appr}) \sum_{k=N}^{\infty} \beta_{k-j} \tag{3.32}$$

where α_k, β_k are defined by (3.14)-(3.15). The relation (3.32) can be after some algebra reduced to

$$p_N(\text{appr}) = \rho p_{c-1}(\text{appr}) - (1-\rho) \sum_{k=c}^{N-1} p_k(\text{appr}). \tag{3.33}$$

By (3.31) and (3.33) we have

$$\rho\pi_{c-1}(\text{appr}) = (1-\rho)\Pi_W(\text{appr}) + (1-p_N(\text{appr}))^{-1} p_N(\text{appr}), \tag{3.34}$$

where $\Pi_W = \sum_{k=c}^{N-1} \pi_k$ denotes the steady-state probability that an entering customer will be delayed. Noting that π_i gives the steady-state probability that at a service completion epoch i customers are left behind and using (3.34), an examination of the proof of Theorem 3.2 shows that the corresponding approximations for the moments of the interdeparture time T_D between two consecutive departures are given by

$$ET_D^m(\text{appr}) = \frac{m!}{\lambda^m} [(1-p_N(\text{appr}))^{-1} - \Pi_W(\text{appr})(\rho - \frac{\lambda^m ES^m}{m!c^m} - (1-\rho)\sum_{i=1}^{m-1} \frac{\lambda^i}{i!}\gamma_i) +$$

$$+ (1-p_N(\text{appr}))^{-1} p_N(\text{appr}) \sum_{i=1}^{m-1} \frac{\lambda^i}{i!}\gamma_i], \quad m \geq 1 \qquad (3.35)$$

Note that for the special case of no waiting room (i.e., $N = c$) we have the remarkable result $p_n(\text{appr}) = C^{-1}(\lambda ES)^n/n!$, $0 \leq n \leq c$ with $C = \Sigma_{k=0}^{c}(\lambda ES)^k/k!$, that is the approximations are equal to the exact values having the famous insensitivity property of depending on the service time only through the first moment.

3.2 The Machine Repair Problem with Multiple Repairmen.

Consider the machine repair problem with N identical machines and c repairmen where $1 < c \leq N$. The machines have independent running times with a common exponential distribution with mean $1/\eta$ and the repair time S of a broken-down machine has a general probability distribution function F. Define the state of the system as the number of machines broken down and let $\lambda_j = (N-j)\eta$ for $0 \leq j \leq N$. Under the approximation assumption with specification (3.10), a generalisation of the proof of Theorem 3.1 as given in the appendix yields

$$p_n(\text{appr}) = \binom{N}{n}(\eta ES)^n p_0(\text{appr}), \quad 0 \leq n \leq c-1$$

$$p_n(\text{appr}) = \lambda_{c-1}p_{c-1}(\text{appr}) \int_0^\infty (1-F_e(t))^{c-1}(1-F(t))\phi_{nc}(t)dt +$$

$$+ \sum_{j=c}^{n} \lambda_j\, p_j(\text{appr}) \int_0^\infty (1-F(ct))\phi_{nj}(t)dt, \quad c \leq n \leq N$$

where

$$\phi_{nj}(t) = \binom{N-j}{n-j}(1-e^{-\eta t})^{n-j} e^{-\eta t(N-n)}, \quad t > 0$$

The above approximations have been derived under the specification (3.10) in the approximation assumption. We now discuss a slightly different specification which in particular yields useful results for the case of deterministic service times. For the M/D/c queue with the service times equal to the constant D, consider the specification in which F_j^* for $1 \le j \le c-2$ and F^* are the same as in (3.10) but the "boundary" distribution function F_{c-1}^* is chosen as

$$F_{c-1}^*(t) = \begin{cases} 1, & t \ge D/c, \\ 0, & t < D/c. \end{cases} \tag{3.36}$$

Denoting the corresponding approximations by a bar, a minor modification of the proof of Theorem 3.1 gives the following approximations for the M/D/c queue

$$p_n\overline{(appr)} = \frac{(\lambda D)^n}{n!} p_0\overline{(appr)}, \quad 0 \le n \le c-2, \tag{3.37}$$

$$p_n\overline{(appr)} = \lambda p_{c-2}\overline{(appr)} \int_0^D (1-\frac{t}{D})^{c-2} \frac{(\lambda t)^{n-c+1}}{(n-c+1)!} e^{-\lambda t} dt + $$
$$+ \lambda \sum_{j=c-1}^{n} p_j\overline{(appr)} \int_0^{D/c} \frac{(\lambda t)^{n-j}}{(n-j)!} e^{-\lambda t} dt, \quad n \ge c-1. \tag{3.38}$$

with $p_0\overline{(appr)} = \Omega$. In particular we find

$$P_W\overline{(appr)} = P_W(exp) - (\frac{\eta_1}{\eta_2} - 1) \frac{(\lambda D)^{c-1}}{(c-1)!} \Omega, \tag{3.39}$$

where

$$\eta_1 = \frac{c-1}{D} \int_0^D (1-\frac{t}{D})^{c-2} e^{-\lambda t} dt \text{ and } \eta_2 = e^{-\lambda D/c} .$$

Note that $P_W\overline{(appr)} < P_W(exp)$ since $\eta_1 > \eta_2$ as is readily verified. Numerical results show that $P_W\overline{(appr)}$ is a very accurate approximation for the delay probability in the M/D/c queue and considerably improves in almost all cases the good Erlang delay probability approximation $P_W(exp)$. Also we find after some algebra

$$EL_q\overline{(appr)} = \frac{(\lambda D)^{c-1}}{(c-1)!} \Omega \left\{\frac{\rho^2}{2(1-\rho)^2} + (\frac{\eta_1}{\eta_2} - 1)\right\}, \tag{3.40}$$

$$EL_q^2\overline{(appr)} = \frac{(\lambda D)^{c-1}}{(c-1)!} \Omega \left\{\frac{9\rho^2-11\rho^3+7\rho^4}{6(1-\rho)^3} - \frac{2\rho^2}{(1-\rho)(c+1)} - (\frac{\eta_1}{\eta_2} - 1)\right\}. \tag{3.41}$$

Further $ET_D^m\overline{(appr)}$ for $m \ge 1$ can be easily derived. Although $ET_D\overline{(appr)} = 1/\lambda$ is exact, our numerical results indicate that $ET_D^m\overline{(appr)}$ is less good than $ET_D^m(appr)$ for $m \ge 2$.

We conclude this section by presenting some numerical results. In table 3.1 we consider the M/D/c queue for several values of ρ and c and we give the delay probability P_W, the mean queue size EL_q and the coefficient of variation cvL_q of the queue size. The top numbers in table 3.1 correspond to the exact values, the second top numbers correspond to the approximate values of (3.16)-(3.18) and the third top numbers correspond to the approximate values (3.39)-(3.41). The exact values were taken from [18]. In the tables 3.2-3.3 we deal with the following three phase-type densities represented by an Erlang density, a mixture of Erlang densities and a hyperexponential density,

Case i $\quad f(t) = \mu^2 t e^{-\mu t} (c_S^2 = 0.5)$,

Case ii $\quad f(t) = p\mu e^{-\mu t} + (1/2)(1-p)\mu^3 t^2 e^{-\mu t}$ with $p = 2/3 - (1/6)\sqrt{7}$ $(c_S^2 = 0.5)$

Case iii $\quad f(t) = p\mu_1 e^{-\mu_1 t} + (1-p)\mu_2 e^{-\mu_2 t}$ with $\frac{p}{\mu_1} = \frac{1-p}{\mu_2}$, $p = (1+1\sqrt{5})/2$ $(c_S^2 = 1.5)$.

For these phase-type service times we give in table 3.2 the delay probability P_W, the mean queue size EL_q and the coefficient of variation cvL_q of the queue size for several values of c with a traffic intensity $\rho = 0.8$, where the top numbers in table 3.2 correspond to the exact values. The exact values were computed by using the decomposition method of [29] to solve equilibrium state equations. Finally, table 3.3 concerns the coefficient of variation c_D of the interdeparture time T_D for deterministic service times $(c_S^2 = 0)$ and the above three phase-type service times for several values of c where the traffic intensity $\rho = 0.8$. The top numbers in table 3.3 give the simulated actual values of c_D with a 95% percent confidence interval and the second top numbers give the approximate value of c_D corresponding to (3.21).

4. Appendix

Proof of Theorem 3.1. Under the condition that the system is empty

Table 3.1. Delay Probability and Queue Size for M/D/C Queue (exact, appr. $\overline{\text{appr.}}$)

	$\rho=0.5$			$\rho=0.8$			$\rho=0.9$			$\rho=0.95$		
	P_W	EL_q	cvL_q	P_W	EL_q	cvL_q	P_W	EL_q	cvL_q	P_W	EL_q	cvL_q
c=2	.3233	.1767	3.110	.7091	1.445	1.515	.8471	3.865	1.231	.9227	8.824	1.110
	.3333	.1944	2.986	.7111	1.517	1.471	.8526	3.965	1.207	.9256	8.940	1.097
	.3193	.1807	3.102	.6915	1.442	1.523	.8393	3.850	1.236	.9180	8.801	1.113
c=3	.2253	.1308	3.717	.6325	1.329	1.614	.8077	3.721	1.271	.9018	8.665	1.129
	.2368	.1480	3.531	.6472	1.424	1.551	.8171	3.861	1.237	.9070	8.832	1.110
	.2227	.1326	3.730	.6226	1.319	1.630	.7996	3.694	1.280	.8968	8.627	1.133
c=5	.1213	.0766	5.051	.5336	1.156	1.787	.7478	3.495	1.339	.8692	8.411	1.159
	.1304	.0869	4.778	.5541	1.256	1.706	.7625	3.660	1.296	.8778	8.617	1.136
	.1206	.0750	5.124	.5283	1.134	1.814	.7427	3.451	1.353	.8658	8.351	1.167
c=10	.0331	.0237	9.612	.3847	.8786	2.163	.6469	3.101	1.474	.8116	7.952	1.218
	.0361	.0254	9.152	.4092	.9523	2.057	.6687	3.256	1.423	.8256	8.164	1.192
	.0330	.0212	9.929	.3875	.8400	2.200	.6491	3.029	1.494	.8128	7.856	1.230
c=15	.0104	.0080	17.08	.2955	.7012	2.510	.5771	2.820	1.587	.7695	7.610	1.266
	.0113	.0081	16.40	.3192	.7501	2.384	.6026	2.949	1.532	.7870	7.803	1.239
	.0104	.0067	17.87	.3016	.6559	2.550	.5842	2.730	1.609	.7743	7.489	1.279
c=25				.1900	.4774	3.196	.4793	2.412	1.782	.7063	7.088	1.344
				.2091	.4954	3.029	.5079	2.497	1.721	.7284	7.239	1.317
				.1973	.4301	3.236	.4920	2.302	1.806	.7164	6.931	1.359
c=50				.0776	.2142	5.123	.3355	1.779	2.203	.6012	6.195	1.497
				.0870	.2073	4.840	.3639	1.795	2.125	.6291	6.264	1.470
				.0819	.1789	5.161	.3522	1.649	2.225	.6185	5.987	1.516
c=100				.0176	.0541	10.94	.1953	1.110	2.989	.4751	5.077	1.740
				.0196	.0470	10.34	.2169	1.072	2.871	.5065	5.047	1.710
				.0185	.0404	11.02	.2099	.9833	2.998	.4979	4.820	1.760
c=200				.0013	.0043	41.14	.0837	.5194	4.707	.3351	3.766	2.149
				.0014	.0033	39.24	.0945	.4672	4.497	.3653	3.642	2.107
				.0013	.0028	41.79	.0914	.4282	4.683	.3590	3.476	2.163

Table 3.2. Delay Prob. and Queue Size for Phase-type Services (exact, appr.)

	case i (c_S^2=0.5)			case ii (c_S^2=0.5)			case iii (c_S^2=1.5)		
	P_W	EL_q	cvL_q	P_W	EL_q	cvL_q	P_W	EL_q	cvL_q
c=2	.7087 .7111	2.148 2.169	1.485 1.475	.7083 .7111	2.151 2.132	1.478 1.487	.7131 .7111	3.522 3.484	1.490 1.501
c=3	.6432 .6472	1.964 1.992	1.585 1.570	.6426 .6472	1.969 1.951	1.577 1.585	.6503 .6472	3.183 3.148	1.595 1.610
c=4	.5914 .5964	1.816 1.847	1.675 1.657	.5907 .5964	1.823 1.804	1.667 1.674	.6003 .5964	2.917 2.890	1.689 1.705
c=5	.5484 .5541	1.693 1.723	1.758 1.737	.5477 .5541	1.700 1.679	1.750 1.758	.5584 .5541	2.697 2.680	1.775 1.792
c=6	.5116 .5178	1.586 1.615	1.837 1.814	.5108 .5178	1.594 1.571	1.829 1.836	.5224 .5178	2.510 2.501	1.855 1.874
c=7	.4794 .4859	1.493 1.520	1.913 1.888	.4786 .4859	1.501 1.476	1.904 1.912	.4908 .4859	2.346 2.345	1.933 1.952
c=8	.4508 .4576	1.409 1.434	1.986 1.959	.4501 .4576	1.417 1.391	1.977 1.986	.4627 .4576	2.202 2.207	2.007 2.028
c=9	.4253 .4322	1.333 1.357	2.057 2.029	.4245 .4322	1.342 1.314	2.049 2.057	.4373 .4322	2.073 2.083	2.079 2.101
c=10	.4021 .4092	1.265 1.286	2.127 2.098	.4014 .4092	1.274 1.245	2.119 2.127	.4143 .4092	1.956 1.971	2.150 2.173
c=15	.3122 .3192	.9955 1.008	2.466 2.430	.3116 .3192	1.004 .9719	2.457 2.466	.3241 .3192	1.507 1.536	2.490 2.518

Table 3.3. The Coefficient of Variation of the Output Process (sim., appr.)

	c_S^2 = 0	case i(c_S^2=0.5)	case ii(c_S^2=0.5)	case iii(c_S^2=1.5)
c=2	.7438 (± .0072) .6849	.8836 (± .0058) .8543	.8979 (± .0064) .8455	1.065 (± .0077) 1.121
c=3	.8074 (± .0073) .7308	.9136 (± .0069) .8725	.9294 (± .0064) .8632	1.043 (± .0077) 1.106
c=4	.8418 (± .0080) .7617	.9321 (± .0063) .8856	.9502 (± .0046) .8760	1.030 (± .0072) 1.096
c=5	.8644 (± .0068) .7847	.9474 (± .0051) .8959	.9635 (± .0068) .8861	1.021 (± .0064) 1.089
c=10	.9303 (± .0040) .8522	.9734 (± .0029) .9273	.9859 (± .0041) .9182	1.007 (± .0038) 1.065
c=15	.9527 (± .0054) .8884	.9849 (± .0047) .9273	.9921 (± .0041) .9370	1.006 (± .0039) 1.051

at epoch 0, define $M_{nk}(t)$ as the joint probability that exactly n customers arrive in (0,t) and that the service completions of the first k customers all occur beyond time t, $1 \leq k \leq c$ and $n \geq k$. In the same way as (3.23), we derive for $1 \leq k \leq c$ and $n \geq k$,

$$\frac{d}{dt} M_{nk}(t) = -\lambda M_{nk}(t) + \lambda(1-F(t))M_{n-1,k-1}(t), \quad t > 0, \tag{A.1}$$

where $M_{n0}(t)$ is defined by

$$M_{n0}(t) = e^{-\lambda t} \frac{(\lambda t)^n}{n!}, \quad n \geq 0, \; t \geq 0. \tag{A.2}$$

Using the approximation assumption with specification (3.10) and using the argument below (2.13), it follows that

$$A_{nj} = \int_0^\infty (1-F_e(t))^j M_{n-j,n-j}(t)dt, \quad 0 \leq j \leq n \leq c-1, \tag{A.3}$$

$$A_{nj} = \int_0^\infty (1-F_e(t))^j M_{n-j,c-j}(t)dt, \quad 0 \leq j \leq c-1, \; n \geq c, \tag{A.4}$$

$$A_{nj} = \int_0^\infty (1-F(ct))e^{-\lambda t} \frac{(\lambda t)^{n-j}}{(n-j)!} dt, \quad c \leq j \leq n. \tag{A.5}$$

By (A.1) and (3.25) we can rewrite (A.3) for $0 \leq j < n$ as

$$A_{nj} = \int_0^\infty (1-F_e(t))^j (1-F(t))M_{n-j-1,n-j-1}(t)dt + \tag{A.6}$$

$$- \frac{j}{\lambda ES} \int_0^\infty (1-F_e(t))^{j-1}(1-F(t))M_{n-j,n-j}(t)dt.$$

where for $j = 0$ the second term in the right side of (A.6) vanishes. Now we first derive from (3.8) that

$$q_k = \frac{1}{\lambda} \frac{(\lambda ES)^k}{k!}, \quad 0 \leq k \leq c-1. \tag{A.7}$$

By (A.7) and (3.9) we get (3.12). We prove (A.7) by induction on n. Clearly, by (3.8), (A.7) holds for $k = 0$. Fix $1 \leq n \leq c-1$. Assume that (A.7) holds for $k = 0,\ldots,n-1$. Using this induction assumption and (A.6), it is readily verified from (3.8) that

$$(1-\lambda A_{nn})q_n = \frac{(\lambda ES)^{n-1}}{(n-1)!} \int_0^\infty (1-F_e(t))^{n-1}(1-F(t))M_{00}(t)dt. \tag{A.8}$$

Further, by partial integration, we get from (A.3) that

$$1-\lambda A_{nn} = \frac{n}{ES} \int_0^\infty (1-F_e(t))^{n-1}(1-F(t))e^{-\lambda t}dt. \tag{A.9}$$

By (A.8)-(A.9), we get (A.7) for $k = n$. Next, we verify (3.13). By rewriting (A.4) in the same way as (A.3) and using (A.7), we find for $n \geq c$

$$\sum_{j=0}^{c-1} \lambda q_j A_{nj} = \lambda q_{c-1} \int_0^\infty (1-F_e(t))^{c-1}(1-F(t))M_{n-c,0}(t)dt. \tag{A.10}$$

By (3.8)-(3.9), (A.5) and (A.10) we get (3.13). Finally, by (3.12)-(3.13) and $\Sigma_0^\infty p_n(\text{appr}) = 1$, we verify $p_0(\text{appr}) = \Omega$.

5. Acknowledgements

We would like to thank R. Corver and H. Groenevelt for their help in writing computer programs.

6. References

[1] Boxma, O. J., Cohen, J. W. & Huffels, N. (1979), Approximations of the mean waiting time in an M/G/c queueing system. Oper. Res., 27, 1115-1127.

[2] Burman, D. Y. & Smith, D. R. (1981), A light-traffic theorem for multi-server queues, Report, Bell Laboratories, Holmdel, New Jersey.

[3] Bux, W. and Herzog, U. (1977), The phase-concept, approximation of measured data and performance analysis, in: K. M. Chandy and M. Reiser (eds.), Proceedings of the International Symposium on Computer Performance, Modeling, Measurement and Evaluation, North-Holland, Amsterdam, pp. 23-38.

[4] Bux, W. (1979), Single server queues with general interarrival and phase-type service time distributions-computational algorithms, Proceedings of the 9th International Teletraffic Congress (Torremolinos).

[5] Cohen, J. W. (1977), On up- and downcrossings, J. Appl. Probab., 14, 405-410.

[6] Crommelin, C. D. (1932), Delay probability formulae when the holding times are constant. P. O. Elec. Engrs. J., 25, 41-50.

[7] Federgruen, A. & Tijms, H. C. (1980), Computation of the stationary distribution of the queue size in an M/G/1 queueing system with variable service rate, J. Appl. Probab., 17, 515-522.

[8] Halachmi, B. & Franta, W. R. (1978), A diffusion process approximation to the multi-server queue, Management Sci., 24, 522-529.

[9] Heffer, J. C. (1969), Steady-state solution of the $M/E_k/c(\infty,FIFO)$ queueing system, J. Canadian Oper. Res. Soc., 7, 16-30.

[10] Hillier, F. S. & Lo, F. D. (1971), Tables for multiple-server queueing systems involving Erlang distributions, Technical Report no. 31, Dept. of Operations Research, Stanford Univ., Stanford.

[11] Hokstad, P. (1978), Approximations for the M/G/m queue, Oper. Res., 26, 511-523.

[12] Hokstad, P. (1980), The steady-state solution of the $M/K_2/m$ queue, Adv. in Appl. Probab., 12, 799-823.

[13] Ishikawa, A. (1979), On the equilibrium solution for the queueing system $GI/E_k/m$, TRU Math., Vol. 15, 47-66.

[14] Kampe, G. & Kühn, P. J. (1976), Graded delay systems with infinite or finite source traffic and exponential or constant holding time, Proceedings of the 8th International Teletraffic Congress (Melbourne), 256-1/10.

[15] Kobayashi, H. (1976), Modelling and Analysis: an Introduction to System Performance and Evaluation Methodology, Addison-Wesley, Reading.

[16] Köllerström, J. (1974), Heavy traffic theory for queues with several servers, J. Appl. Probab., 11, 544-552.

[17] Kühn, P. J. (1972), On the Calculation of Waiting Times in Switching and Computer Systems, 15th Report on Studies in Congestion Theory, Institute of Switching and Data Technics, University of Stuttgart.

[18] Kuhn, P. J. (1976), Tables on Delay Systems, Institute of Switching and Data Technics, University of Stuttgart.

[19] Lavenberg, S. S. (1975), The steady-state queueing time distribution for the M/G/1 finite capacity queue. Management Sci., 25, 501-506.

[20] Marshall, K. T. & Wolff, R. W. (1971), Customer average and the time average queue lengths and waiting times, J. Appl. Probab., 8, 535-542.

[21] Neuts, M. F. (1981), Matrix-Geometric Solutions in Stochastic Models - An algorithmic Approach, The Johns Hopkins University Press.

[22] Newell, G. (1973), Approximate Stochastic Behaviour of n-Server Service Systems with Large n, Lecture Notes in Economics and Mathematical Systems, 87, Springer-Verlag, Berlin.

[23] Nozaki, S. A. & Ross, S. M. (1978), Approximations in finite-capacity multi-server queues with Poisson arrivals, J. Appl. Probab., 15, 826-834.

[24] Pack, C. D. (1978), The output of a multiserver queueing system, Oper. Res., 26, 492-509.

[25] Ross, S. M. (1970), Applied Probability Models with Optimization Applications, Holden-Day, San Francisco.

[26] Sakasegawa, H. (1977), An approximation formula $L_q \simeq \alpha\rho^{\beta}/(1-\rho)$, Ann. Inst. Statist. Math., 29, Part A, 67-75.

[27] Stidham, S., Jr. (1972), Regenerative processes in the theory of queues with applications to the alternating priority queue, Adv. in Appl. Probab., 4, 542-557.

[28] Takacs, L. (1962), Introduction to the Theory of Queues, Oxford University Press.

[29] Takahashi, Y. & Takami, Y. (1976), A numerical method for the steady-state probabilities of a GI/G/c queueing system in a general class, J. Oper. Res. Soc. Japan, 19, 147-157.

[30] Takahashi, Y. (1977), An approximation formula for the mean waiting time of an M/G/c queue, J. Oper. Res. Soc. Japan, 20, 150-163.

[31] Tijms, H. C., Van Hoorn, M. H. & Federgruen, A. (1981), Approximations for the steady-state probabilities in the M/G/c queue, Adv. in Appl. Probab., 13, 186-206.

[32] Tijms, H. C. & Van Hoorn, M. H. (1981), Algorithms for the state probabilities and waiting times in single server queueing systems with random and quasi-random input and phase-type service times, OR Spektrum 2, 145-152.

[33] Van Hoorn, M. H. (1981), Algorithms for the state probabilities in a general class of single server queueing systems with group arrivals, (to appear in Management Sci.).

[34] Van Hoorn, M. H. & Tijms, H. C. (1982), Approximations for the waiting times in the M/G/c queue, (to appear in Performance Evaluation, Vol. 2).

Vrije Universiteit, Amsterdam

Discussant's Report on
"Computational Methods for Single-Server and Multi-Server Queues with Random and Quasirandom Input,"
by H. C. Tijms and M. H. Van Hoorn

This paper presents two recursive schemes: the first computes the exact values of the steady-state probability density function for the M/G/1 system where the arrival rate is allowed to be dependent on the number of customers present, and the second computes an approximation of the steady-state probability density function for the M/G/c system where the arrival rate is not state-dependent. In the M/G/c, corresponding approximations for the probability of delay, the moments of the waiting-time distribution, and the moments of the interdeparture-time distribution can also be computed. My comments and questions are related to this M/G/c approximation.

The approximation is based on a smoothening assumption about the distributions of the time until the next service completion at service completion epochs (see APPROXIMATION ASSUMPTION and (3.10)). To motivate their assumption, the authors make complimentary comparisons of the M/G/c system with light congestion to the M/G/∞ system and of the M/G/c system with heavy congestion, to the M/G/1 system (where the service time is replaced by S/c). It is also interesting to think of treating the M/G/c system in light traffic as an M/G/c/c blocking system and remark that at arbitrary epochs with $j \leq c$ customers present in the M/G/c/c, the time until the next service completion is distributed as the minimum of j independent equilibrium excess service-time distributions. (The authors' approximation is to employ such a distribution at service completion epochs with j customers left behind.) Indeed, the approximation for the expected queueing time proposed in (3.17) and (3.20) can be viewed as a linear combination of the expected queueing time for M/G/c in heavy traffic as shown in Nozaki and Ross

(1978) and the expected queueing time for M/G/c in light traffic as conjectured in Boxma, Cohen and Huffels (1979) and recently proven in Burman and Smith (unpublished). Specifically, (3.17) and (3.20) imply that

$$EW_q(\text{appr}) = \rho \frac{ES^2}{2(ES)^2} EW_q(\text{exp}) + (1-\rho) \frac{c\gamma_1}{ES} EW_q(\text{exp}),$$

and it can be noted that $\frac{ES^2}{2(ES)^2} EW_q(\text{exp})$ is the heavy-traffic delay approximation and $\frac{c\gamma_1}{ES} EW_q(\text{exp})$ is the light-traffic expected delay approximation.

The difficulty in testing approximate solutions is, of course, the availability of exact solutions for various M/G/c systems against which comparisons can be made. On the basis of numerical work can the quality of the approximations be related to the coefficient of variation of the service-time distribution? Is their behavior monotone in the quality as a function of the coefficient of variation?

Putting the question of accuracy aside, the usefulness of the approximation is in its tractability. How costly is the execution of the approximation for large queueing systems, i.e., large numbers of servers and phase-type service distributions with large numbers of phases?

Finally, perhaps this approximation can be extended to the GI/G/c system by making a similar smoothening assumption about the distributions of the time until the next arrival at service completion epochs.

Author's Response to Discussant's Report

We are indebted for pointing out that our approximation for the mean waiting time in the M/G/c queue is a linear combination of exact mean waiting time results for light and heavy traffic. This observation is in support of the quality of our approximations.

With regard to the accuracy of the approximations for the M/G/c queue, we have compared the approximate results with exact results for deterministic service times and for special phase-type service times. For deterministic services tables have been given in Kühn (1976) for c values up to 250. For phase-type services it is only computationally feasible to calculate exact values for the state probabilities for smaller values of c, because of dimensionality problems. Using the decomposition algorithm of Takahashi and Takami (1976), we have computed tables both for mixtures of Erlang-1 and Erlang-2 service distributions with the same scale parameter for c values up to 50 and for hyper-exponential service distributions with coefficient of variation up to 3 for c values up to 25, cf. reference 1 below. For these cases with a coefficient of variation of the service time between 0 and 3 we found that the approximations are of a good quality. For large values of the coefficient of variation our numerical experiments indicate that the approximations may become less good when the traffic intensity is not close enough to 1. A more detailed discussion of the validation of the approximations will be given in reference 2.

In reply to the question of the applicability of the approximation algorithms for practical purposes, it can be said that these algorithms are both easy to understand and easy to program. The only nontrivial aspect is the evaluation of the integrals in the recursive scheme (3.12)-(3.13). For special phase-type distributions as the hyper-exponential distribution the integrals can be reduced to simple sums, otherwise numerical integration procedures have to be used. For constant service times we have used quadrature formulae such as Simpson's rule, while for Erlangian service times we recommend Gauss-Laguerre integration. In reference 2 these numerical aspects are discussed in detail and rules of thumb are given for the accuracy and the tractability of the numerical procedures. The computing time for the recursive approximation scheme does not crucially depend on c; even for very large values of c (say c = 250) the total computing time for the approximate state probabilities is of the order of a few seconds CPU-time. The computing time for the associated approximate waiting time probabilities is in fact independent of c; the average computing time for the waiting time distribution is below 1 second CPU-time. By the accuracy and the low computing time, the approximate methods are well suited for quick use in practice and are superior to simulation. For practical applications in which only the first two moments of the service time are available, we suggest to fit a hyperexponential distribution to these first two moments when the coefficient of variation is above 1 and to fit a mixture of an Erlang-1 and an Erlang-k distribution with the same scale parameters otherwise. Our numerical investigations indicate that for "reasonable" service time densities the cumulative state and waiting time probabilities are rather insensitive to more than the first two moments of the service time.

References

1. Groenevelt, H., Van Hoorn, M. H. and Tijms, H. C. (1982), "Tables for M/G/c Queues with Special Phase-type Service Times," Research Report, Dept. of Actuarial Sciences and Econometrice, Free University, Amsterdam.

2. Van Hoorn, M. H. (1982), "Algorithms and Approximations for Queueing Systems," Ph.D. Thesis, Free University, Amsterdam, (forthcoming).

3. Van Hoorn, M. H. and Tijms, H. C. (1982), "Approximations for the Waiting Times in the M/G/c Queue," Performance Evaluation, 2.

Discussant: Dr. Diane Sheng, Bell Laboratories, Holmdel, New Jersey 07733.

THE TIME FOR A ROUND-TRIP IN A
CYCLE OF EXPONENTIAL QUEUES

H. Daduna
and
R. Schassberger

Abstract

In [1] Chow obtained the steady-state distribution of the cycle time of a customer in a closed tandem queue composed of two exponential FIFO servers and populated by an arbitrary fixed number of customers. The present paper gives the generalization of Chow's result (by means of a different proof) to the case of arbitrarily many exponential servers. Precisely, the model is this (Figure 1):

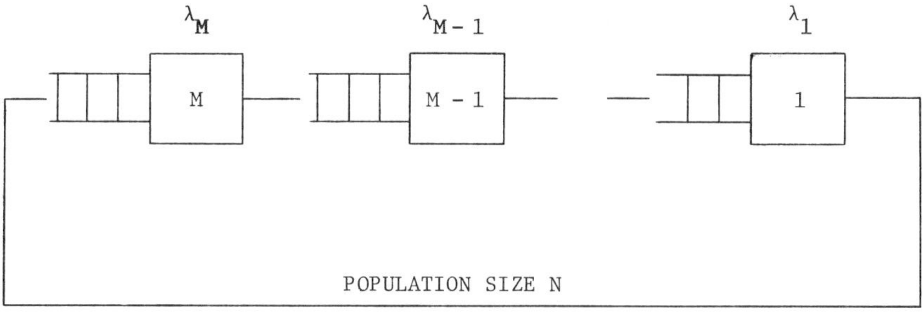

Figure 1

There is a string of M, $M \geq 2$, single servers connected in such a way that a customer, upon departing from server i, joins the queue at server $i-1$, $i = 2,\dots,M$ at no loss of time, and upon departing from server 1 immediately joins the queue at server M. This closed system is populated by a total of N, $N \geq 1$, customers. The customers present at any one of the servers are served in the order of their arrivals. The servers are providing service whenever customers are present. The sequence of service times at server i is a sequence of independent random variables distributed exponentially with mean λ_i^{-1}, $i = 1,\ldots,M$, and these M sequences are mutually independent.

Denote by $Y_i(t)$ the number of customers present at server i at time t, $i = 1,\ldots,M$, $t \geq 0$. Then the stochastic process $\{Y(t);\ t \geq 0\}$, $Y(t) = (Y_M(t),\ldots,Y_1(t))$, is a homogeneous Markov chain on the state space $Z_N^M = \{(j_M,\ldots,j_1):\ 0 \leq j_i \leq N \text{ for } i = 1,\ldots,M,\ j_1+\ldots+j_M = N\}$. Its (unique) stationary probability distribution is well known to be given by

$$P_N^M(j_M,\ldots,j_1) = (K_N^M)^{-1} \prod_{i=1}^{M} \lambda_i^{-j_i}, \quad (j_M,\ldots,j_1) \in Z_N^M,$$

where $(K_N^M)^{-1}$ is a normalizing constant.

The tandem queue shall be said to be in steady state if the population process $\{Y(t);\ t \geq 0\}$ is stationary.

The question to be considered is this: in steady state what amount of time will pass between two successive departures from node 1 of the same customer? Calling this time the steady-state cycle time of a customer, the above question is answered by the following

Theorem. The steady-state cycle time of a customer has the distribution function F_N^M given by

$$F_N^M(t) = \sum_{(j_M,\ldots,j_1)\in Z_{N-1}^M} P_{N-1}^M(j_M,\ldots,j_1)\,(E_{\lambda_M}^{j_M+1} * \ldots * E_{\lambda_1}^{j_1+1})(t), \quad t \geq 0,$$

where $*$ denotes convolution, E_λ^k is the distribution function of the k-fold convolution of the exponential distribution with mean λ^{-1}, $P_{N-1}^M(\cdot)$ denotes the stationary distribution of the population process belonging to population size $N-1$ and having state space Z_{N-1}^M.

The proof of this result starts from the well-known fact that, in steady state, the probability is $P^M_{N-1}(j_M,\ldots,j_1)$ for j_i customers to be found in node i, i = 1,...,M, immediately after the departure from some server, and not counting the departing customer. Thus, if $f^M_N(j_M,\ldots,j_1)$ is the Laplace-Stieltjes transform (LST) of the distribution function of the corresponding conditional travel time of the departing customer until his next departure from the same server, then the LST f^M_n of F^M_N is given by

$$f^M_N(s) = \sum_{(j_M,\ldots,j_1)\varepsilon Z^M_{N-1}} P^M_{N-1}(j_M,\ldots,j_1) f^M_N(j_M,\ldots,j_1)(s).$$

Recursive arguments involving these transforms constitute the remainder of the proof.

The result of the above theorem can be applied to yield a well-known result of Reich ([2]).

When cutting the link between servers 1 and M and feeding, instead, server M by a Poisson arrival stream of intensity λ, $\lambda < \lambda_1,\ldots,\lambda_M$, then a familiar open tandem queue emerges. The population process is ergodic and, when stationary, shows a total population of N, $N \geq 0$, customers with probability

$$P_N = \sum_{(j_M,\ldots,j_1)\varepsilon Z^M_N} \prod_{i=1}^{M} \left(1 - \frac{\lambda}{\lambda_i}\right)\left(\frac{\lambda}{\lambda_i}\right)^{j_i} .$$

Hence the steady-state travel time distribution function for a job through this queue has the LST $f^M(s)$ given by

$$f^M(s) = \sum_{N=0}^{\infty} P_N \, f^M_{N+1}(s) = \prod_{i=1}^{M} \frac{\lambda_i - \lambda}{\lambda_i - \lambda + s} ,$$

which is Reich's result.

1. References

[1] Chow, W. M., The cycle time distribution of exponential cyclic queues, J. Assoc. Comput. Mach., 27, 281-286, 1981.

[2] Reich, E., Waiting times when queues are in tandem. Ann. Math. Stat. 28, 768-773, 1957.

Technische Universitat Berlin, Fachbereich 3 - Mathematik, West Berlin

COMPUTATIONAL ASPECTS OF APPLIED PROBABILITY
Narayan Bhat, Chairman

J. Keilson & U. Sumita
H. Schwetman
V. Ramaswami & D. M. Lucantoni
D. R. Miller

WAITING TIME DISTRIBUTION RESPONSE TO TRAFFIC SURGES VIA THE LAGUERRE TRANSFORM

J. Keilson

and

U. Sumita

1. Introduction and Summary

The Laguerre transform, described in detail elsewhere [A] [B], is a novel tool for mechanizing numerically the operations of convolution, differentiation, integration and polynomial multiplication required for applied probability evaluation. In the M/G/1 queue for example the busy period cumulative distribution function is known [6] to have the real time representation

$$S(\tau) = \int_0^{\tau} \sum_{k=0}^{\infty} e^{-\lambda\tau'} \frac{(\lambda\tau')^k}{(k+1)!} \, dA^{(k+1)}(\tau') d\tau' \tag{1.1}$$

where λ is the arrival rate, $A(\tau)$ is the service time c.d.f., and $A^{(k)}(\tau)$ is its k-fold convolution. Evaluation of (1.1) numerically by discretization and Simpson's rule procedure is clumsy, costly and of uncertain accuracy.

A similar set of difficulties arise in the context of the waiting-time distribution for the G/G/1 queue described by the Lindley process [2]. This Markov process states that

$$W_{k+1} = \max[0, W_k + \xi_{k+1}] \tag{1.2}$$

where (ξ_k) is a sequence of i.i.d. variates, and $W_o = 0$. The stationary distribution of W_k is governed by a Weiner-Hopf type equation with awkward factorization difficulties, necessitating artificial assumption on the form of the underlying service time and interarrival time distributions. A more direct and convenient evaluation of the stationary distribution of W_k is provided by the Laguerre transform method, which evaluates the distributions of W_k sequentially through its ability to mechanize the convolutions required. A description of this procedure and numerical results are given in [B], for distributions of ξ_k for which Weiner-Hopf factorization is intractable.

The power of the Laguerre transform method is seen more vividly in the context of time-inhomogeneous processes where analysis breaks down completely, and seems quite hopeless. Suppose for example one wishes to model the delay in a server system when the input process has a period of peak intensity during which a queue builds and subsides subsequently when arrivals slow down. Consider specifically a single server system governed by Poisson arrivals at rate $\lambda(t)$, a general service time distribution and first come-first served discipline, i.e., $M(\lambda(t))/G/1$. The server-backlog process is quite nasty analytically, and to study system performance simulation would normally be tried. Such simulation provides crude moment information and cruder empirical delay time distribution. The Laguerre transform offers a substitute for simulation of the following character. By mechanizing the operations of convolution differentiation, tail integration and polynomial multiplication, it is able to generate a sequence of distributions indexed by a real time parameter describing the evolution of some system variate of interest e.g., the waiting time at the service facility above. Such distributions, in so far as they capture events of low probability with far greater accuracy and conviction than from simulation, may have correspondingly greater appeal.

Of great theoretical and practical importance for the Laguerre transform method, is the bounding of error due to truncation of the sequence representing a function of interest. A concluding section addresses itself to this need and obtains simple useful bounds for a family of functions of key importance.

2. The Laguerre Transform

In two previous papers [A], [B], the Laguerre transform has been developed as an algorithmic procedure for evaluation of probability distributions requiring multiple convolutions in conjunction with other algebraic and summation operations. The original paper [A[dealt with positive random variables and their distributions. A sequel [B] has extended the formalism to handle distributions and processes on the full continuum. In this section we give a concise summary of the two basic papers for the reader's convenience.

(A) The Laguerre Transform for $f(x) \in L_2(0,\infty)$

The Laguerre polynomials, defined by the Rodrigues formula

$$L_n(x) = \frac{e^x}{n!}\left(\frac{d}{dx}\right)^n (x^n e^{-x}),\ n = 0,1,2,..., \tag{2.1}$$

form a set of orthonormal polynomials with weighting function e^{-x} on $(0,\infty)$. The associated Laguerre functions

$$\ell_n(x) = e^{-\frac{1}{2}x} L_n(x),\ n = 0,1,2,..., \tag{2.2}$$

then provide an orthonormal basis in $L_2(0,\infty)$. For any $f(x) \in L_2(0,\infty)$, one has the Fourier-Laguerre expansion†

$$f(x) = \sum_{n=0}^{\infty} f_n^\dagger \ell_n(x),\ x \geq 0;\ f_n^\dagger = \int_0^\infty f(x)\ell_n(x)dx,\ n = 0,1,2,... \tag{2.3}$$

Let $\phi_f(s) = \int_0^\infty e^{-sx} f(x)dx$. It is known that $\ell_n(x)$ has the Laplace transform

$$\lambda_n(s) = \int_0^\infty e^{-sx} L_n(x)\,dx = \frac{1}{s+\frac{1}{2}}\left(\frac{s-\frac{1}{2}}{s+\frac{1}{2}}\right)^n, \quad \mathrm{Re}(s) > -\frac{1}{2},\ n = 0,1,2,\ldots. \tag{2.4}$$

When $(f_n^\dagger)^\infty \varepsilon \ell_1$, one has formally, from (2.3) and (2.4), that

$$\phi_f(s) = \sum_{n=0}^\infty f_n^\dagger \frac{1}{s+\frac{1}{2}}\left(\frac{s-\frac{1}{2}}{s+\frac{1}{2}}\right)^n \tag{2.5}$$

which is valid at least for s pure imaginary. Let

$$u = \frac{s-\frac{1}{2}}{s+\frac{1}{2}}$$

and define

$$T_f^\dagger(u) = \sum_{n=0}^\infty f_n^\dagger u^n; \quad T_f^\#(u) = \sum_{n=0}^\infty f_n^\# u^n = (1-u)T_f^\dagger(u). \tag{2.6}$$

Corresponding to (2.5) one has, on the unit circle in the complex u-plane, that

$$\phi_f\left(\frac{1}{2}\,\frac{1-u}{1+u}\right) = (1-u)T_f^\dagger(u) = T_f^\#(u). \tag{2.7}$$

We note, from (2.6), that

$$f_o^\dagger = f_o^\#; \quad f_n^\# = f_n^\dagger - f_{n-1}^\dagger,\ n \geq 1; \quad f_n^\dagger = \sum_{j=0}^n f_j^\#. \tag{2.8}$$

Let $f(x)*g(x) = \int_0^x f(x-y)g(y)\,dy$. Since $\phi_{f*g}(s) = \phi_f(s)\phi_g(s)$, one has from (2.7) that

$$T_{f*g}^\#(u) = T_f^\#(u)\,T_g^\#(u). \tag{2.9}$$

The transformations $T_f^\#(u)$ and $T_g^\#(u)$ in (2.7) map functions f(x), g(x) into sequences $(f_n^\#)$, $(g_n^\#)$. Their continuum convolution f(x)*g(x) is mapped into a lattice convolution $(f_n^\#)*(g_n^\#)$, and then back onto the continuum via (2.8) and the representation (2.3). The values of $\ell_n(x)$

are computed quickly by a recurrence formula [A] and this Laguerre transform procedure provides an efficient algorithmic basis for numerical evaluation of convolutions.

(B) The Bilateral Laguerre Transform for $f(x) \varepsilon L_2(-\infty,\infty)$

The Laguerre transform described in (A) can be extended in a natural manner to two-sided functions on the full continuum. Let $U(x) = 1,\ x \geq 0$ and $U(x) = 0,\ x < 0$. Then

$$f(x) = f_+(x) + f_-(x);\ f_+(x) = f(x)U(x);\ f_-(x) = f(x)U(-x) \tag{2.10}$$

where the discrepancy at $x=0$ may be ignored. Clearly $f(x) \varepsilon L_2(-\infty,\infty)$ implies that $f_+(x) \varepsilon L_2(0,\infty)$ and $f_-(x) \varepsilon L_2(-\infty,0)$ and one may write (see endnote for Equation (2.3) on page 4)

$$f_+(x) = \sum_{n=0}^{\infty} f^{\dagger}_{n+} \ell_n(x)U(x);\ f_-(x) = \sum_{n=0}^{\infty} f^{\dagger}_{n-} \ell_n(-x)U(-x). \tag{2.11}$$

Let

$$h_n(x) = \begin{cases} \ell_n(x)U(x) & n \geq 0 \\ -\ell_{-n-1}(-x)U(-x), & n < 0. \end{cases} \tag{2.12}$$

From (2.10) and (2.11) one then has

$$f(x) = \sum_{n=-\infty}^{\infty} f^{\dagger}_n h_n(x), \tag{2.13}$$

where

$$f^{\dagger}_n = f^{\dagger}_{n+},\ n \geq 0;\ f^{\dagger}_n = -f^{\dagger}_{(-n-1)-},\ n < 0. \tag{2.14}$$

It can be readily seen that the set of functions $(h_n(x))_{-\infty}^{\infty}$ provides a complete orthonormal basis for $L_2(-\infty,\infty)$. Specifically one has

$$f^{\dagger}_n = \int_{-\infty}^{\infty} f(x)h_n(x)dx;\ \sum_{n=-\infty}^{\infty} f^{\dagger 2}_n = \int_{-\infty}^{\infty} f^2(x)dx. \tag{2.15}$$

The bilateral Laplace transform of $h_n(x)$ is given by

$$\chi_n(x) = \int_{-\infty}^{\infty} e^{-sx}\, h_n(x)\,dx = \frac{1}{s+\frac{1}{2}}\left(\frac{s-\frac{1}{2}}{s+\frac{1}{2}}\right)^n, \quad -\frac{1}{2} < \mathrm{Re}(s) < \frac{1}{2},$$

$$-\infty < n < \infty. \qquad (2.16)$$

Let $\phi_{Bf}(s) = \int_{-\infty}^{\infty} e^{-sx}\, f(x)\,dx$. When $(f_n^{\dagger})_{-\infty}^{\infty} \in \ell_1$, one then has

$$\phi_{Bf}\left(\frac{1}{2}\cdot\frac{1+u}{1-u}\right) = (1-u)\; T_f^{\dagger}(u) = T_f^{\#}(u) \qquad (2.17)$$

which is valid at least on the unit circle in the complex u-plane where

$$T_f^{\dagger}(u) = \sum_{n=-\infty}^{\infty} f_n^{\dagger} u^n, \quad T_f^{\#}(u) = \sum_{n=-\infty}^{\infty} f_n^{\#} u^n = (1-u)\,T_f^{\dagger}(u).$$

We note that

$$f_n^{\#} = f_n^{\dagger} - f_{n-1}^{\dagger}; \quad f_n^{\dagger} = \sum_{j=-\infty}^{n} f_j^{\#} = -\sum_{j=n+1}^{\infty} f_j^{\#}. \qquad (2.18)$$

For the bilateral Laplace transforms one has $\phi_{Bf*g}(s) = \phi_{Bf}(s)\phi_{Bg}(s)$ where $f(x)*g(x) = \int_{-\infty}^{\infty} f(x-y)g(y)\,dy$. Hence one has again,

$$T_{f*g}^{\#}(u) = T_f^{\#}(u)\,T_g^{\#}(u). \qquad (2.19)$$

The two-sided convolution can then be evaluated precisely as before.

For both unilateral and bilateral Laguerre transform, the sharp coefficients for simple combinations of elementary probability density functions are available analytically [A], [B], [3], [4], [5].

3. The Lindley Process With Independent But Nonidentical Increments

The Lindley process [2] is a random walk on the nonnegative half line defined by

$$W_{k+1} = [W_k + \xi_{k+1}]^+, \quad k = 0,1,2,\ldots \qquad (3.1)$$

where $[x]^+ = \max\{0,x\}$ and the increments ξ_k are i.i.d. W_o is a random variable with known distribution independent of ξ_k. An iterative schema is developed in [B] via the bilateral Laguerre transform for computing the distribution of W_k, when the common distribution of ξ_k is absolutely continuous with density function $a(x) \in L_2(-\infty,\infty)$. As an example the waiting time distribution of the k-th customer for a G/G/1 queueing system is evaluated there.

The iterative schema can be extended easily to the case where ξ_k are independent but not identically distributed. Even though this extension is straightforward, it is of practical importance and we describe it here for completeness. Let

$$A_k(x) = P[\xi_k \leq x] = \int_{-\infty}^{x} a_k(y)dy; \; a_k(x) \in L_2(-\infty,\infty), \tag{3.2}$$

$$E_k = P[W_k = 0], \tag{3.3}$$

and

$$F_k(x) = P[W_k \leq x] = \begin{cases} 0 & x < 0 \\ E_k + \int_0^x f_k(y)dy & x \geq 0 . \end{cases} \tag{3.4}$$

If we define $W_{k+1}^H = W_k + \xi_{k+1}$, then the density of W_{k+1}^H is given by

$$f_{k+1}^H(x) = E_k \, a_k(x) + f_k(x)*a_k(x). \tag{3.5}$$

Let $(a_n^{\#}(k))_{-\infty}^{\infty}$, $(f_n^{\#}(k))_o^{\infty}$ and $(f_n^{H\#}(k))_{-\infty}^{\infty}$ be the Laguerre sharp coefficients of $a_k(x)$, $f_k(x)$ and $f_k^H(x)$ respectively. When E_k, $(a_n^{\#}(k))_{-\infty}^{\infty}$ and $f_n^{\#}(k)_o^{\infty}$ are known, one has from (3.5) that

$$f_n^{H\#}(k+1) = E_k \, a_n^{\#}(k) + \sum_{m=0}^{\infty} a_{n-m}^{\#}(k) \, f_m^{\#}(k), \; -\infty < n < \infty . \tag{3.6}$$

Since $f_{k+1}(x) = f_{k+1}^H(x)U(x)$ and $\sum_{n=-\infty}^{\infty} f_n^{H\#} = 0$ from (2.18), one obtains that

$$f_o^{\#}(k+1) = \sum_{n=-\infty}^{o} f_n^{H\#}(k+1); \; f_n^{\#}(k+1) = f_n^{H\#}(k+1), \; n \geq 1. \tag{3.7}$$

Finally

$$E_{k+1} = 1 - \int_0^{\infty} f_{k+1}(y)dy = 1 - 2 \sum_{n=0}^{\infty} (-1)^n \, f_n^{\dagger}(k+1)$$

since $\int_0^\infty \ell_n(x)dx = (-1)^n\, 2$ (cf. Appendix A of [A]). From (2.8) this leads to

$$E_{k+1} = 1 + 2 \sum_{n=0}^{\infty} f^{\#}_{2n+1}(k+1) \; . \tag{3.8}$$

Hence if E_o, $f_o(x)$ and $a_k(x)$ are known, we again have an iterative procedure for evaluating E_k and $f_k(x)$ via (3.6), (3.7) and (3.8). Applying a simple operational property of the Laguerre transform, $F_k(x)$ can also be computed easily. The first two moments are also available via the Laguerre sharp coefficients. [B].

4. Sequence of Waiting Times for Traffic Surges

We now illustrate the procedure described in Section 3 in a queueing context. Consider an $M(\lambda(t))/M(\mu)/1$ queueing system. Specifically the interarrival times of customers are independently and exponentially distributed with time dependent parameter $\lambda(t)$. There is one server in the system and the service time of each customer is independent and exponentially distributed with constant parameter μ. Let $M(t)$ be the number of customers who arrived at the system in $(0,t)$. Let

$$p_k(t) = P[M(t) = k]; \; g(t,u) = \sum_{k=0}^{\infty} p_k(t)u^k . \tag{4.1}$$

From the forward equations, one then easily finds that

$$\frac{\partial}{\partial t} g(t,u) = -\lambda(t)(1-u)g(t,u). \tag{4.2}$$

When the system starts empty, one has $g(o+, u) = 1$ so that the differential equation (4.2) has the solution

$$g(t,u) = e^{-L(t)(1-u)} \tag{4.3}$$

where $L(t) = \int_0^t \lambda(y)dy = \mu \int_0^t \rho(y)dy$ and $\rho(t) = \lambda(t)/\mu$ is the traffic intensity at t. It can be readily seen that

$$E[M(t)] = \frac{\partial}{\partial u} g(t,u)\Big|_{u=1} = L(t). \tag{4.4}$$

Let t_k be the epoch determined by $E[M(t_k)] = L(t_k) = k$, i.e., let the mean number of customers having arrived at the system in $(0,t_k)$ be k. The sequence t_k is machine generated by solving the equation $L(t_k) = k$. A hybrid description, in the sense given in Section 1, approximating the system behavior is then obtained from the following assumption.

> The k-th customer has the interarrival time T_k exponentially distributed with parameter $\lambda(t_k)$ where $E[M(t_k)] = L(t_k) = k$. The random variables T_k are independent. (4.5)

Under this assumption, which supplants the $M(\lambda(t))/M(\mu)/1$ model but preserves its basic character, the waiting time W_k of the k-th customer can be described by

$$W_{k+1} = [W_k + \xi_{k+1}]^+; \ W_o = 0, \ k = 0,1,2,\ldots \tag{4.6}$$

where the increments $\xi_{k+1} = S - T_{k+1}$ are independent with density function

$$a_{k+1}(x) = \{\mu e^{-\mu x} U(x)\} * \{\lambda(t_{k+1}) e^{\lambda(t_{k+1})x} U(-x)\}. \tag{4.7}$$

Hence the procedure given in Section 3 is applicable and the distributions of W_k can be obtained recursively.

As a concrete numerical example, let

$$\lambda(t) = \frac{5}{54}\left(1 + \frac{\pi}{2}\sin\frac{\pi t}{540}\right), \ 0 \leq t \leq 540, \tag{4.8}$$

with time unit of a minute. Let

$$\mu = \frac{25}{108} \approx 0.2315. \tag{4.9}$$

The traffic intensity $\rho(t)$ is depicted in Figure 4.1.

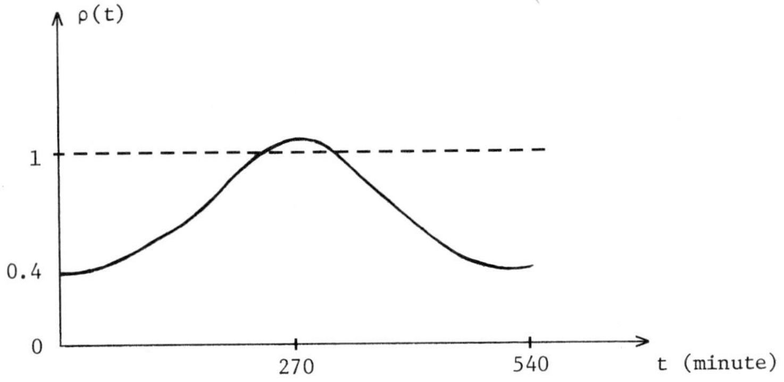

Figure 4.1

For $L(t) = E[M(t)]$ in (4.4), one has

$$L(t) = 50[\frac{t}{540} + \frac{1}{2}(1 - \cos\frac{\pi t}{540})], \quad 0 \leq t \leq 540. \tag{4.10}$$

We note that $L(540) = 100$ and thus 100 customers are expected in 9 hours. The mean traffic intensity $\bar{\rho}$ is given by

$$\bar{\rho} = \frac{1}{540}\int_0^{540} \rho(y)\,dy = 0.8. \tag{4.11}$$

The epoch t_k for determining the interarrival time parameter $\lambda(t_k)$ of the k-th customer under the assumption (4.5) is obtained as a unique zero of

$$f(z_k) = z_k - \pi(\frac{k}{50} - \frac{1}{2}) - \frac{\pi}{2}\cos z_k = 0, \quad 0 \leq z_k \leq \pi \tag{4.12}$$

where $t_k = 540\, z_k/\pi$.

Table 1 gives $\lambda(t_k)$, $\rho(t_k)$, E_k and the mean and standard derivation of W_k for $1 \leq k \leq 100$. The traffic intensity $\rho(t)$ takes the maximum at $k = 50$, while the idle probability E_k attains the minimum at $k = 70$ and the mean waiting time attains the maximum $k = 80$. Hence one observes the lag time among the entities, $\rho(t)$, E_k and $E[W_k]$. When

$\lambda(t)$ is constant with the same average traffic intensity $\bar{\rho} = 0.8$, (i.e., $\lambda(t) = \lambda$ with $\lambda/\mu = \bar{\rho} = 0.8$), one has the mean waiting time of 17.28 at ergodicity. We see in our time inhomogeneous example that $E[W_k]$ exceeds 17.28 for $k \geq 25$. In Table 2, the survival functions of W_k are exhibited with E_k in the first row for k = 1, 10, 20, 50 and 100.

Table 1

Period (Minutes) : 540
Num. of Cust. in the Period: 100
Service Rate : .2315
Average Traffic Intensity : .8000

K	LAM(K)	RO(K)	MASS(K)	E[W(K)]	STD[W(K)]
1	0.101	0.438	0.670	1.603	3.608
2	0.109	0.472	0.578	2.606	4.893
3	0.117	0.504	0.524	3.386	5.816
4	0.124	0.534	0.483	4.064	6.571
5	0.130	0.562	0.449	4.692	7.233
7	0.142	0.613	0.394	5.882	8.407
10	0.158	0.681	0.328	7.648	9.993
15	0.179	0.773	0.245	10.746	12.477
20	0.196	0.847	0.185	14.134	14.906
25	0.210	0.905	0.141	17.822	17.319
30	0.220	0.951	0.108	21.776	19.713
35	0.228	0.985	0.085	25.941	22.076
40	0.234	1.009	0.067	30.243	24.394
45	0.237	1.024	0.054	34.594	26.654
50	0.238	1.028	0.046	38.893	28.848
55	0.237	1.024	0.039	43.023	30.968
60	0.234	1.009	0.036	46.846	33.004
65	0.228	0.985	0.034	50.198	34.944
70	0.220	0.951	0.034	52.879	36.771
75	0.210	0.905	0.037	54.633	38.452
80	0.196	0.847	0.044	55.121	39.930
85	0.179	0.773	0.057	53.876	41.087
90	0.158	0.681	0.083	50.213	41.665
95	0.130	0.562	0.136	43.058	41.027
97	0.117	0.504	0.174	38.848	40.104
98	0.109	0.472	0.199	36.376	39.405
99	0.101	0.438	0.229	33.625	38.494
100	0.093	0.400	0.266	30.561	37.309

Table 2. The Survival Function of the Waiting Time Distribution For the K-th Customer in a Time Inhomogeneous G/G/1 System

X/K	1	25	50	75	100
	0.670	0.141	0.046	0.037	0.266
0.0	0.330	0.859	0.954	0.963	0.734
0.5	0.298	0.844	0.949	0.959	0.722
1.0	0.269	0.830	0.943	0.955	0.710
1.5	0.242	0.815	0.938	0.951	0.699
2.0	0.219	0.800	0.932	0.947	0.688
2.5	0.197	0.786	0.927	0.943	0.678
3.0	0.178	0.772	0.921	0.939	0.668
3.5	0.161	0.757	0.915	0.935	0.658
4.0	0.145	0.743	0.909	0.931	0.649
4.5	0.131	0.729	0.903	0.927	0.640
5.0	0.118	0.715	0.897	0.923	0.632
5.5	0.106	0.701	0.891	0.918	0.623
6.0	0.096	0.688	0.885	0.914	0.615
6.5	0.087	0.674	0.879	0.910	0.607
7.0	0.078	0.661	0.872	0.906	0.600
7.5	0.071	0.647	0.866	0.902	0.592
8.0	0.064	0.634	0.860	0.898	0.585
8.5	0.057	0.621	0.853	0.893	0.578
9.0	0.052	0.608	0.847	0.889	0.572
9.5	0.047	0.595	0.840	0.885	0.565
10.0	0.042	0.583	0.834	0.881	0.559
20.0	0.005	0.364	0.697	0.790	0.454
30.0	0.001	0.211	0.557	0.694	0.374
40.0	0.000	0.114	0.426	0.594	0.307
50.0	0.000	0.058	0.311	0.495	0.248
60.0	0.000	0.028	0.218	0.401	0.197

5. Error Analysis of the Laguerre Transform

The accuracy of the algorithm based on the Laguerre transform depends on one's ability to represent the functions present compactly, i.e., with vectors $(f_n^{\dagger})_{-N}^{N}$ and $(f_n^{\#})_{-N}^{N}$ of reasonable length. Hence it is of crucial importance to know how many terms one requires for a given accuracy. In the original paper [A], some necessary conditions and some sufficient conditions are given for $|f_n^{\dagger}|$ to be $o(n^{-k})$ in terms of differentiability, boundedness etc. of the function. In the second paper [B], the extent $\sum_{-\infty}^{\infty} |n| \, f_n^{\dagger 2} / \sum_{-\infty}^{\infty} f_n^{\dagger 2}$ of the Laguerre coefficients is related to the extent $\int_{-\infty}^{\infty} |x| \, f^2(x)dx/\int_{-\infty}^{\infty} f^2(x)dx$ of the function.

These results, however, describe structural relationships and do not supply a clear error bound.

In this section we study the absolute tail sum of $f_n^{\dagger}$ for the functions $f(x) \in L_2(0,\infty)$ satisfying certain conditions and obtain simple error bounds for such functions. The results for similar two-sided functions $f(x) \in L_2(-\infty,\infty)$ are obtained straightforwardly from (2.10). For $f(x) \in L_2(0,\infty)$ with $(f_n^{\dagger}) \in \ell_1$, let

$$S_N(x) = \sum_{n=0}^{N} f_n^{\dagger} \ell_n(x); \quad R_N(x) = |f(x) - S_N(x)|. \tag{5.1}$$

Since $|\ell_n(x)| \leq 1$ for any x, $x \geq 0$, one sees immediately that

$$R_N(x) \leq \sum_{n=N+1}^{\infty} |f_n^{\dagger}| . \tag{5.2}$$

We wish to find upper bounds for $\sum_{n=N+1}^{\infty} |f_n^{\dagger}|$. We will deal with function $f(x) \in L_2(0,\infty)$ meeting the following assumptions.

(A1) f(x) has the Laplace transform $\phi(s)$, which is a single-valued function with analytic continuation to the entire complex plane apart from singularities.

(A2) Let Q be the set of singular points of $\phi(s)$ in the complex plane. Then Q is bounded.

(A3) $\sup_{s \in Q} \mathrm{Re}(s) < 0$.

(A4) $\phi(s)$ is regular at $s = \infty$ and $\phi(\infty) = 0$.

A referee has pointed out (see Appendix A) that assumptions (A1)-(A4) imply that $f(x) = 0(e^{-cx})$, where $-c > \sup_{s \in Q} \mathrm{Re}(s)$, $c > 0$. Hence under these assumptions, one has $f(x) \in L_2(0,\infty)$.

We recall from (2.7) that $T_f^{\#}(u) = \phi(\frac{1}{2}\frac{1+u}{1-u})$. After a little algebra (see Appendix B), one finds under (A1), (A2), (A3), (A4) that $T_f^{\#}(u)$ is regular in $A_\rho = \{u: |u| < \rho\}$ for $0 < \rho < R_f$, where

$$R_f = \sqrt{\frac{v - \cos\theta^*}{v + \cos\theta^*}} > 1 \tag{5.3}$$

and

$$v = \sup_{s=Re^{i\theta}\in Q} \{R + \frac{1}{4R}\}; \quad \theta^* = \inf_{s=Re^{i\theta}\in Q} \{\theta: \frac{\pi}{2} < \theta < \frac{3}{2}\pi\}. \tag{5.4}$$

The last inequality in (5.3) holds since $\frac{\pi}{2} < \theta^* < \frac{3}{2}\pi$ from (A3).

Theorem 5.1. Let $\phi(s)$ satisfy (A1), (A2), (A3), (A4). Let R_f be as in (5.3) and define $M_f^{\#}(\rho) = \sup_{|u|=\rho} |T_f^{\#}(u)|$, $1 < \rho < R_f$. Then

$$\text{(a)} \quad |f_n^{\dagger}| \le \frac{M_f^{\#}(\rho)}{\rho^n(\rho-1)} \qquad \text{(b)} \quad \sum_{n=N+1}^{\infty} |f_n^{\dagger}| \le \frac{M_f^{\#}(\rho)}{\rho^N(\rho-1)^2} \,.$$

Proof. We note from (2.6) that $T_f^{\dagger}(u) = \frac{1}{1-u} T_f^{\#}(u)$ and both $T_f^{\dagger}(u)$ and $T_f^{\#}(u)$ have the same radius of convergence (cf. P2.3b of [B]). One sees immediately that, for $1 < \rho < R_f$,

$$\max_{|u|=\rho} |T_f^{\dagger}(u)| \le \max_{|u|=\rho} \frac{1}{|1-u|} \cdot \max_{|u|=\rho} |T_f^{\#}(u)| = \frac{1}{\rho-1} M_f^{\#}(\rho). \tag{5.5}$$

Hence by the maximum principle [1], one has

$$|T_f^{\dagger}(u)| \le \frac{1}{\rho-1} M_f^{\#}(\rho), \quad |u| \le \rho \,. \tag{5.6}$$

The Cauchy inequality then leads to

$$|f_n^{\dagger}| = |(\frac{d}{du})^n \, T_f^{\dagger}(u)|_{u=0} \frac{1}{n!} \le \frac{M_f^{\#}(\rho)}{\rho^n(\rho-1)} \,, \tag{5.7}$$

proving statement (a). Statement (b) follows directly from (a). □

Since the Laguerre sharp transform $T_f^{\#}(u)$ is a linear transform, Theorem 5.1 can be easily extended to mixtures of functions satisfying (A1), (A2), (A3), (A4). For convolution $f(x)*g(x) = \int_0^x f(y)g(x-y)dy$, one has $T_{f*g}^{\#}(u) = T_f^{\#}(u)T_g^{\#}(u)$ so that

$$M_{f*g}^{\#}(\rho) \le M_f^{\#}(\rho)\, M_g^{\#}(\rho). \tag{5.8}$$

These observations lead to the following theorem of practical importance.

Theorem 5.2. Let f(x) and g(x) satisfy (A1), (A2), (A3), (A4). Let R_f and R_g be determined as in (5.3) for f(x) and g(x) respectively. Let $1 < \rho < \min\{R_f, R_g\}$. Then the following statements hold.

(a) $r(x) = a\, f(x) + b\, g(x) \Longrightarrow \sum_{n=N+1}^{\infty} |r_n^{\dagger}| \leq \dfrac{|a| M_f^{\#}(\rho) + |b| M_g^{\#}(\rho)}{\rho^N (\rho-1)^2}$

a, b real.

(b) $r(x) = f(x)*g(x) \Longrightarrow \sum_{n=N+1}^{\infty} |r_n^{\dagger}| \leq \dfrac{M_f^{\#}(\rho) M_g^{\#}(P)}{\rho^N (\rho-1)^2}$

Proof. The theorem follows from Theorem 5.1, linearity of $T_f^{\#}(u)$ and (5.8). □

As a special case of Theorem 5.2(b), one has:

Corollary 5.3. Let f(x) satisfy (A1), (A2), (A3), (A4). Let $f^{(k)}(x)$ denote the k-fold convolution of f(x) with itself. Then

$$\sum_{n=N+1}^{\infty} |f_n^{(k)\dagger}| \leq \frac{\{M_f^{\#}(\rho)\}^k}{\rho^N (\rho-1)^2}$$

where $1 < \rho < R_f$.

We now apply the results developed in this section to various probability density functions. Our first example is the simplest one.

(A) Exponential Densities

Let $f(x) = \gamma e^{-\gamma x} U(x)$, $\gamma > 0$. The Laplace transform of f(x) is given by $\phi(s) = \gamma/(s+\gamma)$. Hence $Q = \{-\gamma\}$, and (A1), (A2), (A3), and (A4) are satisfied. One has $\theta^* = \pi$ and $v = \gamma + \frac{1}{4\gamma}$ so that

$$R_f = \sqrt{\frac{\gamma + \frac{1}{4\gamma} + 1}{\gamma + \frac{1}{4\gamma} - 1}} = \left|\frac{\gamma + \frac{1}{2}}{\gamma - \frac{1}{2}}\right| > 1 . \tag{5.9}$$

Since $T_f^{\#}(u) = \phi(\frac{1}{2}\frac{1+u}{1-u})$, we immediately see that

$$T_f^{\#}(u) = \frac{\gamma}{\gamma - \frac{1}{2}} \cdot \frac{u-1}{u-\xi}; \quad \xi = \frac{\gamma + \frac{1}{2}}{\gamma - \frac{1}{2}}. \tag{5.10}$$

Hence in this case R_f agrees with the radius of convergence of $T_f^{\#}(u)$. We note that $|\xi| > 1$ and

$$\begin{cases} 0 < \gamma < 1/2 \Longrightarrow \xi < -1 \\ 1/2 < \gamma \Longrightarrow \xi > 1. \end{cases} \tag{5.11}$$

Let $u = \rho e^{i\theta}$, $1 < \rho < R_f$ and define

$$P(\theta) = \left|\frac{u-1}{u-\xi}\right|_{u=\rho e^{i\theta}} = \frac{\rho^2-2\rho\cos\theta+1}{\rho^2-2\rho\xi\cos\theta+\xi^2}, \quad 0 \le \theta < 2\pi. \tag{5.12}$$

After a little algebra one finds that

$$(\rho^2-2\rho\xi\cos\theta+\xi^2)^2 \frac{d}{d\theta} P(\theta) = 2\rho(1-\xi)(\rho^2-\xi)\sin\theta. \tag{5.13}$$

From (5.11), (5.12) and (5.13), one then concludes that

$$M_f^{\#}(\rho) = \begin{cases} |T_f^{\#}(-\rho)|, & (0 < \gamma < 1/2) \text{ or } (\gamma > 1/2 \text{ and } 1 < \rho < \sqrt{R_f}), \\ |T_f^{\#}(\rho)|, & 1/2 < \gamma \text{ and } \sqrt{R_f} < \rho < R_f. \end{cases} \tag{5.14}$$

Hence we have established the following theorem.

Theorem 5.4. Let $f(x) = \gamma e^{-\gamma x} U(x)$, $\gamma > 0$. Then

$$R_N(x) \le \sum_{n=N+1}^{\infty} |f_n^{\dagger}| \le e_N(\rho) \stackrel{\text{def}}{\equiv} \frac{M_f^{\#}(\rho)}{\rho^N(\rho-1)^2}, \quad 1 < \rho < R_f = \left|\frac{\gamma+1/2}{\gamma-1/2}\right|$$

where

$$M_f^{\#}(\rho) = \begin{cases} \left|\dfrac{\gamma(1+\rho)}{(\gamma+1/2)+(\gamma-1/2)\rho}\right|, & (0<\gamma<1/2) \text{ or } (\gamma<1/2 \text{ and } 1<\rho<\sqrt{R_f}), \\ \left|\dfrac{\gamma(1-\rho)}{(\gamma+1/2)-(\gamma-1/2)\rho}\right|, & 1/2<\gamma \text{ and } \sqrt{R_f}<\rho<R_f. \end{cases}$$

One sees that $e_N(\rho)$ goes to $+\infty$ as $\rho \to 1$ or $\rho \to R_f$. Since $e_N(\rho) > 0$, $1 < \rho < R_f$, one may choose the optimal ρ^* so that $e_N(\rho^*) \leq e_N(\rho)$ for $\rho \in (1,R_f)$. Such ρ^* can be obtained numerically by solving $\frac{d}{d\rho} \log e_N(\rho) = 0$ in the interval $(1,R_f)$. Alternatively one may choose $\tilde{\rho} = \frac{1}{2}(1+R_f)$ to have an upper bound $e_N(\tilde{\rho})$, which has a simple closed form.

Corollary 5.5. Let $f(x) = \gamma e^{-\gamma x} U(x)$, $\gamma > 0$. Let $\tilde{\rho} = \frac{1}{2}(1+R_f)$ where R_f is as in Theorem 5.4. Then

$$R_N(x) \leq \sum_{n=N+1}^{\infty} |f_n^\dagger| \leq e_N(\tilde{\rho}) = \begin{cases} \dfrac{(1-2\gamma)^{N+1}(1-\gamma)}{2\gamma^2}, & 0 < \gamma < 1/2, \\[2ex] \dfrac{(2\gamma-1)^{N+1}}{(2\gamma)^{N-1}}, & 1/2 < \gamma . \end{cases}$$

Proof. The proof is immediate by substituting $\tilde{\rho} = \frac{1}{2}(1+R_f) > \sqrt{R_f}$ into $e_N(\rho)$ in Theorem 5.4. □

We note that $e_N(\tilde{\rho})$ is monotone decreasing in γ for $0 < \gamma < 1/2$ and monotone increasing in γ for $\gamma > 1/2$. Furthermore $e_N(\tilde{\rho}) \to +\infty$ as $\gamma \to 0$ or $\gamma \to +\infty$. This corresponds to the fact that the Laguerre transform method cannot tolerate functions f(x) too closely concentrated at zero or functions f(x) too great in extent. (cf. Section of [B]). When $\gamma = 1/2$, $e_N(\tilde{\rho}) = 0$ and no error will be introduced. This is so because $f(x) = \frac{1}{2} e^{-1/2\, x}\, U(x) = \ell_o(x)U(x)$ and $f_o^\dagger = \frac{1}{2}$ and $f_n^\dagger = 0$, $n \neq 0$. If we scale $f(x) = \gamma e^{-\gamma x}U(x)$ by c so that $c\gamma = 1/2$, one can avoid truncation error.

Remark 5.6. For $f(x) = \gamma e^{-\gamma x}U(x)$, the dagger coefficients are analytically available (cf. Appendix [A]). One has

$$f_n^\dagger = \frac{\gamma}{\gamma + 1/2} \left(\frac{\gamma - \frac{1}{2}}{\gamma + \frac{1}{2}}\right)^n, \quad n = 0,1,2,\ldots \,. \tag{5.15}$$

Hence an upper bound for $R_N(x)$ is immediately obtained as

$$R_N(x) \le \sum_{n=N+1}^{\infty} |f_n^{\dagger}| \le e_N^*(\gamma) = \gamma \left|\frac{\gamma - \frac{1}{2}}{\gamma + \frac{1}{2}}\right|^{N+1} . \tag{5.16}$$

This error bound $e_N^*(\gamma)$ has the geometric decay rate of $1/R_f$, while $e_N(\rho)$ in Theorem 5.4 has $1/\rho$ with $1 < \rho < R_f$. Hence $e_N^*(\gamma)$ is a better bound than $e_N(\rho)$ is. However $e_N^*(\gamma)$ cannot provide clear error bounds for convolution, while $e_N(\rho)$ can via Theorem 5.2.

The results for exponential densities in Theorem 5.4 and Corollary 5.5 combined with Theorem 5.2 provide error bounds for the following families of functions, and functions obtained by convolutions or linear combinations thereof.

(B) Gamma Densities of Integral Order

$$f(x) = \frac{1}{k!} \gamma^{k+1} x^k e^{-\gamma x} U(x), \quad \begin{cases} \gamma > 0, \\ k = 0,1,2\ldots;\ \phi(s) = (\frac{\gamma}{s+\gamma})^{k+1} \end{cases}$$

(C) Completely Monotone Densities

$$f(x) = \sum_{j=1}^{k} p_j \gamma_j e^{-\gamma_j x} U(x);\ \phi(s) = \sum_{j=1}^{k} p_j \frac{\gamma_j}{s+\gamma_j}$$

where $p_j \ge 0$, $\sum_{j=1}^{k} p_j = 1$ and $\gamma_j > 0$.

(D) A Subfamily of PF_∞ Densities

$$f(x) = f_1(x)*f_2(x)*\ldots*f_k(x)$$

where

$$f_j(x) = \gamma_j e^{-\gamma_j x} U(x),\ \gamma_j > 0.$$

$$\phi(s) = \prod_{j=1}^{k} \frac{\gamma_j}{s+\gamma_j} .$$

Remark 5.7. It is known numerically [4] [5] that the probability density functions of Normal, Logistic and Rayleigh distributions have Laguerre coefficients which fall off quite rapidly. The Laplace transforms of those densities, however, are not regular at $s = \infty$ and do not satisfy (A4). These difficulties are under study.

6. Acknowledgement

The authors wish to thank D. R. Miller, R. Lal and J. Wellner for their helpful comments.

7. Appendix A

A referee has pointed out that when $\phi(s) = \int_0^\infty e^{-sx} f(x)dx$ satisfies assumptions (A1), (A2), (A3), (A4), the function $f(x)$ is of exponential type, i.e., $f(x) = 0(e^{-cx})$ for some $c > 0$. The outline of his argument is as follows.

Let $\phi(s)$ be a function whose singularities are contained in some disc $|s| < \rho < \infty$, and suppose $\phi(\infty) = 0$. Then $\phi(s) = \sum_{n=0}^{\infty} a_n s^{-n-1}$, where the radius of convergence of $\phi(1/s)$ is at least ρ^{-1}, i.e., $\limsup_{n\to\infty} |a_n|^{1/n} \leq \rho$. Let $f(z) = \sum_{n=0}^{\infty} a_n z^n/n!$. Then $f(z)$ is an entire function and

$$|f(z)| \leq \sum_{n=0}^{\infty} |a_n|\cdot|z|^n/n! \leq M \sum_{n=0}^{\infty} |z(\rho+\varepsilon)|^n/n!$$

where $\varepsilon > 0$ and $M > 0$ are constants. Hence one has

$$|f(z)| \leq M\, e^{|z(\rho+\varepsilon)|} .$$

We note that $\phi(s) = \int_0^\infty e^{-sx}f(x)dx$. It now follows that, under (A1), (A2), (A3), (A4), $f(x)$ is the restriction to $(0,\infty)$ of an entire function of exponential type. Using the indicator diagram $h(\theta)$ of Boas and Buck ("Polynomial Expansions of Analytic Functions") one has

$$\limsup_{r\to\infty} \frac{1}{r} \log|f(r)| = \sup_{s\varepsilon Q} \mathrm{Re}(s) .$$

This means that $f(x) = O(e^{-cx})$ where $-c > \sup_{s\varepsilon Q} \mathrm{Re}(s)$, $c > 0$.

8. Appendix B

Theorem. Let $\phi(s) = T_f^{\#}(\frac{2s-1}{2s+1})$. Let $\phi(s)$ be regular in $D_\delta = \{s: \delta < \mathrm{Re}(s)\}$ for some $\delta < 0$, as well as at $s = \infty$ with $\phi(\infty) = 0$. Then $T_f^{\#}(u)$ is regular in $A_{R_f} = \{u: |u| < R_f\}$ where

$$R_f = \sqrt{\frac{V - \cos\theta^*}{V + \cos\theta^*}} > 1; \qquad V = \sup_{s=Re^{i\theta}\varepsilon Q} \{R + \frac{1}{4R}\}$$

$$\theta^* = \inf_{s=Re^{i\theta}\varepsilon Q} \{\theta: \frac{\pi}{2} \leq \theta < \frac{3}{2}\pi\}$$

and Q is the set of singular points of $\phi(s)$.

Proof. Let $R^* = \inf_{s\varepsilon Q} \left|\frac{2s-1}{2s+1}\right|$. Then $T_f^{\#}(u)$ is regular in $A_{R^*} = \{u: |u| < R^*\}$. One sees that

$$R^{*2} = \inf_{s\varepsilon Q} \left|\frac{2s-1}{2s+1}\right|^2 = \inf_{s=Re^{i\theta}\varepsilon Q} \{\frac{4R^2 + 1 - 4R\cos\theta}{4R^2 + 1 + 4R\cos\theta}\}$$

$$= \inf_{s=Re^{i\theta}\varepsilon Q} \left\{\frac{1 - \dfrac{\cos\theta}{R + \frac{1}{4R}}}{1 + \dfrac{\cos\theta}{R + \frac{1}{4R}}}\right\}$$

Since $\phi(s)$ is regular in D_δ, one has $\cos\theta < 0$ for $s = Re^{i\theta} \varepsilon Q$. Hence

$$R^{*2} \geq \frac{1 - \sup\{\dfrac{\cos\theta}{R + \frac{1}{4R}}\}}{1 + \sup\{\dfrac{\cos\theta}{R + \frac{1}{4R}}\}}$$

where the supremum is taken over $s = Re^{i\theta}$ in Q. We note that

$$\frac{1-x}{1+x} = -1 + \frac{2}{1+x} \quad ,$$

which is monotone decreasing for x, $x \neq -1$. Let R_f be as given. Since $\phi(s)$ is regular at $s = \infty$, one has $V < \infty$ and $\frac{\pi}{2} < \theta^* < \frac{3}{2}\pi$. Hence

$$-1 \leq \sup_{s=Re^{i\theta}\varepsilon Q} \left\{ \frac{\cos\theta}{R + \frac{1}{4R}} \right\} \leq \frac{\cos\theta^*}{V} < 0 \; .$$

One then concludes that $R^* \geq R_f > 1$, completing the proof. □

9. References

[A] Keilson, J. and Nunn, W. R. (1979), "Laguerre Transformation as a tool for the numerical solution of integral equations of convolution type", Appl. Math. Comput., 5, pp. 313-359.

[B] Keilson, J., Nunn, W. R., and Sumita, U. (1981), "The bilateral Laguerre transform", Appl. Math. Comput., 8, pp. 137-174.

[1] Levinson, N. and Redheffer, R. M. (1970), Complex Variables, Holden-Day, Inc., San Francisco, California.

[2] Lindley, D. V. (1952), "The theory of queues with a single server", Proc. Cambridge Phil. Soc., 48, pp. 277-289.

[3] Keilson, J., Petrondas, D., Sumita, U. and Wellner, J., (1980), "Significance points for some tests of uniformity on the sphere", submitted for publication.

[4] Sumita, U. (1979), "Numerical evaluation of multiple convolutions, survival functions, and renewal functions for the one-sided normal distribution and the Rayleigh distribution via the Laguerre transformation", Working Paper Series No. 7912 Graduate School of Management, University of Rochester.

[5] Sumita, U. (1980), "On sums of independent logistic and folded logistic variants", Working Paper Series No. 8001, Graduate School of Management, University of Rochester.

[6] Takács, L. (1962), Introduction to the Theory of Queues, Oxford University Press, New York.

10. Endnotes

† In general, the expansion converges in the L_2-sense. Throughout this paper we consider functions f(x) with $(f_n^\dagger) \in \ell_1$ so that point-wise convergence is guaranteed. For sufficient conditions for f(x) to have

$(f_n^{\dagger}) \in \ell_1$, see, e.g., [A], [B] and R. V. Churchill, "Operational Mathematics", p. 452.

This research was conducted at the M.I.T. Laboratory for Information and Decision Systems, with partial support provided by the United States Air Force Office of Scientific Research, Grant No. AFOSR-79-0043 and National Science Foundation Grant No. NSF/ECS 79-19880.

University of Rochester, Rochester, New York; and Visiting Massachusetts Institute of Technology, Laboratory for Information and Decision Systems, Cambridge, Massachusetts 02139.

Discussant's Report on
"The Equilibrium Waiting Time Distribution of G/G/1 Via the Bilateral Laguerre Transform, and Related Results,"
by Julian Keilson

Professor Keilson has introduced an interesting new transform technique to applied probability. The Laguerre transform maps densities on a continuum into countable dimension vectors while preserving the convolution operation. Thus it has obvious application to numerical convolution algorithms for continuous densities. This transform has the very interesting feature that it is apparently conceived specifically for its computational advantages. The alternate numerical method for dealing with convolutions of functions on a continuum is to discretize the continuum, approximate the functions by the appropriate vector, and then perform discrete convolution. The Laguerre transform approach looks like it has the potential for achieving comparable accuracy using a vector of much fewer components than the alternate approach. It would be very interesting to perform a numerical study comparing the efficiency of the two approaches.

The Laguerre transform has one very outstanding advantage over other transform techniques such as Laplace or Fourier. There is no inversion problem!

A difficulty encountered in the application of this technique is the truncation of the transform vector. Keilson and Nunn have shown certain qualitative properties concerning the speed that the components of the vector go to zero. However in actual application it seems that one still must do something naive like truncating after 30 terms and also after 60 terms, say, computing the answer for both and comparing them; if they are within the accuracy desired then 30 terms sufficed, if not then another computation must be made with 100 terms, say, etc. Is there any better, more efficient way to achieve a desired degree of accuracy in the final answer?

The truncated Laguerre expansion of a density is essentially a smoothing. When faced with the problem of statistical density estimation, might it be possible to estimate the first n Laguerre coefficients and then use the truncated expansion with estimated coefficients as a smoothed density estimate?

Author's Response to Discussant's Report

Although the Laguerre Transform Method has not been compared with other numerical procedures in any systematic way, numerical accuracy of the method has been tested for the normal density and the exponential density, where direct comparison is possible. For the case of the standard normal density, for example, use of 250 Laguerre coefficients provides more than 10 digit accuracy uniformly in the range $0 \leq x \leq 10$ after convolving 20 times. The result is even better for the exponential density.

In the Laguerre Transform Method, inversion difficulty for Laplace or Fourier transforms is replaced by the problem of finding the Laguerre coefficients. Fortunately, for a wide range of probability density functions, the Laguerre coefficients are available either analytically or through simple numerical procedures.

The question of truncation error is partially answered in the body of the paper. For Gamma densities of integral order, completely monotone densities with finite and discrete spectral span, and sums of scaled exponential variates, the exact number of terms required to satisfy a given accuracy can be found. For other functions not in this family, one may calculate $\int_{-\infty}^{\infty} x^k f(x)dx$ for $k = 0,1,2$ using the Laguerre coefficients [B]. When the length of the Laguerre coefficients is chosen to satisfy a given accuracy of these moments, one finds that the function values f(x) satisfy the same accuracy. This empirical observation is supported extensively.

Discussant: Dr. Douglas R. Miller, Dept. of Operations Research, School of Engineering, George Washington University, Washington, D.C. 20052.

SOME COMPUTATIONAL ASPECTS OF QUEUEING NETWORK MODELS

Herb Schwetman

Abstract

Queueing network models have emerged as useful models of computer systems. While such models have been known and studied for some time in the area of operations research, their application to the study of computers has stimulated further research, both into extending the basic results and into developing efficient techniques for their solution. Currently, there does not appear to be any "best" technique. This paper surveys three well-known methods for solving queueing network models and discusses their complexity, size and numerical properties.

1. Introduction

Queueing network models (QNM's) have been studied and used for some time in the area of operations research. Since around 1968, these models have been used as models of computer and communications systems. This new use has stimulated research into enlarging the class of models which can be solved and into developing new solution techniques [KiSe79]. These new techniques demonstrate improvements with respect to either complexity, size (when implemented in computer programs) and/or numerical properties.

Because of the complexity of networks of interest and of the solution algorithms, these techniques are often implemented as computer programs. In fact, several packages have been developed to solve fairly general classes of QNM's [Buze78]. Each of these packages has a network definition language, a solution portion implementing some solution algorithm, and produces output, in the form of tables. The topic, computational aspects, is cast in the framework of these solution programs.

Developers and users of QNM solution programs are concerned with several issues, including:

- What are the types of models solved?
- Is the program giving correct answers?
- Is the program susceptible to problem-dependent numerical error?
- What are the time and space requirements for the program?

This paper begins by introducing QNM's and the associated notation. Their use as models of computer systems is illustrated with an example. Three solution techniques, all developed since 1970, are described. The single class version of each technique is presented in detail, using standard and consistent notation. This presentation, which in a sense repeats earlier work, is in a form which will permit the reader to understand, implement and analyze each algorithm. The paper concludes by discussing each technique with respect to the issues mentioned above. It will be seen that there is currently no technique which is "best" with respect to all desirable attributes.

2. QNM's: Notation and an Example

In its simplest form, a QNM consists of a collection of K devices (also called facilities); jobs (tasks or customers) travel from one device to another, selecting the next device to visit according to the transition probabilities, q_{ij}, where q_{ij} is the probability that a job at device i will next visit device j. The subscript 0 denotes arrivals from or departures to "outside of the system". When jobs arrive at a device, they wait until the single server is idle; waiting jobs are selected for service using a first-come, first-served (FCFS) scheduling discipline. The time spent in service at device i is governed by a negative exponential distribution with mean S_i. Queueing networks can be either open or closed. In a closed model, there are N jobs circulating amongst the K devices (i.e., no external arrivals). In an open model, jobs arrive at the network at rate X_0. Since equilibrium must be maintained, these jobs also depart at the same rate.

A QNM of this form can be solved, to obtain values for the following performance variables: for device i

X_i - throughput rate (jobs per unit time).

U_i - utilization.

$\bar{n}_i$ - mean queue length, and

R_i - mean response time.

The example shown in Figure 1 is a very simple model; it serves to illustrates the method of identifying resources and jobs which use these resources. The use of more elaborate models of actual computer and communication systems has been reported in the literature; e.g., [KiSe79], [Rose78] and [Schw80c].

QNM's can be solved even when the descriptions of the models are enlarged to include:

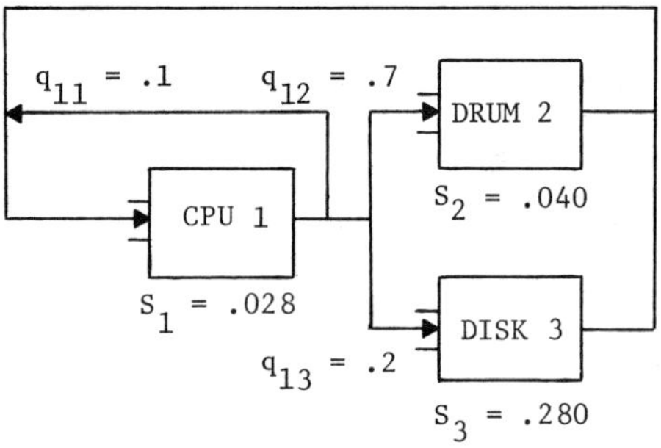

Figure 1. Example of Single Class, Closed QNM [Buze73]

(1) a total of four different scheduling disciplines at the facilities: FCFS - first-come, first-served, IS - infinite servers (no delay), PS - processor sharing, and LCFSPR - last-come, first-served preempt;

(2) "general" service time distributions (distributions with rational Laplace transforms) for stations which are not FCFS;

(3) multiple classes of jobs, where a job class is described in terms of its transition probabilities, mean service intervals, population and whether it is open or closed;

(4) for facilities which are not FCFS, jobs in different classes can have different mean service times;

(5) class switching - jobs can switch between classes, subject to constraints involving flow conservation (e.g., jobs which are external arrivals must eventually depart the network); and

(6) the service time at a device can depend on the number of jobs at that device (in particular devices with multiple servers can be modeled).

The restrictions mentioned in (2) and (4) limit the range of solvable models to those with resources which, in some sense, preserve the arrival process. These restrictions are explicitly mentioned in

[BCMP75].

3. Solution Techniques

While QNM's with multiple classes can be solved, the notation becomes very cumbersome. Thus, except as noted, the remainder of this paper will consider only closed, single class QNM's. The landmark paper by Baskett, Chandy, Muntz and Palacios [BCMP75] gives the closed form solution for the general class of QNM's. The paper by Bruell, Balbo and Schwetman [BaBS77], among others, gives details about solving multiple class models.

Closed, single class QNM's with K devices are completely characterized by the following parameters:

N = number of jobs,

$S_i(n)$ = the reciprocal of the service rate of device i, when there are n jobs present, and

either

V_i = the expected number of visits to device i per job,

or

q_{ij} = the probability that a job at device i next visits device j.

V_0, the expected number of visits to the system per job, is used to "normalize" the visits per job to the devices of the system. It can be shown that the $\{V_i\}$ can be obtained as the solution to the set of equations:

$$V_0 = 1$$

$$V_j = \sum_{i=0}^{K} V_i \, q_{ij} \quad \text{for } j = 1,\ldots,K.$$

For closed models, the V_i are unique up to a multiplicative constant. If the model is open, then X_0 is the arrival rate to the model; in this case,

$$X_j = \sum_{i=0}^{K} X_i \, q_{ij}$$

are uniquely determined, and

$$V_i = \frac{X_i}{X_0} \, .$$

Let

$$\underline{n} = (n_1, \ldots, n_K)$$

be a state of the model, where n_i is the number of jobs at device i. Then

$$S(N,K) = \{\underline{n} = (n_1, \ldots, n_K) \mid n_i \geq 0 \text{ and } \sum_{i=1}^{K} n_i = N\}$$

is the set of feasible states for a closed QNM. The probability of finding the model in state $\underline{n}$ is

$$p(\underline{n}) = \frac{1}{G(N)} \prod_{i=1}^{K} f_i(n_i)$$

where

$$f_i(n) = \begin{cases} 1 & n = 0 \\ V_i \, S_i(n) \, f_i(n-1) & n > 0 \end{cases}$$

and G(N) is a normalizing constant. For devices with IS scheduling, $f_i(n)$ can be given as

$$f_i(n) = \frac{(V_i S_i)^n}{n!}$$

Because the $p(\underline{n})$ must sum to 1.0, the G(N) can be calculated as

$$G(N) = \sum_{\underline{n} \in S(N,K)} \prod_{i=1}^{K} f_i(n_i)$$

While this is a solution, the number of terms in the sum render it impractical. Denning and Buzen [DeBu78] (among others) give the number of feasible states as

$$\binom{N + K - 1}{K - 1}$$

If K = 8 and N = 20, then the calculation of G(N) requires the summation of 888030 terms, each the product of 8 factors.

4. Convolution Algorithm

Beginning with Buzen [Buze73], several researchers [ReKo75] [BaBS77] have shown that G(N) can be computed without recomputing each factor of the sum. To do this, define

$$G_k(n) = \sum_{\underline{n} \in S(n,k)} \prod_{i=1}^{k} f_i(n_i).$$

Then $G_K(N) = G(N)$. We can factor $G_k(n)$ as follows:

$$G_k(n) = \sum_{j=0}^{n} \Big[\sum_{\underline{n} \in S(n,k) \text{ and } n_k = j} \prod_{i=1}^{k} f_i(n_i) \Big]$$

$$= \sum_{j=0}^{n} f_k(j) \Big[\sum_{\underline{n} \in S(n-j,k-1)} \prod_{i=1}^{k-1} f_i(n_i) \Big]$$

$$= \sum_{j=0}^{n} f_k(j)\, G_{k-1}(n-j)$$

If we define the N + 1 dimensional vectors

$$\underline{f}_k = (f_k(0), \ldots f_k(N))$$

and

$$\underline{G}_0 = (1,0,\ldots,0)$$

then the vector

$$\underline{G}_k = (G_k(0),\ldots,G_k(N))$$

with $G_k(n)$ as defined above is the vector convolution

$$\underline{G}_k = \underline{f}_k * \underline{G}_{k-1}$$

If a device k has a single load independent server, then the elements $G_k(n)$ can be computed as

$$G_k(n) = V_k\ S_k\ G_k(n-1) + G_{k-1}(n).$$

This is Buzen's algorithm as given in [Buze73]. Bruell and Balbo [BrBa80] give a very complete treatment of the convolution algorithm. In addition to the solution for the single class model given here, they also present the convolution algorithm for multiple classes.

The required performance values can be calculated using the normalizing constants and auxiliary functions; the throughput rate at device i for load N is

$$X_i(N) = V_i\ \frac{G(N-1)}{G(N)}$$

For devices with load independent servers, the utilization is

$$U_i(N) = S_i\ X_i(N)$$

and the mean queue length is

$$\bar{n}_i(N) = \sum_{j=1}^{N} (V_i\ S_i)^j\ \frac{G(N-1)}{G(N)} = \frac{1}{G(N)} \sum_{j=1}^{N} (V_i\ S_i)^j G(N-j)$$

For devices with load dependent servers, define $P_i(n,N)$, the probability of having n jobs at device i when the load is N as

$$P_i(n,N) = \frac{f_i(n)}{G(N)}\ g_K^i(N-n)$$

where

$$g_K^i(n) = \begin{cases} G_K(0) = 1.0 & n = 0 \\ G(N) - \sum_{j=1}^{n} f_i(j)\ g_K^i(N-j) & n > 0 \end{cases}$$

Then, for these load dependent devices,

$$U_i(N) = \begin{cases} 1 - \dfrac{g_K^i(N)}{G(N)} & i < K \\ \\ 1 - \dfrac{G_{K-1}(N)}{G(N)} & i = K \end{cases}$$

and

$$\bar{n}_i(N) = \begin{cases} \sum_{j=1}^{N} j\ P_i(j,N) & i < K \\ \\ \dfrac{1}{G(N)} \sum_{j=1}^{N} j\ f_K(j)\ G_{K-1}(n - j) & i = K \end{cases}$$

For all devices, the mean response time is

$$R_i(N) = \frac{\bar{n}_i(N)}{X_i(N)}$$

If device K (the last device) is a load dependent device, then the auxiliary functions $g_K^i(n)$ and $P_i(n,N)$ are not required for that device, as the alternative forms of the equations for $U_i(N)$ and $\bar{n}_i(N)$ can be used. Buzen [Buze78] used this feature to develop a variation of the convolution algorithm which does not require the auxiliary functions; rather, the model is repetitively solved with each load dependent device moved to the K-th position. This procedure, called "reindexing", while requiring more computational effort, has better numerical properties than the standard convolution technique.

5. Mean Value Analysis

Another technique has been developed by Reiser and Lavenberg [ReLa80]. This technique, called Mean Value Analysis (MVA), does away with the need for calculating the normalization constant. The critical formulae, for load independent devices, are [BuDe80]:

$$R_i(n) = \begin{cases} S_i[1 + \bar{n}_i(n-1)] & \text{if FCFS, PS, or LCFSPR} \\ S_i & \text{if IS} \end{cases}$$

Note, the $\bar{n}_i(0)$ are initialized to 0. The overall throughput rate then is

$$X_0(n) = \frac{n}{\sum_{j=1}^{K} V_j R_j(n)}$$

and the device throughput rate and mean queue length are

$$X_i(n) = X_0(n)\, V_i$$

$$\bar{n}_i(n) = X_i(n)\, R_i(n).$$

For devices with load dependent service rates, the equation for the mean response time is modified to be

$$R_i(n) = \sum_{\ell=1}^{n} \ell\, S_i(\ell)\, P_i(\ell-1, n-1),$$

and $P_i(\ell,n)$ is given by

$$P_i(\ell,n) = X_i(n)\, S_i(\ell)\, P_i(\ell-1,n-1) \quad \text{for } \ell = 1,\ldots,n,$$

and

$$P_i(0,n) = 1 - \sum_{\ell=1}^{n} P_i(\ell,n) .$$

These formulae lead to an algorithm which builds up the values of the desired performance variables from $n = 0$ to $n = N$. For load independent devices, the utilization is given by

$$U_i(N) = X_i(N)\, S_i,$$

and for load dependent devices,

$$U_i(N) = 1 - P_i(0,N).$$

As presented in [ReLa80], the MVA algorithm has a slightly different form, which can result in an increase in the storage and a decrease in the number of operations required. Briefly, this modified MVA algorithm is as follows:

$$W_i(n) = V_i R_i(n) = \begin{cases} V_i \; S_i[1+\bar{n}_i(n-1)] & \text{if FCFS, PS, or LCFSPR} \\ V_i \; S_i & \text{if IS} \end{cases}$$

$$X_0(n) = \frac{n}{\sum_i V_i \; R_i(n)} = \frac{n}{\sum_i W_i(n)}$$

$$\bar{n}_i(n) = X_i(n) \; R_i(n) = X_0(n) \; V_i \; R_i(n) = X_0(n) \; W_i(n)$$

For load dependent devices,

$$W_i(n) = V_i \; R_i(n) = \sum_{\ell=0}^{n} \ell \; V_i \; S_i(\ell) \; P_i(\ell-1, n-1)$$

Also, Reiser and Lavenberg present a special form of the load dependent device equations for multiple server devices which reduces the number of factors in these equations.

The MVA algorithm can be extended to handle models with multiple job classes. Also, both open and closed classes can be accommodated. A technical report by Schwetman [Schw80b] describes computational procedures for all of these extensions, plus efficient procedures for managing storage for these.

6. LBANC Algorithm

Chandy and Sauer have developed a new algorithm [ChSa80], called Local Balance Algorithm for Normalizing Constants (LBANC). LBANC is similar to the convolution algorithm in that it produces normalizing constants, G(n); it differs by using a technique similar to MVA to do this. Let $\bar{m}_i(n)$ be the unnormalized mean queue length at device i, i.e.,

$$\bar{m}_i(n) = \bar{n}_i(n)\ G(n).$$

Since

$$\sum_{i=1}^{K} \bar{n}_i(n) = n,$$

we have

$$G(n) = \frac{\sum_{i=1}^{K} \bar{m}_i(n)}{n} \quad .$$

To obtain the $\bar{m}_i(n)$, apply Little's Law to the MVA equation for $R_i(n)$, to get for load independent devices with FCFS, PS or LCFSPR scheduling

$$\bar{n}_i(n) = X_i(n)\ R_i(n) = X_i(n)\ S_i[1 + \bar{n}_i(n-1)],$$

or

$$\bar{m}_i(n) = G_i(n)\ X_i(n)\ S_i[1 + \bar{n}_i(n-1)]$$

$$= V_i\ S_i[G(n-1) + \bar{m}_i(n-1)].$$

The last equation follows from the definition of $X_i(n)$ from the convolution algorithm. For devices with IS scheduling,

$$\bar{m}_i(n) = V_i\ S_i\ G(n-1).$$

By initializing

$$G(0) = 1$$

and

$$\bar{m}_i(0) = 0 \quad i = 1,\ldots,K$$

and then successively computing the $\bar{m}_i(n)$ and $G(n)$ for $n = 1,\ldots,N$, $\bar{m}_i(N)$ and $G(N)$ are obtained. The performance variables are then calculated as

$$\bar{n}_i(N) = \frac{\bar{m}_i(N)}{G(N)}$$

$$X_i(N) = V_i \frac{G(N-1)}{G(N)}$$

$$U_i(N) = X_i(N) S_i$$

$$R_i(N) = \frac{\bar{n}_i(N)}{X_i(N)}$$

For devices with load dependent servers, the calculation of $\bar{m}_i(n)$ is modified as follows:

$$\bar{m}_i(n) = \sum_{\ell=1}^{n} \ell \; p_i(\ell,n)$$

where $p_i(\ell,n)$, the unnormalized queue length probability, is given by

$$p_i(\ell,n) = V_i \; S_i(n) \; p_i(\ell-1,n-1)$$

and

$$p_i(0,n) = G(n) - \sum_{\ell=1}^{n} p_i(\ell,n).$$

For these devices, the utilization is given by

$$U_i(N) = 1 - \frac{p_i(0,N)}{G(N)} \quad .$$

7. Computational Aspects

As these algorithms are implemented as programs, several issues arise. One is the amount of resources (the amount of storage and the amount of computational effort) required to solve QNM's. Thus, it is of interest to evaluate each algorithm in these terms. In order to do this, it is necessary to state our chosen measures of resourcc usage. In this paper, these will be the size of arrays and the number of arithmetic operations (adds, subtracts, multiples and divides) required. The evaluation is restricted to single-class, closed models with N jobs and K devices; α of these devices have load independent servers, and β have load dependent servers ($\alpha + \beta = K$).

It is assumed that the implementation of each algorithm will require the same input values; an estimate of the space required to store these is given in Table 1. Also assume that each solution will consist of $U_i(N)$, $X_i(N)$, $\bar{n}_i(N)$ and $R_i(N)$ for each of the K devices; these will be computed but not stored.

The amount of storage and the number of operations for the convolution algorithm are estimated in Table 2. Tables 3 and 4 present the estimated storage requirements and operation counts for the MVA and the modified MVA algorithms. Table 5 gives the estimated storage and number of operations for the LBANC Algorithm. These estimates are obtained directly from examination of the formulae presented in the previous sections. While they do not give absolute estimates of resources requirements, they are good indicators of the relative requirements for the different techniques. This type of estimation is similar to the methods used by Zahorjan [Zaho80] and others.

Table 1. Storage for Inputs

item	storage
V_i	$\alpha + \beta$
S_i	α
$S_i(N)$	βN
Load indep.	2α
Load dep.	$\beta(1 + N)$

In Tables 2, 3, 4 and 5, the total amount of storage and the estimated total number of operations are obtained by summing the load independent and load dependent estimates. Table 6 gives an example of a model, along with the storage and operation counts for the three algorithms.

Table 2. Convolution Method - Storage and Operations

item	storage	no. operations
V_iS_i	α	α
$f_i(N)$	$\beta(N + 1)$	$2\beta N$
$G(N)$	$N + 1$	$2\alpha N + 2\beta(N^2 + N)$
$g_K^i(N)$	$\beta(N + 1)$	$2\beta N$
$X_i(N)$		$\alpha + \beta$
$U_i(N)$		$\alpha + 2\beta$
$\bar{n}_i(N)$		$3\alpha N + 3\beta N$
$R_i(N)$		$\alpha + \beta$
Load indep.	$\alpha + N + 1$	$\alpha(5N + 4)$
Load dep.	$2\beta(N + 1)$	$\beta(2N^2 + 9N + 1)$

Table 3. MVA Method - Storage and Operations

item	storage	no. operations
$R_i(N)$	$\alpha + \beta$	$2\alpha N + \frac{3\beta}{2}(N^2 + N)$
$P_i(\ell,N)$	βN	$2\beta(N^2 + N)$
$P_i(0,N)$	β	$\frac{\beta}{2}(N^2 + N)$
$X_0(N)$		$2\alpha N + 2\beta N$
$X_i(N)$	$\alpha + \beta$	$\alpha N + \beta N$
$\bar{n}_i(N)$	$\alpha + \beta$	$\alpha N + \beta N$
$U_i(N)$		$\alpha + \beta$
Load indep.	3α	$2\alpha(3N + 1)$
Load dep.	$\beta(N + 4)$	$2\beta(2N^2 + 4N + 1)$

Table 4. Modified MVA - Storage and Operations

item	storage	no. operations
V_iS_i	α	α
$V_iS_i(N)$	βN	βN
$W_i(n)$	$\alpha + \beta$	$2\alpha N + \frac{3\beta}{2}(N^2 + N)$
$P_i(\ell,n)$	βN	$2\beta(N^2 + N)$
$P_i(0,n)$	β	$\frac{\beta}{2}(N^2 + N)$
$X_0(n)$		$\alpha N + \beta N$
$\bar{n}_i(n)$	$\alpha + \beta$	$\alpha N + \beta N$
$X_i(N)$	$\alpha + \beta$	$\alpha + \beta$
$U_i(N)$		$\alpha + \beta$
$R_i(N)$		$\alpha + \beta$
Load indep.	4α	$4\alpha(N + 1)$
Load dep.	$\beta(2N + 4)$	$\beta(4N^2 + 9N + 3)$

Table 5. LBANC Method - Storage and Operations

item	storage	no. operations
V_iS_i	α	α
$\bar{m}_i(N)$	$\alpha + \beta$	$2\alpha N + \beta(N^2 + N)$
$G(N)$		$\alpha N + \beta N$
$p_i(\ell,N)$	βN	$\beta(N^2 + N)$
$p_i(0,N)$	β	$\frac{\beta}{2}(N^2 + N)$
$X_i(N)$		$\alpha + \beta$
$U_i(N)$		$\alpha + 2\beta$
$\bar{n}_i(N)$		$\alpha + \beta$
$R_i(N)$		$\alpha + \beta$
Load indep.	2α	$\alpha(3N + 5)$
Load dep.	$\beta(N + 2)$	$\frac{\beta}{2}(5N^2 + 7N + 10)$

Table 6. Examples

N	α	β	method	storage	no. operations
7	5	2	con.	45	519
			MVA	37	760
			MVA mod.	56	684
			LBANC	28	434
10	10	0	conv.	21	540
			MVA	30	620
			MVA mod.	40	440
			LBANC	20	350
10	0	10	conv.	231	2910
			MVA	140	8820
			MVA mod.	240	4930
			LBANC	120	2900
10	5	5	conv.	126	1725
			MVA	85	4410
			MVA mod.	120	2685
			LBANC	60	1625

It can be noted that all of the presented algorithms require only modest amounts of storage. However, these storage requirements expand rapidly as multiple job classes are introduced. A similar comment can also be made about the execution times.

Also to be considered is the range of applicability of the different algorithms. Chandy and Sauer [ChSa80] discuss the generality of these three algorithms, stating that "convolution is the only algorithm which has been applied to the full class of product form networks". Schwetman [Schw80b] has shown how to apply MVA to most forms of product form network; the exceptions are external arrival rates which depend on the population and load dependent servers whose service rate depends on the number of jobs in each class at the device (as opposed to the total number of jobs at the device). Chandy and Sauer [ChSa81] have given the LBANC algorithm for all product form networks except for these two types. Recently, this has been extended to include jobs from open classes at load dependent devices [Saue81].

Another implementation issue is the ease of understanding and the ease of implementation. Here, MVA is the winner, as it is based on some very intuitive arguments [SaCh80]. The convolution method is the most complex. This difference is especially true when programs are extended to include multiple job classes.

A remaining issue is that of verifying programs are producing correct solutions. Related to this is the issue of problem-dependent errors, usually due to errors caused by the arithmetic properties of computers (numerical stability). While the issue of numerical stability has been discussed in several places, the issue of verifying correct results has received almost no attention. Schwetman [Schw80a] has presented a preliminary approach to developing procedures for testing QNM software. This approach is based on solving a succession of models, each with solutions which can be verified by alternative methods.

Each of the three algorithms is known to have a potential for two kinds of numerical difficulties [ChSa80]. Briefly, the convolution algorithm and LBANC may compute normalizing constants G(n) which exceed the range of numbers representable in a computer word. The effects of this problem can be minimized by a judicious choice of the $\{V_i\}$ (which are unique up to a multiplicative constant). Reiser and Sauer [ReSa78] detail a procedure for scaling the V_i, as do Bruell and Balbo [BrBa80].

The second type of difficulty occurs in the calculation of queue length distributions. This is required in both MVA and LBANC, for devices with load dependent servers. In some cases, the calculation of $P_i(0,n-1)$ becomes inaccurate; because $P_i(n,N)$ is computed using $P_i(n-1,N-1)$, these errors propagate, leading to "chaotic behavior" [ChSa80]. Similar problems can occur in the calculation of the auxiliary functions $g_K^i(n)$ in the convolution algorithm for load dependent devices.

8. Discussion

This paper has presented three solution algorithms for single class, closed queueing network models. The resource usage for each of them was estimated, in terms of the amount of storage and the number of operations. Other issues relevant to implementing each solution were discussed. The conclusion is that there is currently no single algorithm which is "best" in all situations. Fortunately, the difficulties which do arise do so only for extreme values of the input parameters. Models of actual systems are usually well-behaved and can be reliably solved by these algorithms.

This paper has avoided discussing models with multiple classes of jobs. As would be expected, these cause a great increase in the complexity of the solution programs, the resources required, and the evaluation of each algorithm. Some of these issues are discussed in [BrBa80] and [Zaho80] and are the subject of current research.

This paper has also avoided the issue of developing realistic and useful models of actual systems. In addition to requiring reliable solution techniques, this activity requires a detailed knowledge of the system being modeled and of a variety of modeling techniques, to handle critical features of models. These considerations have led to the development of approximate solution techniques, for obtaining solutions to extended forms of QNM's. Many of these are based on the MVA algorithm presented above.

9. References

[BaBS77] Balbo, G. and S. Bruell, and H. Schwetman, "Customer classes and closed network models - a solution technique," Proceedings IFIPS 77, North Holland Pub., 1977, p. 559.

[BCMP75] Baskett, F., Chandy, K., Muntz, R. and F. Palacios, "Open, Closed, and Mixed Networks of Queues with Different Classes of Customers," J. Assoc. Comput. Mach., 22, April 1975, p. 248.

[BrBa80] Bruell, S. and G. Balbo, Computational Algorithms for Closed Queueing Networks, North Holland Pub., 1980.

[Buze73] Buzen, J., "Computational Algorithms for Closed Queueing Networks with Exponential Servers," Comm. ACM, 16 Sept. 1973, p. 527.

[Buze78] Buzen, J. at.al., "Best/1 - Design of a Tool for Computer System Capacity Planning," Proceedings NCC 1978, AFIPS, 1978, p. 447.

[BuDe80] Buzen, J. and P. Denning, "Operational Treatment of Queue Distributions and Mean Value Analysis," Comput. Perform., 1, June 1980, p. 6.

[ChSa80] Chandy, K. and C. Sauer, "Computational Algorithms for Product Form Queueing Networks," Comm. ACM, 23, Oct. 1980. p.

[DeBu78] Denning, P. and J. Buzen, "The Operational Analysis of Queueing Network Models," Comput. Surveys, 10, Sept. 1978, p. 225.

[KiSe79] Kienzle, M. and K. Sevcik, "Survey of Analytic Queueing Network Models of Computer Systems," Proc. Conference on Simulation, Measurement and Modeling of Computer Systems, ACM/SIGMETRICS, August 1979, p. 113.

[ReKo75] Riser, M. and H. Kobayashi, "Queueing Networks with Multiple Closed Chains: Theory and Computational Algorithms," IBM J. Res. Develop., 19, May 1975, p. 283.

[ReLa80] Reiser, M. and S. Lavenberg, "Mean Value Analysis of Closed Multichain Queueing Networks," J. Assoc. Comput. Mach., 27, April 1980, p. 313.

[Rose78] Rose, C., "A Measurement Procedure for Queueing Network Models of Computer Systems," Comput. Surveys, 10, September 1978, p. 263.

[SaCh81] Sauer, C. and K. Chandy, Computer System Performance Modeling, Prentice-Hall, 1981.

[Saue81] Sauer, C., "Computational Methods for Product Form Queueing Networks," these Proceedings.

[Schw80a] Schwetman, H., "Testing Network-of-queues Software," CSD-TR 330, Dept. Computer Sciences, Purdue Univ., 1980.

[Schw80b] Schwetman, H., "Implementing the Mean Value Algorithm for the Solution of Queueing Network Models," CSD-TR 355, Dept. Computer Sciences, Purdue Univ., 1980.

[Schw80c] Schwetman, H., "Modeling Performance of the B6700: A Case Study," CMG Transactions, No. 28, June 1980, p. 5-3.

[Zaho80] Zahorjan, J., "The Approximate Solution of Large Queueing Network Models," Ph.D. Thesis, Dept. of Computer Science, Univ. of Toronto, 1980; available as Technical Report CSRG-122, Computer Systems.

Department of Computer Sciences, Purdue University, West Lafayette, IN 47907

ALGORITHMIC ANALYSIS OF A DYNAMIC PRIORITY QUEUE

V. Ramaswami[1]

D. M. Lucantoni[2]

Abstract

Implementable algorithms are developed for a single server queue serving two types of customers and operating under a dynamic non-preemptive priority rule which assigns higher priority to type 1 customers when their queue length exceeds a pre-assigned theshold N, and to type 2 customers otherwise. The model is discussed under the assumption of a finite waiting room of size K-1 for type 2 customers and for the case of general service times and Poisson arrivals. The recursive computational schemes, which are obtained here using the function-analytic methods recently introduced by M. F. Neuts, can be used interactively in the design of such systems for arriving at desirable values of N and K. A self-contained section of numerical examples illustrates some interesting qualitative aspects of such a model.

1. Introduction

Often one encounters queueing systems whose smooth functioning is hampered by certain customers who require much longer service times than most others. Such unusually long services, infrequent though they may be, have the effect of introducing wide oscillations in the sample paths, and in practice result in large build-ups of the queue and longer waiting times. The FIFO discipline does not appear to be desirable in designing such systems. An effective method of alleviating the undesirable effects of these occasionally long services is to assign a lower priority to the class of customers needing such long services, permitting them to enter service only when the queue length of the higher priority class falls below a certain pre-assigned threshold; one may also impose a finite waiting room for the lower priority class. This paper is an attempt to model such systems under fairly general assumptions and in a computationally tractable form, and to obtain the queue characteristics by fast algorithms which may be used interactively in arriving at desirable values for the threshold and the waiting room size. It may be stressed that the highly complex dependence of the queue characteristics on the parameters is a serious deterrent to the use of analytic methods such as Markov decision theory. Further, in a model such as the one discussed here, where one expects considerable variability in sample paths, it is indeed doubtful if the optimization of an 'expected cost function' would suffice or would even be appropriate.

From a mathematical standpoint, the model here falls under the class of models "of the M/G/1 type" studied by M. F. Neuts [11] (see also [8], [12], [16]). However, certain irreducibility conditions assumed in these papers fail to hold here, and this calls for certain *ad hoc* methods. The extra effort needed in developing the algorithms is nevertheless richly rewarded by major savings in computational effort. While the algorithms are developed for general service times, we point out that the prohibitive overhead computations required to set up the transition matrix alone make the algorithms impractical except in some special cases if general service times were to be used. We then show that if one assumes that the service time distributions are of phase type [10], [13], then practical recursive schemes can be developed and the algorithms made implementable. The well-known versatility of the phase type distributions should render our models sufficiently general for most practical purposes.

2. The Model and the Embedded Semi-Markov Process

Our model is a single server facility into which two independent Poisson streams of customers - henceforth to be called 1-customers and 2-customers-arrive and form separate queues; the rates of these streams are denoted by λ_1 and λ_2 respectively. The successive customers have independent service times, the c.d.f. of the service time of an i-customer being $\tilde{H}_i(\cdot)$ which is assumed to have a finite mean μ_i', $i = 1,2$. We shall assume that a finite waiting room of size K-1, where $K \geq 1$ is fixed, is imposed on the 2-customers, and that any 2-customer arriving when the waiting room is full is lost. A dynamic non-preemptive priority is specified in terms of a pre-assigned number $N \geq 0$, called the threshold, which operates as follows: At a service completion if only one type of customers is present, one of the waiting customers immediately enters service. However, if at the end of a service completion both queues are non-empty, then a 1-customer is served next if the number of 1-customers is at least $N+1$, whereas a 2-customer is served next if the number of 1-customers is at most N. An arrival to the empty system, irrespective of its type, enters service immediately. While in most practical applications $\lambda_1 >> \lambda_2$ and $\mu_1' << \mu_2'$, no such restrictions will be assumed in the ensuing analysis. Also for the purpose of studying queue lengths, the order of service within each class is immaterial.

Previous literature related to a special case of our model may be found in [2], [7], [9], and [18] which discuss M/G/1 type queues under a priority rule assigning fixed non-preemptive priority to customers whose service times do not exceed a given positive number ϕ. Under the additional assumption of a finite waiting room for customers whose service times exceed ϕ, such a model employing a "short job-long job split" is easily seen to be a special case of ours with $N = 0$, $\lambda_1 = \tilde{H}(\phi)\lambda$, $\lambda_2 = \lambda - \lambda_1$,

$$\tilde{H}_1(x) = \begin{cases} 0 & , \quad x < 0 \\ \tilde{H}(x)/\tilde{H}(\phi) & , \quad 0 \leqq x < \phi \\ 1 & , \quad \phi \leqq x < \infty \quad , \end{cases}$$

and

$$\tilde{H}_2(x) = \begin{cases} 0 & , \quad x < \phi \\ \dfrac{\tilde{H}(x)-\tilde{H}(\phi)}{1-\tilde{H}(\phi)} & , \quad \phi \leqq x < \infty \quad , \end{cases}$$

where λ is the combined arrival rate and $\tilde{H}(\cdot)$ is the service time c.d.f. of a customer. It is claimed in [9] that the "short job-long job split" is preferable to the "shortest processing time discipline" in that the former does not need as much information by the server regarding the service times of waiting customers as the latter does. Note that the queue discipline considered here does not require that the server know the exact service requirements of individual customers beforehand, but all that needs to be known is to which class the customer belongs; often such information is available from other considerations such as for example in computing facilities where batch jobs and jobs requiring the use of auxiliary devices like tapes, plotter, etc., may routinely be assigned lower priority, irrespective of the duration of service. In the general situation considered here we do not require any ordering between the service times of the two classes of customers; $\tilde{H}_1$ and $\tilde{H}_2$ are arbitrary. Finally, the imposition of a finite waiting room for

2-customers permits us to consider a wider set of interesting practical situations which, without such assumption, may lead to an unstable queueing model.

For later use we now define the following quantities:

$$p_n^{(i)}(t) = e^{-\lambda_i t}(\lambda_i t)^n/n!;\ t \in [0,\infty),\ n \geq 0,\ i = 1,2. \tag{2.1}$$

$$\tilde{\alpha}_{jk}^{(i)}(x) = \int_0^x p_j^{(1)}(t)p_k^{(2)}(t)\,d\tilde{H}_i(t);\ x \in [0,\infty),\ j,k \geq 0,\ i = 1,2. \tag{2.2}$$

$$\alpha_{jk}^{(i)}(s) = \int_0^\infty e^{-sx}\,d\tilde{\alpha}_{jk}^{(i)}(x);\ \text{Re } s \in [0,\infty),\ j,k \geq 0,\ i = 1,2. \tag{2.3}$$

$$\alpha_{jk}^{(i)} = \alpha_{jk}^{(i)}(0) = \tilde{\alpha}_{jk}^{(i)}(\infty);\ j,k \geq 0,\ i = 1,2. \tag{2.4}$$

Note that $\tilde{\alpha}_{jk}^{(i)}(x)$ is the probability that the service time of an i-customer lasts at most x units of time and that during such a service time exactly j 1-customers and k 2-customers arrive.

The notational convention embodied in equations (2.2)-(2.4), viz., that of denoting a distribution function on $[0,\infty)$ by a letter superscripted by a tilde, its Laplace-Stieltjes transform (LST) by the same letter without the tilde, and finally the value of the LST at $s = 0$ by the letter representing the LST without the function argument, will be followed throughout the paper. Thus e.g., given a matrix $\tilde{A}(\cdot)$ of d.f.s. on $[0,\infty)$, we have, $A(s) = \int_0^\infty e^{-sx}d\tilde{A}(x)$ and $A = A(0) = \tilde{A}(\infty)$. Also, we shall denote by $\underline{e}$ a column vector of 1's of dimension appropriate to the formula in which it appears.

Given a nonnegative sequence $\{a_j: j \geq 0\}$, with $\sum_0^\infty a_j < \infty$, and $m \leq n$, let $\Delta(m,n;\ \{a_j: j \geq 0\})$ denote the $m \times n$ matrix

$$\begin{bmatrix} a_0 & a_1 & a_2 & \cdots & \cdots & a_{n-2} & \sum_{j=n-1}^{\infty} a_j \\ 0 & a_0 & a_1 & \cdots & \cdots & a_{n-3} & \sum_{j=n-2}^{\infty} a_j \\ 0 & 0 & a_0 & \cdots & \cdots & a_{n-4} & \sum_{j=n-3}^{\infty} a_j \\ \vdots & \vdots & \vdots & \vdots & \vdots & \vdots & \vdots \\ 0 & 0 & 0 & \cdots & \cdots & \cdots & \sum_{j=n-m}^{\infty} a_j \end{bmatrix} .$$

A number of matrices appearing in the sequel are of this type, and concerning such matrices we have the following results which are easy to establish.

Lemma 2.5.

$$\Delta(m,n;\ \{a_j:\ j\geqq 0\}) + \Delta(m,n;\ \{b_j:\ j\geqq 0\}) = \Delta(m,n;\ \{a_j+b_j:\ j\geqq 0\}). \tag{2.6}$$

$$\Delta(m,n;\ \{a_j:\ j\geqq 0\}) \cdot \Delta(n,k;\ \{b_j:\ j\geqq 0\}) = \Delta(m,k;\ \{(a*b)_j:\ j\geqq 0\}), \tag{2.7}$$

where

$$(a*b)_j = \sum_{\nu=0}^{j} a_\nu\, b_{j-\nu}\ , \quad j \geqq 0,$$

is the convolution of the sequences $\{a_\nu\}$ and $\{b_\nu\}$.

Having established the above notations we return to the queueing model and define τ_n to be the epoch of the n-th departure, $\tau_0 = 0$. Letting X_n and J_n denote respectively the number of 1-customers and the number of 2-customers in the system at τ_n+, we have the following result.

Theorem 2.8. $\{(X_n,\ J_n,\ \tau_{n+1}-\tau_n):\ n \geqq 0\}$ is a semi-Markov process (SMP) with state space $\{0,1,\ldots\} \times \{0,\ldots,K-1\}$. After writing the states in lexicographic order, the transition function $\tilde{Q}(x)$, $x \geqq 0$, of this SMP is of the form given in Figure 1 when $N = 0$ and in Figure 2 when $N \geqq 1$.

$$\tilde{Q} = \begin{bmatrix} \tilde{D}_0 & \tilde{D}_1 & \tilde{D}_2 & \tilde{D}_3 & \cdots & \cdots \\ \tilde{A}_0 & \tilde{A}_1 & \tilde{A}_2 & \tilde{A}_3 & \cdots & \cdots \\ 0 & \tilde{A}_0 & \tilde{A}_1 & \tilde{A}_2 & \cdots & \cdots \\ 0 & 0 & \tilde{A}_0 & \tilde{A}_1 & \cdots & \cdots \\ 0 & 0 & 0 & \tilde{A}_0 & \cdots & \cdots \\ \cdots & \cdots & \cdots & \cdots & \cdots & \cdots \end{bmatrix}$$

Figure 1

$$\tilde{Q} = \left[\begin{array}{ccccccc|ccccc} \tilde{D}_0 & \tilde{D}_1 & \tilde{D}_2 & \cdots & \cdots & \tilde{D}_{N-1} & \tilde{D}_N & \tilde{D}_{N+1} & \tilde{D}_{N+2} & \tilde{D}_{N+3} & \cdots & \cdots \\ \tilde{B}_0 & \tilde{B}_1 & \tilde{B}_2 & \cdots & \cdots & \tilde{B}_{N-1} & \tilde{B}_N & \tilde{B}_{N+1} & \tilde{B}_{N+2} & \tilde{B}_{N+3} & \cdots & \cdots \\ 0 & \tilde{B}_0 & \tilde{B}_1 & \cdots & \cdots & \tilde{B}_{N-2} & \tilde{B}_{N-1} & \tilde{B}_N & \tilde{B}_{N+1} & \tilde{B}_{N+2} & \cdots & \cdots \\ 0 & 0 & \tilde{B}_0 & \cdots & \cdots & \tilde{B}_{N-3} & \tilde{B}_{N-2} & \tilde{B}_{N-1} & \tilde{B}_N & \tilde{B}_{N+1} & \cdots & \cdots \\ \cdots & \cdots & \cdots & \cdots & \cdots & \cdots & \cdots & \cdots & \cdots & \cdots & \cdots & \cdots \\ 0 & 0 & 0 & \cdots & \cdots & \tilde{B}_0 & \tilde{B}_1 & \tilde{B}_2 & \tilde{B}_3 & \tilde{B}_4 & \cdots & \cdots \\ \hline 0 & 0 & 0 & \cdots & \cdots & 0 & \tilde{A}_0 & \tilde{A}_1 & \tilde{A}_2 & \tilde{A}_3 & \cdots & \cdots \\ 0 & 0 & 0 & \cdots & \cdots & 0 & 0 & \tilde{A}_0 & \tilde{A}_1 & \tilde{A}_2 & \cdots & \cdots \\ 0 & 0 & 0 & \cdots & \cdots & 0 & 0 & 0 & \tilde{A}_0 & \tilde{A}_1 & \cdots & \cdots \\ 0 & 0 & 0 & \cdots & \cdots & 0 & 0 & 0 & 0 & \tilde{A}_0 & \cdots & \cdots \\ \cdots & \cdots & \cdots & \cdots & \cdots & \cdots & \cdots & \cdots & \cdots & \cdots & \cdots & \cdots \end{array}\right]$$

Figure 2

The $K \times K$ blocks appearing in these figures are as follows:

$$\tilde{A}_\nu = \Delta(K,K;\ \{\tilde{\alpha}^{(1)}_{\nu j} : j \geqq 0\}),\ \nu \geqq 0\ , \tag{2.9}$$

$$\tilde{B}_0 = \left[\begin{array}{c} \Delta(1,K;\ \{\tilde{\alpha}^{(1)}_{0j} : j \geqq 0\}) \\ \hline 0_{K-1\times K} \end{array}\right], \tag{2.10}$$

$$\tilde{B}_\nu = \left[\begin{array}{c} \Delta(1,K;\ \{\tilde{\alpha}^{(1)}_{\nu j}:\ j \geqq 0\}) \\ \hdashline \Delta(K-1,K;\ \{\tilde{\alpha}^{(2)}_{\nu-1,j}:\ j \geqq 0\}) \end{array}\right], \quad \nu \geqq 1, \tag{2.11}$$

$$\tilde{D}_\nu = \left[\begin{array}{c} \Delta(1,K;\ \{\tilde{d}_{\nu j}:\ j \geqq 0\}) \\ \hdashline \Delta(K-1,K;\ \{\tilde{\alpha}^{(2)}_{\nu j}:\ j \geqq 0\}) \end{array}\right], \quad \nu \geqq 0, \tag{2.12}$$

where,

$$\tilde{d}_{\nu j}(x) = \int_0^x \lambda_1 e^{-\lambda_1 u} e^{-\lambda_2 u} \tilde{\alpha}^{(1)}_{\nu j}(x-u)\, du \tag{2.13}$$

$$+ \int_0^x \lambda_2 e^{-\lambda_2 u} e^{-\lambda_1 u} \tilde{\alpha}^{(2)}_{\nu j}(x-u)\, du ,$$

and the $\tilde{\alpha}^{(i)}_{k\ell}$ are as in (2.2).

Proof. A detailed proof is omitted, and we shall only outline the important facts leading to formulas (2.9)-(2.13). When $X_n \geqq N + 1$, irrespective of the value of J_n, a 1-customer is served next, and this leads to (2.9). When $1 \leqq X_n \leqq N$, the queue length of 1-customers can decrease only if a 1-customer is served next, and the latter happens iff $J_n = 0$; this leads to (2.10). When $1 \leqq X_n \leqq N$, a 1-customer is served next if $J_n = 0$, otherwise a 2-customer is served next; these observations lead to (2.11). Finally, when $X_n = 0$ and $J_n > 0$ a 2-customer is served next, whereas when $X_n = J_n = 0$ an idle time results and the customer who is the first to arrive is served next; these considerations lead to (2.12) and (2.13).

Remarks. (i) The structures of the blocks described above play a significant role in obtaining a number of simplifications in the algorithms in the sequel, and the reader is advised to write out these matrices in expanded form before proceeding further.

(ii) The transition function $\tilde{Q}$ in Figure 1 has the structure examined by M. F. Neuts in [11], and that in Figure 2 has such a structure for the "non-boundary" states N+1,N+2,... . For a survey of results pertaining to stochastic models with an embedded SMP of these forms, we refer the reader to D. M. Lucantoni and M. F. Neuts [8]. For a specific example, refer to M. F. Neuts [12] or to V. Ramaswami [16].

(iii) In the analysis developed in [11], it is assumed that the matrix $A = \sum_{\nu=0}^{\infty} A_\nu$, (recall $A_\nu = \tilde{A}_\nu(\infty)$), is irreducible. In our case, A, being upper triangular, is reducible. The implications of this will be explored in detail in the following sections.

(iv) It is easy to verify that the stochastic matrix $Q = \tilde{Q}(\infty)$ is irreducible and áperiodic. It will be verified later that the corresponding Markov chain is positive recurrent, recurrent null or transient according as $\rho_1 = \lambda_1\mu_1'$ is less than, equal to or greater than one.

(v) The computation of the entries $\alpha_{k\ell}^{(i)}$ which are needed to set up Q is no easy task when $\tilde{H}_1$ and $\tilde{H}_2$ are general distributions. In Section 6, we shall provide efficient methods for computing these recursively without any numerical integrations when $\tilde{H}_1$ and $\tilde{H}_2$ are probability distributions of phase type [10], [13].

3. The Fundamental Period and the Non-linear Matrix Equation

For $i \geq 0$, by level $\underline{i}$ we denote the set of states $\{(i,0),\ldots,(i,K-1)\}$. Basic to our analysis of the queue are the first passage times of the SMP $\tilde{Q}(\cdot)$ from level $\underline{i+1}$ to level $\underline{i}$, $i \geq N$; these are called "the fundamental periods" by M. F. Neuts. For $i \geq N$, in view of the structure of $\tilde{Q}(\cdot)$, the distribution of the first passage time from $\underline{i+1}$ to $\underline{i}$ does not depend on i. Let $\tilde{G}_{jj'}(k,x)$ denote the conditional probability, given it starts in $(i+1,j)$, $i \geq N$, that the SMP enters $\underline{i}$ for the first time at or before time x and in exactly k steps by visiting the state (i,j'), $0 \leq j, j' \leq K-1$. For the matrices $\tilde{G}(k,x)$ with entries $\tilde{G}_{jj'}(k,x)$, we define the transform

$$G(z,s) = \sum_{k=0}^{\infty} z^k \int_0^{\infty} e^{-sx}\, d\tilde{G}(k,x), \quad |z| \leqq 1, \quad \text{Re } s \geqq 0, \tag{3.1}$$

and let $G(z) = G(z,0)$ and $G = G(1) = G(1,0)$. From the Markov property at these first passage epochs and from the fact that a first passage in $\tilde{Q}(\cdot)$ from $\underline{i+\nu}$ to $\underline{i}$ must occur via a sequence of first passages from $\underline{i+\nu}$ to $\underline{i+\nu-1}, \ldots, \underline{i+1}$ to $\underline{i}$, it is easily seen that $G^{\nu}(z,s)$, the ν-th power of $G(z,s)$, completely describes first passages from $\underline{i+\nu}$ to $\underline{i}$ for $i \geqq N$ and $\nu \geqq 1$. By appealing to the general results obtained in [11] we also have the following theorem.

Theorem 3.2. (i) $G(z,s)$ satisfies the non-linear matrix functional equation

$$G(z,s) = z \sum_{n=0}^{\infty} A_n(s)\, G^n(z,s), \quad |z| \leqq 1, \ \text{Re } s \geqq 0. \tag{3.3}$$

(ii) For $0 \leqq z \leqq 1$, $s > 0$, there exists a unique nonnegative matrix $G(z,s)$ which satisfies (3.3). The entries of $G(z,s)$ are analytic functions of z and s, and $G(z,s)$ may be written in the form (3.1), where $\tilde{G}(k,\cdot)$, $k \geqq 0$, are matrices of (defective) probability mass functions. The matrix $G = G(1-,0+)$, defined by continuity, is substochastic.

(iii) G is the minimal solution, in the class of nonnegative matrices of the equation

$$G = \sum_{n=0}^{\infty} A_n G^n \tag{3.4}$$

obtained by setting $z = 1$, $s = 0$ in (3.1).

(iv) G is the limit as $n \to \infty$ of the entrywise non-decreasing sequence $\{G_n : n \geqq 0\}$ of matrices defined recursively by

$$\left.\begin{aligned} G_0 &= A_0 \\ G_{n+1} &= \sum_{\nu=0}^{\infty} A_{\nu} G_n^{\nu}, \quad n \geqq 0. \end{aligned}\right\} \tag{3.5}$$

The first result we obtain by exploitation of the special structure of our model concerns the structure of the matrix G and results in major simplifications in the computation of G. This is presented as the next theorem.

Theorem 3.6. (i) There exists a sequence $\{g_n: n \geqq 0\}$ of positive numbers with $\sum_{n=0}^{\infty} g_n \leqq 1$ such that $G = \Delta(K,K; \{g_n: n \geqq 0\})$.

(ii) Letting $\underline{g} = (g_0, g_1, \ldots)$ and $\underline{\alpha}_n^{(1)} = (\alpha_{n0}^{(1)}, \alpha_{n1}^{(1)}, \ldots)$, $n \geqq 0$, we have

$$\underline{g} = \sum_{n=0}^{\infty} \underline{\alpha}_n^{(1)} * \underline{g}^{(n)} \tag{3.7}$$

where $\underline{g}^{(n)}$ is the n-fold convolution of $\underline{g}$ with itself and $\underline{g}^{(0)} = (1,0,0,\ldots)$.

(iii) $\underline{g}$ is the minimal non-negative solution of (3.7) and can be computed by successive substitutions in (3.7) starting with the initial iterate $\underline{\alpha}_0^{(1)}$. That is, $\underline{g} = \lim_{\nu\to\infty} \underline{g}_\nu$ where

$$\left.\begin{aligned} \underline{g}_0 &= \underline{\alpha}_0^{(1)} \\ \underline{g}_{\nu+1} &= \sum_{n=0}^{\infty} \underline{\alpha}_n^{(1)} * \underline{g}_\nu^{(n)}, \quad \nu \geqq 0 \end{aligned}\right\} \tag{3.8}$$

(iv) G is stochastic iff $\rho_1 = \lambda_1 \mu_1' \leqq 1$.

Proof. The convergence of the recursive scheme (3.8) to the minimal non-negative solution of (3.7) is obtained as in Theorem 4 in D. Heiman and M. F. Neuts [4]; the lengthy details are omitted. Using Lemma 2.5 and mathematical induction, it is easily established that the successive iterates G_n in (3.5) are given by $G_n = \Delta(K,K;\underline{g}_n)$. Parts (i) and (ii) of the present theorem now follow from Theorem 3.2(iv) which asserts that $G = \lim_{n\to\infty} G_n$.

To prove (iv), let $\gamma = \sum_{n=0}^{\infty} g_n$ and $\alpha_{n\cdot}^{(1)} = \sum_{k=0}^{\infty} \alpha_{nk}^{(1)}$, $n \geqq 0$. It is clear from equation (3.7) and the minimality of $\underline{g}$ that γ is the minimal nonnegative solution of the equation

$$\gamma = \sum_{n=0}^{\infty} \alpha_{n\cdot}^{(1)} \gamma^{n}. \tag{3.9}$$

A familiar argument based on the consideration of the graphs of both sides of (3.9) now yields that $\gamma = 1$ iff $\sum_{n=0}^{\infty} n\,\alpha_{n\cdot}^{(1)} = \rho_1 \leqq 1$. From (i) it is clear that $G\underline{e} = \gamma\underline{e}$. Hence, part (iv) is proved.

Remarks. a) The computational use of the above theorem is enormous. As will be seen later, for computing a large number of quantities of interest one needs the matrix G. What we have shown is that once one computes $\underline{g}$ using (3.8), then for <u>any</u> value of N and K, the associated matrix G can be obtained without any further calculations. It may be recalled from [8], [11], [16] that the major computational effort in solving models of the M/G/1 type is in the solution of the non-linear matrix equation $G = \sum_{\nu=0}^{\infty} A_{\nu} G^{\nu}$. That we can solve several such equations for different values of N and K, all at one stroke, is what makes the model algorithmically very tractable.

b) In practice, for purposes of design, one can choose an upper bound $K_0 - 1$ for the size of the waiting room for the 2-customers. In such situations all one needs are the first $K_0 - 1$ terms of $\underline{g}$. Our computational experience shows that one may routinely compute the first 50 terms of $\underline{g}$ without any difficulty by implementing (3.8) using Horner's method to compute the successive iterates in (3.8).

From its definition, it is clear that the j-th component of $G\underline{e}$ is the probability that the SMP $\tilde{Q}$ eventually visits $\underline{i}$ given that it starts in $(i+1,j)$, $i \geqq N$, $0 \leqq j \leqq K-1$. Thus if $\rho_1 > 1$, then $G\underline{e} \ll \underline{e}$ and therefore $\tilde{Q}$ is transient. We shall from now on assume that $\rho_1 \leqq 1$. To obtain the condition for the positive recurrence of $\tilde{Q}$, we now examine the vector $\underline{\mu} = G'(1)\underline{e}$, whose components give the mean number of service completions in a first passage from $(i+1,j)$ to $\underline{i}$, $0 \leqq j \leqq K-1$, $i \geqq N$.

Lemma 3.10. $\underline{\mu}$ is finite iff $\rho_1 < 1$.

Proof. By differentiating $G(z) = z \sum_{\nu=0}^{\infty} A_\nu G^\nu(z)$, we get upon letting $z \uparrow 1$ the equation

$$G'(1) = \sum_{\nu=0}^{\infty} A_\nu G^\nu + \sum_{\nu=1}^{\infty} A_\nu \sum_{n=0}^{\nu-1} G^n G'(1) G^{\nu-n-1} .$$

Multiplying the above equation by $\underline{e}$, noting that $G\underline{e} = \underline{e}$ and using (3.4), we get

$$\underline{\mu} = \underline{e} + \left(\sum_{\nu=1}^{\infty} A_\nu \sum_{n=0}^{\nu-1} G^n \right) \underline{\mu} . \tag{3.11}$$

Letting $L = \sum_{\nu=1}^{\infty} A_\nu \sum_{n=0}^{\nu-1} G^n$, we note that $L\underline{e} = \sum_{\nu=1}^{\infty} \nu A_\nu \underline{e} = \rho_1 \underline{e}$, and (3.11) now implies by successive substitutions that

$$\underline{\mu} = (1 + \rho_1 + \rho_1^2 + \ldots + \rho_1^{n-1})\underline{e} + L^n \underline{\mu},\ n \geqq 1. \tag{3.12}$$

The lemma follows immediately.

Corollary 3.13. $\tilde{Q}$ is positive recurrent iff $\rho_1 < 1$. Thus the queue is stable iff $\rho_1 < 1$.

Corollary 3.14. Let $\rho_1 < 1$. Then

$$\underline{\mu} = (1-\rho_1)^{-1} \underline{e} \tag{3.15}$$

Proof. This follows easily from (3.12) by letting $n \to \infty$.

Remarks. a) Henceforth we shall assume that $\rho_1 < 1$.

b) Noting that during a first passage from $\underline{i+1}$ to $\underline{i}$, $i \geqq N$, the queue of 1-customers behaves like an ordinary M/G/1 queue, formula (3.15) and the stability condition are very intuitive.

4. Stationary Queue Length Densities at Points of Departure

In this section, we shall be concerned with the computation of the steady state vector $\underline{x}$ defined uniquely by

$$\underline{x}\, Q = \underline{x},\ \underline{x}\, \underline{e} = 1, \tag{4.1}$$

where $Q = \tilde{Q}(\infty)$, and $\tilde{Q}(\cdot)$ is given by Figure 1 or Figure 2 depending upon whether $N = 0$ of $N \geqq 1$. Choosing the indices for the components of $\underline{x}$ in a natural manner, we note that the component $x(i,j)$ of $\underline{x}$ is the stationary probability that a departure leaves behind i 1-customers and j 2-customers, $i \geqq 0$, $0 \leqq j \leqq K-1$. The basic strategy adopted here is the one proposed by M. F. Neuts [11], viz., to obtain an adequate number of components of $\underline{x}$ by a direct Markov renewal argument and then to perform a block Gauss-Seidel iteration on the (suitably truncated) steady state equations. The novelty here is in the computational and theoretical simplifications that result due to the special structure, and these will be pointed out below at appropriate places. In view of the different forms of $\tilde{Q}$ and for certain algorithmic purposes which will become clear later, we shall discuss the cases $N = 0$ and $N \geqq 1$ separately. Before we take up these cases let us recall (2.9)-(2.12) and our convention that $A_\nu = \tilde{A}_\nu(\infty)$ etc., and introduce the following additional notations:

$$\bar{A}(z) = \sum_{\nu=0}^{\infty} z^\nu A_\nu; \quad \bar{A}^{(n)}(z) = \frac{d^n}{dz^n}\bar{A}(z); \quad \underline{a}^{(n)} = \bar{A}^{(n)}(1-)\underline{e}, \quad n \geqq 0 \tag{4.2}$$

$$\bar{B}(z) = \sum_{\nu=0}^{\infty} z^\nu B_\nu; \quad \bar{B}^{(n)}(z) = \frac{d^n}{dz^n}\bar{B}(z); \quad \underline{b}^{(n)} = \bar{B}^{(n)}(1-)\underline{e}, \quad n \geqq 0 \tag{4.3}$$

$$\bar{D}(z) = \sum_{\nu=0}^{\infty} z^\nu D_\nu; \quad \bar{D}^{(n)}(z) = \frac{d^n}{dz^n}\bar{D}(z); \quad \underline{d}^{(n)} = \bar{D}^{(n)}(1-)\underline{e}, \quad n \geqq 0 \tag{4.4}$$

$$A = \bar{A}(1) = \sum_{\nu=0}^{\infty} A_\nu; \quad B = \bar{B}(1) = \sum_{\nu=0}^{\infty} B_\nu; \quad D = \bar{D}(1) = \sum_{\nu=0}^{\infty} D_\nu. \tag{4.5}$$

Further, let

$$\mu_n'(i) = \int_0^\infty x^n \, d\tilde{H}_i(x), \quad n \geqq 0, \quad i = 1,2 \tag{4.6}$$

denote the moments of the service time cdfs. We shall also state the following result for later use.

Lemma 4.7. We have

$$\underline{a}^{(0)} = \underline{b}^{(0)} = \underline{d}^{(0)} = \underline{e} \tag{4.8}$$

$$\underline{a}^{(n)} = \lambda_1^n \mu_n'(1) \, \underline{e}, \quad n \geqq 1 \tag{4.9}$$

$$\underline{b}^{(n)} = \begin{bmatrix} \lambda_1^n \mu_n'(1) \\ \hdashline \{\lambda_1^n \mu_n'(2) + n\lambda_1^{n-1} \mu_{n-1}'(2)\}\underline{e} \end{bmatrix}, \quad n \geqq 1, \tag{4.10}$$

$$\underline{d}^{(n)} = \begin{bmatrix} p\lambda_1^n \mu_n'(1) + (1-p)\lambda_1^n \mu_n'(2) \\ \hdashline \lambda_1^n \mu_n'(2)\underline{e} \end{bmatrix}, \quad n \geqq 1, \tag{4.11}$$

where $p = \lambda_1/(\lambda_1+\lambda_2)$. Also the $\underline{e}$ in (4.10) and (4.11) is of order $K-1$ for $K \geqq 2$; where $K=1$, the quantities $b^{(n)}$ and $d^{(n)}$ are scalars with values given by the first component of the vectors shown above.

Proof. These are obtained by a direct computation using (2.9)-(2.12); we omit the details, and, for illustration, prove only (4.9).

From (2.9), we have $A_\nu \underline{e} = \alpha_{\nu\cdot}^{(1)} \underline{e}$, where $\alpha_{\nu\cdot}^{(1)} = \sum_{j=0}^{\infty} \alpha_{\nu j}^{(1)} = \int_0^\infty e^{-\lambda_1 x} \frac{(\lambda_1 x)^\nu}{\nu!} d\tilde{H}_1(x)$.

Now, from (4.2),

$$\underline{a}^{(n)} = \left\{ \sum_{\nu=n}^{\infty} \nu(\nu-1) \ldots (\nu-n+1)\alpha_{\nu\cdot}^{(1)} \right\} \underline{e} = \lambda_1^n \mu_n'(1)\underline{e}.$$

4A. The Case N = 0

The case N = 0 corresponds to the situation of fixed priority under which no 2-customer can enter service as long as 1-customers are present. In this case we partition $\underline{x}$ as $\underline{x} = (\underline{x}_0, \underline{x}_1, \ldots)$, $\underline{x}_i \in \mathcal{R}^K$, and note that the steady state equations $\underline{x}\, Q = \underline{x}$ are given by

$$\underline{x}_\nu = \underline{x}_0\, D_\nu + \sum_{j=1}^{\nu+1} \underline{x}_j\, A_{\nu+1-j}, \quad \nu \geqq 0. \tag{4.12}$$

Defining

$$\underline{X}(z) = \sum_{\nu=0}^{\infty} z^\nu \underline{x}_\nu, \quad |z| \leqq 1, \tag{4.13}$$

it is easily seen from (4.12) that

$$\underline{X}(z)\, \{zI-\bar{A}(z)\} = \underline{x}_0\, \{z\bar{D}(z)-\bar{A}(z)\}. \tag{4.14}$$

To compute $\underline{x}_0$ we consider the successive return times to level $\underline{0}$ and consider the discrete time Markov renewal process (MRP) at these epochs with states $\{(0,0),(0,1),\ldots,(0,K-1)\}$; the time interval between two successive transitions in this MRP is the number of steps (service completions) in the Markov chain Q between the two successive visits to level $\underline{0}$. This MRP is governed by the (discrete) transform matrix $L(z) \overset{\text{def}}{=} \sum_{\nu=1}^{\infty} z^{\nu} L_{\nu}$, $|z| \leqq 1$, where L_{ν} is the $K \times K$ matrix whose (j,j')th entry is the conditional probability, given that it starts in $(0,j)$, that the Markov chain Q re-enters level $\underline{0}$ in exactly ν steps by visiting $(0,j')$.

Theorem 4.15. The matrix $L(z)$ is given by

$$L(z) = \sum_{\nu=0}^{\infty} z\, D_{\nu}\, G^{\nu}(z), \tag{4.16}$$

where $G(z) = G(z,0)$ is defined in Section 3. Also

$$L(1) = \begin{bmatrix} \Delta(1,K;p\underline{g}+(1-p)\underline{h}) \\ \hline \Delta(K-1,K;\underline{h}) \end{bmatrix}, \tag{4.17}$$

where $\underline{g}$ was defined in Section 3 and

$$\underline{h} = \sum_{\nu=0}^{\infty} \underline{\alpha}_{\nu}^{(2)} * \underline{g}^{(\nu)} \tag{4.18}$$

with

$$\underline{\alpha}_{\nu}^{(2)} = (\alpha_{\nu 0}^{(2)}, \alpha_{\nu 1}^{(2)}, \ldots), \quad \nu \geqq 0.$$

Proof. Equation (4.16) is obtained by considering the first transition in the Markov chain governed by Q. In the right side of (4.16), the first term $z\, D_0$ corresponds to the case where the chain returns to $\underline{0}$ in the very first step; for $\nu \geqq 1$, the term $z\, D_{\nu}\, G^{\nu}(z)$ corresponds to the case where the chain in its first step enters level $\underline{\nu}$ and then makes a first passage from $\underline{\nu}$ to $\underline{0}$, because, by the remark preceding Theorem 3.2, the number of steps in a first passage from $\underline{\nu}$ to $\underline{0}$ is governed by $G^{\nu}(z)$.

Setting $z = 1$ in (4.16), we have $L(1) = \sum_{\nu=0}^{\infty} D_{\nu}\, G^{\nu}$. Since

$$D_{\nu} = \begin{bmatrix} \Delta(1,K;\{d_{\nu j}: j \geqq 0\}) \\ \hline \Delta(K-1,K;\{\alpha_{\nu j}^{(2)}: j \geqq 0\}) \end{bmatrix}$$

and $G = \Delta(K,K;\underline{g})$, as is seen from

(2.12) and Theorem 3.2(i) respectively, by repeated use of Lemma 2.5, we have

$$L(1) = \begin{bmatrix} \Delta(1,K;\ \sum_{\nu=0}^{\infty} \underline{d}_\nu * \underline{g}^{(\nu)}) \\ \hline \Delta(K-1,K;\ \sum_{\nu=0}^{\infty} \underline{\alpha}_\nu^{(2)} * \underline{g}^{(\nu)}) \end{bmatrix}, \text{ where } \underline{d}_\nu = (d_{\nu 0}, d_{\nu 1}, \ldots) \text{ and}$$

$\underline{\alpha}_\nu^{(2)} = (\alpha_{\nu 0}^{(2)}, \alpha_{\nu 1}^{(2)}, \ldots)$. Now from (2.13), $\underline{d}_\nu = p\ \underline{\alpha}_\nu^{(1)} + (1-p)\ \underline{\alpha}_\nu^{(2)}$ whence $\sum_{\nu=0}^{\infty} \underline{d}_\nu * \underline{g}^{(\nu)} = p \sum_{\nu=0}^{\infty} \underline{\alpha}_\nu^{(1)} * \underline{g}^{(\nu)} + (1-p) \sum_{\nu=0}^{\infty} \underline{\alpha}_\nu^{(2)} * \underline{g}^{(\nu)} = p\ \underline{g} + (1-p)\ \underline{h}$ by using (3.7), and the proof is complete.

Corollary 4.19. The matrix L(1) is stochastic, irreducible and áperiodic.

Corollary 4.20. The vector $\underline{d}^*$, whose i-th component is the mean number of service completions during a return to $\underline{0}$ starting in (0,i), is given by

$$\underline{d}^* = L'(1)\ \underline{e} = \underline{e} + (1-\rho_1)^{-1}\ \underline{d}^{(1)}. \tag{4.21}$$

Proof. Differentiating both sides of (4.16) with respect to z yields

$$L'(z) = \sum_{\nu=0}^{\infty} D_\nu\ G^\nu(z) + \sum_{\nu=1}^{\infty} z\ D_\nu \sum_{r=0}^{\nu-1} G^r(z)\ G'(z)\ G^{\nu-r-1}(z).$$

Setting z = 1- and multiplying by $\underline{e}$ we get, by using $D\underline{e} = G\underline{e} = \underline{e}$ and Corollary 3.14,

$$\underline{d}^* = \sum_{\nu=0}^{\infty} D_\nu\ G^\nu \underline{e} + \sum_{\nu=1}^{\infty} D_\nu \sum_{r=0}^{\nu-1} G^r\ G'(1)\ \underline{e}$$

$$= \underline{e} + \sum_{\nu=1}^{\infty} D_\nu \sum_{r=0}^{\nu-1} G^r\ \underline{\mu} = \underline{e} + (1-\rho_1)^{-1} \sum_{\nu=1}^{\infty} \nu\ D_\nu\ \underline{e}$$

$$= \underline{e} + (1-\rho_1)^{-1}\ \underline{d}^{(1)}.$$

We now have the following result which gives the vector $\underline{x}_0$.

Theorem 4.22. The vector $\underline{x}_0$ is given by

$$\underline{x}_0 = c\ \underline{d}, \tag{4.23}$$

where $\underline{d}$ is the invariant probability vector of L(1) uniquely defined by

$$\underline{d}\, L(1) = \underline{d}, \quad \underline{d}\, \underline{e} = 1, \tag{4.24}$$

and

$$c^{-1} = \underline{d}\, \underline{d}^* = 1 + (1-\rho_1)^{-1} \{\lambda_1 \mu_1'(2) + p\, d_0\, \lambda_1 [\mu_1'(1) - \mu_1'(2)]\} \tag{4.25}$$

where d_0 is the first entry of $\underline{d}$.

Proof. Note that x(0,j) is the inverse of the mean recurrence time of the state (0,j) in the Markov chain Q. This mean recurrence time is the same as the mean recurrence time of (0,j) in the MRP governed by L(z). By Theorem 2.11 of Hunter [5], the inverse of this mean recurrence time is given by

$$m^{-1}(0,j) = d_j / \underline{d}\, \underline{d}^*, \quad 0 \leqq j \leqq K-1, \tag{4.26}$$

whence the theorem follows by using (4.21) and (4.11).

Remarks. i) In the classical approach one examines the equation det $(zI-\bar{A}(z)) = 0$ and provides a formula for $\underline{x}_0$ in terms of the solutions to this equation in the unit disk. The disadvantages of this approach based on Rouché's theorem are discussed in [14] and [17]. In the present model, the solutions of det $(zI-\bar{A}(z)) = 0$ are none other than the eigenvalues of G, and thus this equation has two roots g_0 and 1 of which the former is of multiplicity $K-1$.

ii) To compute the vector $\underline{d}$, note that L(1) is almost upper triangular. Thus setting $d_0 = 1$, one can successively solve the equations $\underline{d}\, L(1) = \underline{d}$ and then normalize the resulting vector. Also this entire computation can be carried out without creating the matrix L(1), and the only significant effort is in computing $\underline{h}$ using (4.18). These facts result in savings in computational effort and in storage.

Having computed $\underline{x}_0$ and observing that the matrix A_0 is nonsingular and upper triangular, it is indeed tempting to compute $\underline{x}_\nu$, $\nu \geqq 1$ by rewriting (4.12) as

$$\underline{x}_1 = \underline{x}_0(I - D_0)\, A_0^{-1}, \tag{4.27}$$

$$\underline{x}_{\nu+1} = \left[\underline{x}_\nu - \left\{\underline{x}_0\, D_\nu + \sum_{j=1}^{\nu} \underline{x}_j\, D_{\nu+1-j}\right\}\right] A_0^{-1}, \quad \nu \geq 1. \tag{4.28}$$

Such a procedure is, however, extremely unstable for it essentially involves computing the product of certain vectors by successive powers of A_0^{-1}. We, therefore, recommend a modified block Gauss-Seidel procedure which will be discussed below. Before we do so we shall discuss the vector

$$\underline{X}(1) = \sum_{\nu=0}^{\infty} \underline{x}_\nu \tag{4.29}$$

whose components yield the steady state probability distribution of 2-customers.

Theorem 4.30. We have

$$\underline{X}(1) = \underline{e}_K' + \underline{x}_0\, (D - A)\, (I - A + E)^{-1}, \tag{4.31}$$

where $\underline{e}_K' = (0,\ldots,0,1) \in R^K$ and E is a K × K matrix whose last column is $\underline{e}$ and all other columns are $\underline{0}$.

Proof. Setting z = 1 in (4.14) we get

$$\underline{X}(1)\, [I - A] = \underline{x}_0\, (D - A). \tag{4.32}$$

Adding $\underline{X}(1)\, E = \underline{e}_K'$ to both sides, and noting that $(I - A + E)$ is upper triangular with positive diagonal entries and hence non-singular, we get (4.31) because $\underline{e}_K'(I - A + E)^{-1} = \underline{e}_K'$.

Remarks. i) In models where A is irreducible it is customary [8], [11] to solve equations of the type (4.32) by using the generalized inverse $(I - A + \Pi)^{-1}$, where $\Pi = \underline{e}\ \underline{\pi}$ and $\underline{\pi}$ is the invariant probability vector of A. The reducibility and special structure of A in our case do not permit the use of this procedure. However, Hunter has shown in [6] that any matrix $(I - A + \underline{tu})^{-1}$, where $\underline{u}\ \underline{e} \neq 0$ and $\underline{t} \neq \underline{0}$, is a generalized inverse of $I-A$. In our case, the choice $E = \underline{e}\,\underline{e}_K'$ leads to particularly simple formulas.

ii) Equation (4.31) serves as a powerful accuracy check on the numerical computation of $\underline{x}$.

The following theorem gives the first two moments of the queue length of 1-customers. In the general case of [8] and in [12], [16] for obtaining the moments of $\underline{x}$, it was necessary to obtain the successive derivatives of the Perron Frobenius eigenvalue and eigenvectors of $\bar{A}(z)$. The simplified formulas here are again due to the structure of $\bar{A}(z)$.

Theorem 4.33. Let ξ_1 and ξ_2 denote the first two factorial moments of the stationary queue length density at points of departure for 1-customers. We have

$$\xi_1 = [\underline{x}_0\{2\underline{d}^{(1)} + \underline{d}^{(2)} - \underline{a}^{(2)}\} + \lambda_1^2\mu_2'(1)]/\{2(1-\rho_1)\} \tag{4.34}$$

$$\xi_2 = [\underline{x}_0\{3\underline{d}^{(2)} + \underline{d}^{(3)} - \underline{a}^{(3)}\} + 3\xi_1\lambda_1^2\mu_2'(1) + \lambda_1^3\mu_3'(1)]/\{3(1-\rho_1)\}. \tag{4.35}$$

Proof. Differentiating twice in (4.14), setting $z = 1$ and multiplying by $\underline{e}$ we get

$$\underline{X}''(1)\ [I-A]\ \underline{e} + 2\underline{X}'(1)\ \{\underline{e}-\underline{a}^{(1)}\} - \underline{X}(1)\ \underline{a}^{(2)} = \underline{x}_0\ \{\underline{d}^{(2)}+2\underline{d}^{(1)}-\underline{a}^{(2)}\}.$$

Using (4.9) in the left side of this equation, we get (4.34) by noting that $(I-A)\underline{e} = \underline{0}$, $\xi_1 = \underline{X}'(1)\underline{e}$ and $\underline{X}(1)\underline{e} = 1$. Equation (4.35) is obtained by computing $\xi_2 = \underline{X}''(1)\underline{e}$ in a similar manner by differentiating thrice in (4.14).

Having computed $\underline{x}_0$ we recommend that $\underline{x}_1$ be computed using (4.27). To get $\underline{x}_\nu$, $\nu \geq 2$ we suggest truncating the system (4.12) initially at some ν^* by using a "$\mu + 3\sigma$" rule. This truncated system is solved by a block Gauss-Seidel iterative procedure where the (n+1)st iterate is computed as follows:

$$\underline{x}_\nu(n+1) = \underline{b}_\nu' + \sum_{j=2}^{\nu-1} \underline{x}_j(n+1)\ A_{\nu+1-j}' + \underline{x}_{\nu+1}(n)\ A_0' \quad \text{for } \nu = 2,\dots,\nu^*, \tag{4.36}$$

where $\underline{b}'_\nu = (\underline{x}_0 D_\nu + \underline{x}_1 A_\nu)\ (I - A_1)^{-1}$, $\nu \geqq 0$, is a known vector, $A'_\nu = A_\nu (I - A_1)^{-1}$, $\nu \geqq 0$, and $\underline{x}_{\nu^*+1}(n) = \underline{0}$ for all n.

Note that in computing the vector $\underline{x}_\nu(n+1)$ we always use the most recent estimates of the other vectors $\underline{x}_j$, $j \neq \nu$. This procedure is continued until the successive iterates converge. If the components of the vectors $\underline{x}_\nu$ so computed do not add up to within the desired tolerance to 1, then one can progressively increase the truncation index ν^* to "$\mu + 4\sigma$" etc., using the most recent estimate of $\underline{x}$ as the initial solution. Our computational experience with such a precedure has been extremely satisfactory with respect to the various accuracy checks discussed above. In implementing the Gauss-Seidel iterations of (4.36), one can exploit the structure of the matrices $\{A_\nu\}$ and $\{D_\nu\}$ in computing the right side of (4.36) and write a code which saves a substantial number of operations. Indeed, in a large scale problem, these computations should be done without storing the matrices A_ν, D_ν, $\nu \geqq 0$, but rather by computing the products via the stored vectors $\underline{\alpha}_\nu^{(1)}$ and $\underline{\alpha}_\nu^{(2)}$, $\nu \geqq 0$. These details of programming become absolutely essential if one were to deal with models with non-Poisson arrivals to be discussed elsewhere.

Theorem 4.37. Let π_1 and π_2 denote respectively the long run proportions of 1-customers and 2-customers out of the total number served. Then

$$\pi_2 = 1 - \pi_1 = \underline{x}_0\underline{e} - p\ x(0,0).$$

Proof. Noting that a 2-customer is served iff either a departure leaves the system empty and then a 2-customer arrives first, or a departure leaves only 2-customers behind, we have $\pi_2 = x(0,0)\ (1-p) + \sum_{j=1}^{K-1} x(0,j)$, and the theorem follows.

We conclude the first part of our discussion of the case N = 0 with the following result which will be found useful in the sequel.

Theorem 4.38. For $\nu \geqq 1$, let $x_{\nu.} = \underline{x}_\nu \underline{e}$ and $\eta_\nu = \underline{x}_0 D_\nu \underline{e}$. Also let $\underline{x}^* = (x_{1.}, x_{2.}, \ldots)$ and $\underline{\eta} = (\eta_1, \eta_2, \ldots)$. The vector $\underline{x}^*$ is the unique nonnegative solution of the equation

$$\underline{x}^* = \underline{\eta} + \underline{x}^* P, \tag{4.39}$$

where

$$P = \begin{bmatrix} \alpha_{1.}^{(1)} & \alpha_{2.}^{(1)} & \alpha_{3.}^{(1)} & \alpha_{4.}^{(1)} & \cdots \\ \alpha_{0.}^{(1)} & \alpha_{1.}^{(1)} & \alpha_{2.}^{(1)} & \alpha_{3.}^{(1)} & \cdots \\ 0 & \alpha_{0.}^{(1)} & \alpha_{1.}^{(1)} & \alpha_{2.}^{(1)} & \cdots \\ 0 & 0 & \alpha_{0.}^{(1)} & \alpha_{1.}^{(1)} & \cdots \\ 0 & 0 & 0 & \alpha_{0.}^{(1)} & \cdots \\ \cdots & \cdots & \cdots & \cdots & \cdots \end{bmatrix}.$$

Proof. That $\underline{x}^*$ satisfies (4.39) is easily seen from (4.12) by multiplying both sides of the latter by $\underline{e}$. The sequence of vectors $\{\underline{u}_n\}_0^\infty$ defined by $\underline{u}_0 = \underline{\eta}$, $\underline{u}_{n+1} = \underline{\eta} + \underline{u}_n P$ for $n \geqq 0$, is easily shown (by mathematical induction) to be non-decreasing, and converging to $\underline{x}^*$ as $n \to \infty$. Now, for any other nonnegative solution $\underline{\tilde{x}}$ of (4.39), we have, $\underline{u}_n \leqq \underline{\tilde{x}}$ for all n whence

$$\underline{0} \leqq (\underline{\tilde{x}} - \underline{x}^*) = (\underline{\tilde{x}} - \underline{x}^*)P = \ldots = (\tilde{x} - x^*)P^n \to \underline{0} \quad \text{as } n \to \infty.$$

Thus $\underline{\tilde{x}} = \underline{x}^*$; hence uniqueness.

Varying the waiting room size parameter K.

Having discussed the computations for a given value of K, we now turn our attention to the important problem of computing the steady state vectors corresponding to different values of K and will show that, by proper nesting of the computations, substantial savings can be effected. In particular, our results will show that having computed the steady state vector for a given value of K, most of the components of the steady state vector for the next lower value of K can be obtained by a mere rescaling of the corresponding components of the already computed steady state vector by an easily computable constant. In view of this, we recommend that computations be done first for the largest value of K under consideration and then the results below be used to obtain the steady state vectors for lower values of K.

For the purpose of this discussion (only), all quantities defined earlier in the paper will be denoted by the same symbols used earlier but such symbols will now be augmented when necessary by a function argument K. Thus, for example, the steady state vectors $\{\underline{x}_\nu\}$ discussed earlier will be denoted by $\{\underline{x}_\nu(K)\}$, the matrix L(1) in (4.17) by L(1,K), the vector $\underline{d}$ in Theorem 4.22 by $\underline{d}(K)$ etc. Also given a vector $\underline{u}$, we denote by $\hat{\underline{u}}$ the vector obtained from $\underline{u}$ by deleting the last component of $\underline{u}$. Finally, $\hat{\hat{\underline{u}}} = (\hat{\hat{\underline{u}}})$.

Lemma 4.40. For $K \geq 2$, we have

$$\underline{d}(K-1) = [1-d_{K-1}(K)]^{-1}\, \hat{\underline{d}}(K),$$

where $d_{K-1}(K)$ is the last component of $\underline{d}(K)$.

Proof. Using (4.17) the reader may easily verify that the first K-2 equations of the system $\underline{d}\, L(1,K) = \underline{d}$ are the same as those of the system $\underline{d}\, L(1,K-1) = \underline{d}$. This fact along with the forward substitution technique to solve these systems, discussed in Remark (ii) following Theorem 4.22, immediately yields the lemma.

Theorem 4.41. For $K \geq 2$, we have

$$\underline{x}_0(K-1) = \bar{\theta}_K\, \hat{\underline{x}}_0(K), \tag{4.42}$$

where

$$\bar{\theta}_K = [\theta_K(\bar{\eta}_1+\bar{\eta}_2(K))]/[\bar{\eta}_1+\theta_K\bar{\eta}_2(K)],$$

$$\bar{\eta}_1 = 1 + (1-\rho_1)^{-1}\,\lambda_1\mu_1'(2),$$

$$\bar{\eta}_2(K) = p\,d_0(K)\,(1-\rho_1)^{-1}\,\lambda_1[\mu_1'(1)-\mu_1'(2)],$$

and

$$\theta_K = [1-d_{K-1}(K)]^{-1}.$$

Proof. By Theorem 4.22, we have $\underline{x}_0(\nu) = c_\nu\,\underline{d}(\nu)$, where $c_\nu^{-1} = \bar{\eta}_1 + \bar{\eta}_2(\nu)$. The theorem now follows by comparing the components of $\underline{x}_0(K)$ and $\underline{x}_0(K-1)$ using Lemma 4.40.

Remark. Direct computation of $\underline{x}_0(K-1)$ using Theorem 4.22 is laborious. Having computed $\underline{d}(K)$ and $\underline{x}_0(K)$, computation of $\underline{x}_0(K-1)$ becomes almost trivial in view of the above theorem.

Further exploitation of the structure of the present model permits us also to simplify the computation of the remaining components of $\underline{x}(K-1)$. The next result will show that the first K-2 components of each $\underline{x}_\nu(K-1)$ can be obtained by multiplying the corresponding components of $\underline{x}_\nu(K)$ by the constant $\bar{\theta}_K$. Having thus computed all but the last component of $\underline{x}_\nu(K-1)$, we may compute the last component of $\underline{x}_\nu(K-1)$ by computing $\{\underline{x}_\nu(K-1)\underline{e}\}$ using Theorem 4.38. The substantial reduction (by a factor of K) in the number of linear equations to be solved by using such a procedure reduces the computational burden enormously.

Before we state the next theorem we introduce some additional notation. Let $\underline{X}(z;\ell)$, $\bar{A}(z;\ell)$ and $\bar{D}(z;\ell)$ denote the quantities $\underline{X}(z) = \sum_0^\infty z^\nu\,\underline{x}_\nu$, $\bar{A}(z) = \sum_0^\infty z^\nu\,A_\nu$ and $\bar{D}(z) = \sum_0^\infty z^\nu\,D_\nu$ respectively associated with the parameter value $K = \ell$. Also let $\underline{\alpha}^{(i)}(z) = \sum_{\nu=0}^\infty z^\nu\underline{\alpha}_\nu^{(i)}$, $i = 1,2$ and $\underline{d}(z) = \sum_{\nu=0}^\infty z^\nu\underline{d}_\nu$. Finally, let $A^*(z)$ denote the submatrix formed by the first $K-2$ rows and the first $K-2$ columns of $\bar{A}(z;K)$, and let $D^*(z)$ denote the submatrix formed by the first $K-1$ rows and the first $K-2$ columns of $\bar{D}(z;K)$.

Theorem 4.43. Let $K \geqq 3$. For $\nu \geqq 1$,

$$\underline{\hat{x}}_\nu(K-1) = \bar{\theta}_K \, \underline{\hat{\hat{x}}}_\nu(K)$$

Proof. By (4.14),

$$\underline{X}(z;\ell)\ \{zI_\ell - \bar{A}(z;\ell)\} = \underline{x}_0(\ell)\ \{z\ \bar{D}(z;\ell) - \bar{A}(z;\ell)\},\quad \ell = K-1, K, \tag{4.44}$$

and for $|z| < 1$, in view of the non-singularity of $\{zI - \bar{A}(z;\ell)\}$, the above equation uniquely determines $\underline{X}(z;\ell)$. From (4.44) we have, in view of the structure of $\bar{A}(z;\ell) = \Delta(\ell,\ell;\underline{\alpha}^{(1)}(z))$ and of

$$\bar{D}(z;\ell) = \left[\begin{array}{c} \Delta(1,\ell;\underline{d}(z)) \\ \hline \Delta(\ell-1,\ell;\underline{\alpha}^{(2)}(z)) \end{array}\right],$$ the following equations:

$$\underline{\hat{\hat{X}}}(z;K)\ \{zI_{K-2} - A^*(z)\} = \underline{\hat{\hat{x}}}_0(K)\ \{zD^*(z)\} - \underline{\hat{\hat{x}}}_0(K)\ A^*(z) \tag{4.45}$$

$$\underline{\hat{X}}(z;K-1)\ \{zI_{K-2} - A^*(z)\} = \underline{x}_0(K-1)\ \{zD^*(z)\} - \underline{\hat{x}}_0(K-1)\ A^*(z). \tag{4.46}$$

Equation (4.45) is got by comparing the first $K-2$ columns in both sides of (4.44) for $\ell = K$, and equation (4.46) is got by comparing the first $K-1$ columns in both sides of (4.44) for $\ell = K-1$. Since by (4.42), $\underline{x}_0(K-1) = \bar{\theta}_K\ \underline{\hat{x}}_0(K)$, equations (4.45) and (4.46) yield

$$\underline{\hat{X}}(z;K-1) \equiv \bar{\theta}_K\ \underline{\hat{\hat{X}}}(z;K) \quad \text{for all} \quad |z| < 1,$$

and the theorem follows.

Corollary 4.47. For the marginal distribution of 2-customers, we have

$$\underline{\hat{X}}(1;K-1) = \bar{\theta}_K\ \underline{\hat{\hat{X}}}(1;K).$$

This concludes our discussion of the case $N = 0$. In the next section, we develop the necessary recursions on N for fixed K.

4B. The Case $N \geqq 1$

When $N \geqq 1$, we may partition the matrix $Q = \tilde{Q}(\infty)$ defined by Figure 2 into the form

$$Q = \begin{bmatrix} R_0 & R_1 & R_2 & R_3 & \cdots \\ C_0 & A_1 & A_2 & A_3 & \cdots \\ 0 & A_0 & A_1 & A_2 & \cdots \\ 0 & 0 & A_0 & A_1 & \cdots \\ 0 & 0 & 0 & A_0 & \cdots \\ \vdots & \vdots & \vdots & \vdots & \cdots \end{bmatrix},$$

where the $(N+1)K \times (N+1)K$ matrix R_0 is given by

$$R_0 = \begin{bmatrix} D_0 & D_1 & \cdots & D_N \\ B_0 & B_1 & \cdots & B_N \\ 0 & B_0 & \cdots & B_{N-1} \\ \vdots & & & \\ 0 & 0 & \cdots & B_1 \end{bmatrix},$$

the $(N+1)K \times K$ matrices R_ν, $\nu \geqq 1$, are given by

$$R_\nu = \begin{bmatrix} D_{N+\nu} \\ B_{N+\nu} \\ B_{N+\nu-1} \\ \vdots \\ B_{\nu+1} \end{bmatrix}, \quad \nu \geqq 1,$$

and the $K \times (N+1)K$ matrix

$$C_0 = [0,\ldots,0,A_0].$$

We now define the levels of states in this Markov chain by $\underline{0} = \{(\nu,j)$: $0 \leqq \nu \leqq N$, $0 \leqq j \leqq K-1\}$, and $\underline{i} = \{(N+i,j)$: $0 \leqq j \leqq K-1\}$, $i \geqq 1$. The invariant probability vector $\underline{x}$ of Q defined by the equations $\underline{x}\, Q = \underline{x}$, $\underline{x}\, \underline{e} = 1$ is partitioned into $\underline{x} = (\underline{x}_0,\underline{x}_1,\underline{x}_2,\ldots)$ where $\underline{x}_0 \in R^{(N+1)K}$ and $\underline{x}_i \in R^K$, $i \geqq 1$. Just as in Section 4A, we propose to compute $\underline{x}_0$ directly by a Markov renewal argument, and then to compute the remaining components of $\underline{x}$ by a block Gauss-Seidel iterative scheme. The major significant idea to be developed below is that in computing $\underline{x}_0$ one may perform an efficient recursion on N.

Analogously as in Section 4A if we let L(z) denote the transform matrix which governs the return times to level $\underline{0}$, it may easily be seen by a standard first passage argument that

$$L(z) = z R_0 + \sum_{\nu=1}^{\infty} z R_\nu G^{\nu-1}(z) H(z), \tag{4.48}$$

where the $K \times (N+1)K$ matrix

$$H(z) = [0,\ldots,0,G(z)], \tag{4.49}$$

governs the first entrance times from $\underline{1}$ to $\underline{0}$, and $G(z) = G(z,0)$ is as defined in Section 3. The structure of H(z) given in (4.49) is obvious from the structure of Q and can also be proved directly by considering the equation

$$H(z) = z C_0 + \sum_{\nu=1}^{\infty} z A_\nu G^{\nu-1}(z) H(z),$$

which is obtained by a direct argument conditioning on the first step of the Markov chain Q.

Using the same arguments as in the proof of Theorem 4.22, we see that $\underline{x}_0 = (\underline{d}\,\underline{d}^*)^{-1}\,\underline{d}$, where $\underline{d}$ is the invariant probability vector of L(1) and

$$\underline{d}^* = L'(1)\,\underline{e} = \underline{e} + (1-\rho_1)^{-1} \sum_{\nu=1}^{\infty} \nu R_\nu\,\underline{e}. \tag{4.50}$$

As the first step in establishing the desired recursion on N for computing $\underline{x}_0$, we will show how the vectors $\underline{d}^*$ corresponding to various values of N can be computed recursively on N. To this end, we shall introduce the notational convention that a letter augmented by a function argument N will stand for the quantity denoted by that letter and corresponding to the value N of the threshold. For example, under this convention we would denote the matrix R_0 displayed earlier as

$R_0(N)$, etc. Also we set $\underline{r}_\nu(N) = R_\nu(N)\underline{e}$, $\underline{b}_\nu = B_\nu \underline{e}$, $\underline{d}_\nu = D_\nu \underline{e}$, $\nu \geq 0$.

Theorem 4.51. The sequence $\{\underline{d}^*(N);\ N \geq 1\}$ can be computed from the recurrence formula

$$\underline{d}^*(N+1) = \begin{bmatrix} \underline{d}^*(N)+(1-\rho_1)^{-1}\underline{r}_0(N)-\underline{e} \\ \hdashline (1-\rho_1)^{-1}(\underline{b}_0+\underline{b}^{(1)}) \end{bmatrix} - \rho_1(1-\rho_1)^{-1}\,\underline{e}, \quad N \geq 0, \tag{4.52}$$

where $\underline{d}^*(0)$ is the vector $\underline{d}^*$ given by (4.21) and $\underline{b}^{(1)}$ is defined by (4.10). (Recall that the generic notation $\underline{e}$ stands for a column vector of 1's of appropriate dimension.)

Proof. From (4.50),

$$\underline{d}^*(N+1) = \underline{e} + (1-\rho_1)^{-1} \sum_{\nu=1}^{\infty} \nu\, \underline{r}_\nu(N+1).$$

It may easily be checked that

$$\sum_{\nu=1}^{\infty} \nu\, \underline{r}_\nu(N+1) = \begin{bmatrix} \sum_{\nu=1}^{\infty} \nu\, \underline{r}_{\nu+1}(N) \\ \hdashline \sum_{\nu=1}^{\infty} \nu\, \underline{b}_{\nu+1} \end{bmatrix} = \begin{bmatrix} \sum_{\nu=1}^{\infty} \nu\, \underline{r}_\nu(N)+\underline{r}_0(N)-\underline{e} \\ \hdashline \underline{b}^{(1)} + \underline{b}_0 - \underline{e} \end{bmatrix}.$$

Finally, the fact that $\sum_{\nu=1}^{\infty} \nu\, \underline{r}_\nu(N) = (1-\rho_1)\ \{\underline{d}^*(N)-\underline{e}\}$ and some straightforward algebra yields (4.52).

Remark. Note that the vectors $\underline{r}_0(N)$, $N \geq 0$ appearing in (4.52) can themselves be computed recursively using

$$\begin{aligned} \underline{r}_0(0) &= \underline{d}_0, \\ \underline{r}_0(N+1) &= \begin{bmatrix} \underline{r}_0(N) + \underline{r}_1(N) \\ \hdashline \underline{b}_0 + \underline{b}_1 \end{bmatrix}, \quad N \geq 0, \end{aligned} \tag{4.53}$$

where

$$\underline{r}_1(N) = \begin{bmatrix} \underline{d}_{N+1} \\ \underline{b}_{N+1} \\ \vdots \\ \underline{b}_2 \end{bmatrix}. \tag{4.54}$$

This scheme is obtained easily by computing $R_0(N+1)\underline{e}$ and $R_0(N)\underline{e}$ directly.

Having discussed the recursive schemes for computing $\underline{d}^*(N)$, $N \geqq 1$, we now take up the computation of the matrices $L(1,N)$, $N \geqq 1$ and their respective invariant probability vectors $\underline{d}(N)$, $N \geqq 1$; $L(1,N)$ is the matrix $L(1)$ defined by (4.48) and (4.49) and corresponding to the value N of the threshold. Using (4.48) and (4.49) it is easily seen that for $N \geqq 1$

$$L(1,N+1) = \left[\begin{array}{ccccc|c} & & R_0(N) & & & \displaystyle\sum_{\nu=0}^{\infty} \begin{pmatrix} D_{N+\nu+1} \\ B_{N+\nu+1} \\ B_{N+\nu} \\ \vdots \\ B_{\nu+1} \end{pmatrix} G^{\nu} \\ \hline 0 & 0 & \cdots & 0 & B_0 & \end{array}\right]. \quad (4.55)$$

At first glance, the computation of the last K columns in the right side of (4.55) looks forbidding. But recalling the structures of the matrices $\{D_\nu\}$, $\{B_\nu\}$ and G, it is easy to see that computation of sums such as $\sum_{\nu=0}^{\infty} B_{k+\nu} G^{\nu}$ and $\sum_{\nu=0}^{\infty} D_{k+\nu} G^{\nu}$ essentially involves only the computation of the first K components of the vectors $\sum_{\nu=0}^{\infty} \underline{\alpha}_{k+\nu}^{(1)} * \underline{g}^{(\nu)}$ and $\sum_{\nu=0}^{\infty} \underline{\alpha}_{k+\nu}^{(2)} * \underline{g}^{(\nu)}$ which can be carried out efficiently using Horner's method.

We observe that the matrices $L(1,N+1)$ and $L(1,N)$ are such that the first NK equations in the two systems of linear equations

$$\underline{d}(N+1)\ L(1,N+1) = \underline{d}(N+1) \text{ and } \underline{d}(N)\ L(1,N) = \underline{d}(N)$$

coincide. This suggests that having computed the invariant probability vector $\underline{d}(N)$ of $L(1,N)$, to compute the vector $\underline{d}(N+1)$, one may make successive substitutions in the system $\underline{d}(N+1) = \underline{d}(N+1)\ L(1,N+1)$ starting with the initial iterate $[\underline{d}(N),\underline{0}]$. Indeed, the scheme

$$\underline{v}_0 = [\underline{d}(N), \underline{0}]$$

$$\underline{v}_{j+1} = \underline{v}_j \; L(1,N+1), \quad j \geq 0, \tag{4.56}$$

is such that $\underline{v}_j \underline{e} \equiv 1$, and $\underline{v}_j \to \underline{d}(N+1)$ as $j \to \infty$. To see the latter, note that since L(1,N+1) is irreducible, áperiodic and stochastic, as $j \to \infty$,

$$\begin{aligned} \underline{v}_{j+1} - \underline{v}_j &= \underline{v}_j \; [L(1,N+1)-I] \\ &= \underline{v}_0 \; L^j(1,N+1) \; [L(1,N+1)-I] \\ &\to \underline{v}_0 \; \{\underline{e} \; \underline{d}(N+1)\} \; [L(1,N+1)-I] = \underline{0}. \end{aligned}$$

The multiplications in (4.56) can be carried out without creating and storing the large matrix L(1,N+1) but by directly using the vectors $\underline{\alpha}_n^{(1)}$, $\underline{\alpha}_n^{(2)}$ and $\underline{g}$.

Having thus computed $\underline{x}_0$, we propose that $\underline{x}_1$ be computed by solving the non-singular, upper triangular system

$$\underline{x}_1 \; A_0 = \hat{\underline{x}}_0 - \underline{x}_0 \begin{bmatrix} D_N \\ B_N \\ B_{N-1} \\ \vdots \\ B_1 \end{bmatrix},$$

where $\hat{\underline{x}}_0$ is the last K components of $\underline{x}_0$, which is obtained by comparing the last K components of both sides of the steady-state equation $\underline{x}_0 \; R_0 + \underline{x}_1 \; C_0 = \underline{x}_0$. Then the remaining components of $\underline{x}$ can be computed by a block Gauss-Seidel iteration as in Section 4A. For the truncation of the steady state equations it appears expedient to start the block Gauss-Seidel scheme with the truncation index obtained for the previous value of N and increasing it progressively if it is found necessary to do so. Our computational experience with the procedure outlined here has been very satisfactory.

Defining the generating function

$$\underline{X}(z) = \sum_{\nu=1}^{\infty} z^{N+\nu} \underline{x}_{\nu}, \tag{4.57}$$

it is easily obtained from the steady state equations

$$\underline{x}_{\nu} = \underline{x}_0 R_{\nu} + \sum_{j=1}^{\nu+1} \underline{x}_j A_{\nu+1-j}, \quad \nu \geqq 1,$$

that

$$\underline{X}(z)\{zI-\bar{A}(z)\} = z^{N+1}\{\underline{x}_0\bar{R}(z)-\underline{x}_1A_0\}, \tag{4.58}$$

where

$$\bar{R}(z) = \sum_{\nu=1}^{\infty} z^{\nu} R_{\nu} .$$

Putting $z = 1$ in (4.48) and carrying through the same algebra as in Theorem 4.30, we get

$$\underline{X}(1) = \sum_{\nu=1}^{\infty} \underline{x}_{\nu} = (1-\underline{x}_0\underline{e})\underline{e}_K + \{\underline{x}_0 \sum_{\nu=1}^{\infty} R_{\nu}-\underline{x}_1A_0\}\ (I-A+E)^{-1}, \tag{4.59}$$

where $\underline{e}_K = (0,\ldots,0,1) \in R^K$ and $E = \underline{e}\ \underline{e}_K$. Equation (4.59) serves as a powerful accuracy check on the numerical computation of $\underline{x}$. Partitioning the vector $\underline{x}_0$ as $\underline{x}_0 = (\underline{x}_{00},\ldots,\underline{x}_{0N})$, where $\underline{x}_{0j} \in R^K$, $0 \leqq j \leqq N$, it is clear that the vector $\left(\sum_{j=0}^{N} \underline{x}_{0j}\right) + \underline{X}(1)$ yields the steady state queue length density at points of departures for 2-customers. Letting ξ_1 and ξ_2 denote the first two steady state factorial moments of the queue length of 1-customers, we also have the following result.

Theorem 4.60. We have

$$\xi_1 = \{\underline{X}'(1)\underline{e}\} + \sum_{j=1}^{N} j\ (\underline{x}_{0j}\underline{e}), \tag{4.61}$$

and

$$\xi_2 = \{\underline{X}''(1)\underline{e}\} + \sum_{j=2}^{N} j(j-1)(\underline{x}_{0j}\underline{e}), \tag{4.62}$$

where

$$\underline{X}'(1)\underline{e} = [2(1-\rho_1)]^{-1}[\lambda_1^2\mu_2'(1)\ \{1-\underline{x}_0\underline{e}\} + 2(N+1)\underline{x}_0\bar{R}'(1)\underline{e} + \underline{x}_0\bar{R}''(1)\underline{e}],$$

and

$$\underline{X}''(1)\underline{e} = [3(1-\rho_1)]^{-1}[3\lambda_1^2\mu_2'(1)\underline{X}'(1)\underline{e} + \lambda_1^3\mu_3'(1)\{1-\underline{x}_0\underline{e}\}$$
$$+ 3(N+1)N\underline{x}_0\bar{R}'(1)\underline{e} + 3(N+1)\underline{x}_0\bar{R}''(1)\underline{e} + \underline{x}_0\bar{R}'''(1)\underline{e}].$$

Proof. Letting p_ν denote the steady state probability that a departure leaves behind ν 1-customers, we have

$$p_j = \underline{x}_{0j}\underline{e}, \quad j = 0,\dots,N$$

$$p_{N+j} = \underline{x}_j\underline{e}, \quad j = 1,2,\dots .$$

Thus $\xi_1 = \sum_{j=1}^{N} j\, p_j + \sum_{j=1}^{\infty} (N+j)\, p_{N+j}$, and $\xi_2 = \sum_{j=2}^{N} j(j-1)\, p_j + \sum_{j=1}^{\infty} (N+j)(N+j-1)\, p_{N+j}$. Equations (4.61) and (4.62) now follow from (4.57). The formulas for $\underline{X}'(1)\underline{e}$ and $\underline{X}''(1)\underline{e}$ are got by differentiating in (4.58) twice and thrice respectively, setting $z = 1$ and multiplying by $\underline{e}$. The simplifications using (4.9) are omitted.

Remark. Using the structure of R_ν, $\nu \geqq 1$, one can efficiently compute $\bar{R}'(1)\underline{e}$, $\bar{R}''(1)\underline{e}$ and $\bar{R}'''(1)\underline{e}$; for brevity we shall leave the details to the reader.

5. Queue Lengths at an Arbitrary Epoch

Having obtained the stationary queue length densities at points of departure, we now obtain formulas for the stationary densities at an arbitrary epoch. Let $x(i,j)$ denote the (i,j)th component of the steady state vector $\underline{x}$ discussed in Section 4. Letting X_t, Y_t and Z_t denote respectively the number of 1-customers, the number of 2-customers, and the type of customer (if any) in service at time t+, the quantities to be computed are

$$y(0,0) = \lim_{t\to\infty} \Pr\{X_t=0,\ Y_t=0 \mid X_0,Y_0\}, \tag{5.1}$$

$$y_1(i,j) = \lim_{t\to\infty} \Pr\{X_t=i,\ Y_t=j,\ Z_t=1 \mid X_0,Y_0\},\ i \geqq 1,\ 0 \leqq j \leqq K-1, \tag{5.2}$$

and

$$y_2(i,j) = \lim_{t\to\infty} \Pr\{X_t=i,\ Y_t=j,\ Z_t=2 \mid X_0,Y_0\},\ i \geqq 0,\ 1 \leqq j \leqq K. \qquad (5.3)$$

It is well-known that these limits do not depend on the initial conditions and also that they may be interpreted as the long-run proportion of time spent by the system in the various states. These quantities yield a number of operating characteristics, such as the proportion of time the server is idle, the proportion of time the waiting room for 2-customers is full, the proportion of time the server spends serving each type, etc., which are important from the point of view of design.

Let $m(i,j)$ denote the mean recurrence time of the state (i,j) in the Markov renewal process $\tilde{Q}(\cdot)$, and note that these are given by

$$m^{-1}(i,j) = (\underline{x}\ \underline{\delta})^{-1}\ x(i,j),\ i \geqq 0,\ 0 \leqq j \leqq K-1, \qquad (5.4)$$

where

$$\underline{\delta} = \int_0^\infty x\, d\tilde{Q}(x)\ \underline{e}. \qquad (5.5)$$

After partitioning $\underline{\delta}$ into subvectors $\underline{\delta}_\nu$, $\nu \geqq 0$ of dimension K, we can easily see by a direct computation that

$$\underline{\delta}_0 = \begin{bmatrix} p\mu_1' + (1-p)\mu_2' \\ \hline \mu_2'\ \underline{e} \end{bmatrix} \qquad (5.6)$$

$$\underline{\delta}_\nu = \begin{bmatrix} \mu_1' \\ \hline \mu_2'\ \underline{e} \end{bmatrix},\ 1 \leqq \nu \leqq N, \qquad (5.7)$$

and

$$\underline{\delta}_\nu = \mu_1'\ \underline{e},\ \nu \geqq N+1, \qquad (5.8)$$

where $p = \lambda_1/(\lambda_1+\lambda_2)$. Thus the computation of the quantities in (5.4) is trivial once the vector $\underline{x}$ has been computed. The steady state probabilities in (5.1)-(5.3) are to be expressed in terms of these mean recurrence times and the quantities

$$\gamma_{jk}^{(i)} = \int_0^\infty p_j^{(1)}(t)\, p_k^{(2)}(t)\, \frac{1-\tilde{H}_i(t)}{\mu_i'}\, dt, \quad j,k \geqq 0, \quad i = 1,2 \quad , \tag{5.9}$$

where $p_j^{(i)}(\cdot)$ are defined in (2.1). The computation of the quantities in (5.9) can be done efficiently when both service time cdfs are of phase type (see Section 6). We now state our main result as a theorem.

Theorem 5.10. We have

$$y(0,0) = m^{-1}(0,0)/(\lambda_1+\lambda_2) \tag{5.11}$$

$$y_1(i,j)/\mu_1' = m^{-1}(0,0) \cdot p\, \gamma_{i-1,j}^{(1)} + \sum_{\nu=1}^{N\wedge i} m^{-1}(\nu,0) \cdot \gamma_{i-\nu,j}^{(1)} + \sum_{\nu=N+1}^{i} \sum_{\ell=0}^{j} m^{-1}(\nu,\ell) \cdot \gamma_{i-\nu,j-\ell}^{(1)}, \quad i \geqq 1,\ 0 \leqq j \leqq K-2, \tag{5.12}$$

$$y_1(i,K-1)/\mu_1' = m^{-1}(0,0) \cdot p \sum_{j=K-1}^{\infty} \gamma_{i-1,j}^{(1)} + \sum_{\nu=1}^{N\wedge i} m^{-1}(\nu,0) \cdot \sum_{j=K-1}^{\infty} \gamma_{i-\nu,j}^{(1)} + \sum_{\nu=N+1}^{i} \sum_{\ell=0}^{K-1} m^{-1}(\nu,\ell) \cdot \sum_{j=K-1-\ell}^{\infty} \gamma_{i-\nu,j}^{(1)}, \quad i \geqq 1. \tag{5.13}$$

$$y_2(i,j)/\mu_2' = m^{-1}(0,0) \cdot (1-p)\, \gamma_{i,j-1}^{(2)} + \sum_{\nu=0}^{N\wedge i} \sum_{\ell=1}^{j} m^{-1}(\nu,\ell) \cdot \gamma_{i-\nu,j-\ell}^{(2)}, \quad i \geqq 0,\ 1 \leqq j \leqq K-1 \tag{5.14}$$

$$y_2(i,K)/\mu_2' = m^{-1}(0,0) \cdot (1-p) \sum_{j=K-1}^{\infty} \gamma_{i,j}^{(2)} + \sum_{\nu=0}^{N\wedge i} \sum_{j=1}^{K-1} m^{-1}(\nu,j) \cdot \sum_{\ell=K-j}^{\infty} \gamma_{i-\nu,\ell}^{(2)}, \quad i \geqq 0. \tag{5.15}$$

In the above formulae any sum in which the upper limit of summation is smaller than the lower limit is taken to be zero, and $N \wedge i = \min(N,i)$.

Proof. These formulae are derived routinely by considering the epoch of the last departure before time t and applying the Key Renewal Theorem ([1], Theorem 6.3); we shall omit the arduous details.

It is elementary to show that the quantities computed here are also the stationary probabilities at points of arrival.

6. The Case of Phase Type Services

The first step in the implementation of the algorithms presented in the earlier sections is the computation of the quantities $\alpha_{jk}^{(i)}$ and $\gamma_{jk}^{(i)}$, $j,k \geqq 0$, $i = 1,2$ which were defined by (2.2) and (5.9) respectively. This entails the evaluation of integrals of the form

$$\alpha_{jk} = \int_0^\infty e^{-\lambda_1 x} \frac{(\lambda_1 x)^j}{j!} e^{-\lambda_2 x} \frac{(\lambda_2 x)^k}{k!} d\tilde{H}(x), \quad j,k \geqq 0, \tag{6.1}$$

and

$$\gamma_{jk} = \int_0^\infty e^{-\lambda_1 x} \frac{(\lambda_1 x)^j}{j!} e^{-\lambda_2 x} \frac{(\lambda_2 x)^k}{k!} \frac{1-\tilde{H}(x)}{\mu} dx, \quad j,k \geqq 0, \tag{6.2}$$

where $\tilde{H}(\cdot)$ is a cumulative probability distribution function on $[0,\infty)$ and $\mu < \infty$ is its mean. While, in general, the computation of (6.1) and (6.2) is a formidable task, for the case when $\tilde{H}$ is a probability distribution of phase type (PH-distribution), we will show that these can be computed recursively without numerical integrations. Probability distributions of phase type were introduced by M. F. Neuts [10], [13] and include a number of familiar distributions such as the exponential, generalized Erlang and the hyperexponential as very special cases. Though (only) slightly less general than probability distributions with rational Laplace transforms [3], yet the class of PH-distributions is a versatile class with a number of closure properties, and due to the algorithmic tractability of its members, it occupies an important place in computational probability; hence our interest in this class.

A continuous PH-distribution is obtained as the distribution of the time till absorption in a finite state, continuous time Markov chain (MC) with exactly one absorbing state into which absorption is certain. To be specific, consider a MC with state space $\{1,\ldots,m+1\}$, initial probability vector $(\underline{\beta},0) \in R^{m+1}$, and infinitesimal generator

$$Q = \begin{bmatrix} S & \underline{S}^\circ \\ \underline{0} & 0 \end{bmatrix}$$

where S is an $m \times m$ non-singular matrix with $S_{ii} < 0$ and $S_{ij} \geqq 0$ for $i \neq j$ and $\underline{S}^\circ = -S\,\underline{e} \geqq \underline{0}$. The time till absorption in such a MC has cdf

$$\tilde{H}(x) = 1 - \underline{\beta}\,\exp\,(Sx)\,\underline{e}, \quad x \geqq 0, \tag{6.3}$$

which is called the PH-distribution with representation $(\underline{\beta}, S)$. For certain technical reasons and without loss of generality [10] we shall assume that the representation $(\underline{\beta}, S)$ is so chosen that the matrix $Q^* = S + \underline{S}^\circ\underline{\beta}$ is irreducible. For later use we shall let $\underline{\theta}$ denote the stationary probability vector corresponding to the infinitesimal generator Q^*, i.e., $\underline{\theta}$ is the unique solution of the equations $\underline{\theta}\,Q^* = \underline{0}$, $\underline{\theta}\,\underline{e} = 1$. We recall [13] that $\underline{\theta} = -\mu^{-1}\,\underline{\beta}\,S^{-1}$, where $\mu = -\underline{\beta}\,S^{-1}\underline{e}$ is the mean of $\tilde{H}$.

The analogous definition of a discrete PH-distribution is obtained by considering a discrete MC with one-step transition probability matrix

$$P = \begin{bmatrix} T & \underline{T}^\circ \\ \underline{0} & 1 \end{bmatrix}$$

and initial probability vector $(\underline{\alpha}, \alpha_{m+1}) \in R^{m+1}$. The cdf $\tilde{F}(\cdot)$ of the time till absorption in such a MC, which is the discrete PH-distribution with representation $(\underline{\alpha}, T)$ is given by the probability density function on the nonnegative integers

$$f(\nu) = \begin{cases} \alpha_{m+1}, & \nu = 0 \\ \underline{\alpha}\,T^{\nu-1}\,\underline{T}^\circ, & \nu \geqq 1. \end{cases} \tag{6.4}$$

For examples of PH-distributions and their properties, we refer the reader to [10] and [13]. For our purpose, we shall need the following two important results concerning PH-distributions.

Theorem 6.5. Let $\tilde{H}$ be a continuous PH-distribution with representation $(\underline{\beta},S)$ and let λ be a positive constant. Then the pdf defined by

$$f(\nu) = \int_0^\infty e^{-\lambda x} \frac{(\lambda x)^\nu}{\nu!} \, d\tilde{H}(x), \quad \nu \geqq 0$$

is a discrete PH-distribution with representation $(\underline{\alpha},T)$, where $T = \lambda(\lambda I-S)^{-1}$, and $\underline{\alpha} = \underline{\beta}\, T$.

Proof. This result is proved as Theorem 19 in M. F. Neuts [12].

Theorem 6.6. Let $\tilde{H}$ be a continuous PH-distribution with representation $(\underline{\beta},S)$. Then the cdf $\mu^{-1} \int_0^x \{1-\tilde{H}(u)\}\, du$, where μ is the mean of $\tilde{H}$, is a continuous PH-distribution with representation $(\underline{\theta},S)$, where $\underline{\theta} = -\mu^{-1}\, \underline{\beta}\, S^{-1}$.

Proof. This is Property 8 in M. F. Neuts [10].

Having recalled the essential results concerning PH-distributions we now take up the computation of the quantities defined in (6.1) and (6.2) where $\tilde{H}$ is a PH$(\underline{\beta},S)$. In the sequel, we shall use the suggestive notations $\alpha_{j\cdot}$, $\alpha_{\cdot k}$, $\gamma_{j\cdot}$ and $\gamma_{\cdot k}$ of ANOVA to denote the sums $\sum_k \alpha_{jk}$, $\sum_j \alpha_{jk}$, $\sum_k \gamma_{jk}$ and $\sum_j \gamma_{jk}$ respectively.

Theorem 6.7. Let $\tilde{H}$ be PH$(\underline{\beta},S)$. Then the probability distributions $\{\alpha_{j\cdot}: j \geqq 0\}$, $\{\alpha_{\cdot k}: k \geqq 0\}$, $\{\gamma_{j\cdot}: j \geqq 0\}$, and $\{\gamma_{\cdot k}: k \geqq 0\}$ are all discrete PH-distributions. Denoting their representations by $(\underline{\alpha}_i, T_i)$ $i = 1,2,3,4$ respectively, we have

$$T_1 = T_3 = \lambda_1(\lambda_1 I-S)^{-1}, \quad T_2 = T_4 = \lambda_2(\lambda_2 I-S)^{-1} \tag{6.8}$$

$$\underline{\alpha}_1 = \underline{\beta}\, T_1, \quad \underline{\alpha}_2 = \underline{\beta}\, T_2, \quad \underline{\alpha}_3 = \underline{\theta}\, T_3, \quad \underline{\alpha}_4 = \underline{\theta}\, T_4. \tag{6.9}$$

Proof. These results are easily obtained using Theorem 6.5 and Theorem 6.6.

Thus to compute the two dimensional arrays (α_{jk}) and (γ_{jk}), we first compute the row and column sums of these arrays. The terms of each of these marginal distributions are computed recursively until they add up to $1 - \varepsilon$ where ε is a pre-assigned small positive number. Such a procedure yields a natural way of truncating the infinite arrays. In computing these marginal distributions, note that in view of (6.8), it is efficient to compute $\{\alpha_{j\cdot}\}$ and $\{\gamma_{j\cdot}\}$ side by side, and $\{\alpha_{\cdot k}\}$ and $\{\gamma_{\cdot k}\}$ side by side. For instance, using Theorem 6.7 and equation (6.4) we have

$$\alpha_{0\cdot} = 1 - \underline{\alpha}_1\underline{e}, \qquad \gamma_{0\cdot} = 1 - \underline{\alpha}_3\underline{e}$$

and for $\nu \geqq 1$,

$$\alpha_{\nu\cdot} = \underline{\alpha}_1\underline{g}_{\nu-1}, \qquad \gamma_{\nu\cdot} = \underline{\alpha}_3\underline{g}_{\nu-1},$$

where

$$\underline{g}_0 = \underline{T}_1^\circ = \underline{e} - T_1\underline{e}$$

and

$$\underline{g}_\nu = T_1\underline{g}_{\nu-1}, \quad \nu \geqq 1,$$

and the side by side computation of $\{\alpha_{j\cdot}\}$ and $\{\gamma_{j\cdot}\}$ results in considerable savings, for it obviates the need to compute the sequence $\{\underline{g}_\nu\}$ twice. Also note that in implementing the above scheme, only the latest vector $\underline{g}_\nu$ has to be stored.

We now present the following theorem which yields the necessary recursive schemes to compute the entries $\{\alpha_{jk}\}$ and $\{\gamma_{jk}\}$.

Theorem 6.10. Let $\tilde{H}$ be a continuous PH-distribution with representation $(\underline{\beta}, S)$. Then we have the following recurrence relations for the quantities defined by (6.1) and (6.2):

$$\begin{aligned} \alpha_{00} &= \underline{\beta}\,\underline{\eta}_0\,, & \gamma_{00} &= \underline{\theta}\,\underline{\eta}_0\,, \\ \alpha_{j,k+1} &= \frac{\lambda_2}{k+1}\,\xi_{j+k+1}\,\alpha_{jk}\,, & \gamma_{j,k+1} &= \frac{\lambda_2}{k+1}\,\bar{\xi}_{j+k+1}\,\gamma_{jk}\,, \\ \alpha_{j+1,k} &= \frac{\lambda_1}{j+1}\,\xi_{j+k+1}\,\alpha_{jk}\,, & \gamma_{j+1,k} &= \frac{\lambda_1}{j+1}\,\bar{\xi}_{j+k+1}\,\gamma_{jk}\,, \end{aligned} \tag{6.11}$$

where

$$\underline{\eta}_0 = [(\lambda_1+\lambda_2)I-S]^{-1}\ \underline{S}^\circ$$
$$\underline{\eta}_\nu = \nu[(\lambda_1+\lambda_2)I-S]^{-1}\ \underline{\eta}_{\nu-1}, \quad \nu \geqq 1, \tag{6.12}$$

and

$$\xi_\nu = (\underline{\beta}\ \underline{\eta}_\nu)/(\underline{\beta}\ \underline{\eta}_{\nu-1}), \quad \nu \geqq 1$$
$$\bar{\xi}_\nu = (\underline{\theta}\ \underline{\eta}_\nu)/(\underline{\theta}\ \underline{\eta}_{\nu-1}), \quad \nu \geqq 1\ . \tag{6.13}$$

Proof. Letting

$$\underline{\eta}_\nu = \int_0^\infty e^{-[(\lambda_1+\lambda_2)I-S]x}\ x^\nu\ dx\ \underline{S}^\circ, \quad \nu \geqq 0,$$

it is easily verified that (6.12) holds. Using (6.3) in (6.1) we get

$$\alpha_{jk} = \frac{\lambda_1^j\lambda_2^k}{j!k!}\ \underline{\beta}\ \underline{\eta}_{j+k}\ , \quad j,k \geqq 0. \tag{6.14}$$

Using Theorem 6.6 and (6.2) we also have

$$\gamma_{jk} = \frac{\lambda_1^j\lambda_2^k}{j!k!}\ \underline{\theta}\ \underline{\eta}_{j+k}\ , \quad j,k \geqq 0. \tag{6.15}$$

The recurrence formulas (6.11), (6.13) are now immediate from (6.14) and (6.15).

A number of detailed comments pertaining to the implementation of Theorem 6.10 are entirely in order. Having computed the row and column sums of the array (α_{jk}), we compute α_{00} directly. Then the remaining elements in 'row 0' and 'column 0' are computed using the recursive schemes till they add up close enough to the appropriate marginal sums $\alpha_{0\cdot}$ and $\alpha_{\cdot 0}$ computed earlier. Then one computes α_{11} using α_{01} or α_{10}. In general, having computed α_{ii} (using $\alpha_{i,i-1}$ or $\alpha_{i-1,i}$) the remaining entries in the i-th row and i-th column are computed until they add close enough to the directly computed i-th row and i-th column sums respectively. In doing so note from (6.12) and (6.13) that one has to store only the latest two $\underline{\eta}$-vectors. Also as one computes the entries

α_{jk}, one computes both the constant ξ_ν and $\bar{\xi}_\nu$ as they become available and stores them. The significant savings in storage and arithmetic operations that result from such an arrangement of the computational scheme should be clear from (6.13). Having completed the computation of (α_{jk}), we compute γ_{00} directly and then implement the recursive scheme for (γ_{jk}) in a similar manner. The stored values of $\bar{\xi}_\nu$ are now used whenever needed, and when a value of $\bar{\xi}_\nu$ not already stored is required (this could happen, for the truncation indices of the two arrays are in general different), these are computed using the last two stored $\underline{\eta}$-vectors, and (6.12) and (6.13).

If α_{00} or γ_{00} is very small one could run into difficulties in directly implementing the scheme above, but in such a case one can rescale the quantities of interest and implement the scheme successfully. It can be shown that the entries of $\underline{\eta}_\nu$ grow as the ν-th moment of $\tilde{H}$. Being so, if a large number of vectors $\underline{\eta}_\nu$ are needed, then one could encounter problems of overflow. (This did occur in several examples during our computations.) Luckily, however, this problem can be overcome by rescaling both the stored $\underline{\eta}$-vectors whenever either gets "too large"; this procedure leaves ξ_ν and $\bar{\xi}_\nu$ unaltered (see (6.13)) and the rescaled $\underline{\eta}$-vectors also satisfy the original recursive scheme (see (6.12)).

The schemes described above have been implemented by us for various PH-densities with as many as ten phases, and despite the disadvantages of using an APL code, our experience has been extremely satisfactory. For the case when $\tilde{H}$ is the exponential distribution,

$$\alpha_{jk} \equiv \gamma_{jk} = \binom{j+k}{j} \left[\frac{\lambda_1}{\lambda_1+\lambda_2+\mu}\right]^j \left[\frac{\lambda_2}{\lambda_1+\lambda_2+\mu}\right]^k \left[\frac{\mu}{\lambda_1+\lambda_2+\mu}\right]$$

as is easily verified. By representing the exponential distribution in different ways as a PH-distribution, one can test the correctness of the program; this and several other tests of checking the computations are now routinely used by those working with PH-distributions and for want of space cannot be described here even briefly. (See e.g., the discussion in [15].)

7. Numerical Examples

The algorithms presented in this paper have been programmed in APL, and we shall, in this section, present some numerical examples and illustrate how one may use interactive computing to obtain useful information on the sensitivity of the model to input characteristics and the model parameters. For brevity, only output pertaining to a few selected characteristics of the steady state distribution $\underline{x}$ of the queue lengths at departure epochs will be discussed.

For the specific examples discussed here we assumed that 1-customers have an arrival rate $\lambda_1 = 0.8$ and have i.i.d. service times which are distributed as

$$PH(\underline{\beta}_1, S_1), \text{ where } \underline{\beta}_1 = (1,0,0) \text{ and } S_1 = \begin{bmatrix} -2.5 & 2.5 & 0 \\ 0 & -5 & 5 \\ 0 & 0 & -2.5 \end{bmatrix}. \text{ This}$$

PH-distribution is the generalized Erlang obtained as the convolution of three exponential distributions with means 0.4, 0.2 and 0.4. It is trivial to verify that its mean is 1. The arrival rate of 2-customers, λ_2, was taken to be 0.2, and for their service time distribution $\tilde{H}_2$, a number of distributions each with mean 4 were considered, the intent being to examine the effect of the variability in $\tilde{H}_2$ on the system. Thus, in our models, "on the average" 2-customers require four times as much service as 1-customers, but arrive four times less frequently. The specific distributions considered for $\tilde{H}_2$ were:

1) H_3: The hyperexponential given by $PH(\underline{\beta}_2, S_2)$, where $\underline{\beta}_2 = (.25,.25,.5)$, and $S_2 = \text{diag}(-.125,-.25,-.5)$.

2) E(2,.5): The Erlang with two phases each with mean duration 2.

3) E(8,2): The Erlang with eight phases each with mean duration 0.5.

4) D: Deterministic with value 4.

The standard deviations of the above four distributions are 5.292, 2.828, 1.414 and 0 respectively.

The following quantities were computed for N = 0, and K = 1,2,4 and 6:

ν_i = mean queue length of i-customers, i = 1,2

σ_i = s.d. of the queue length of i-customers, i = 1,2

e_{0i} = probability of emptiness for i-customers, i = 1,2

π_2 = proportion of services rendered to 2-customers

f = probability of a full waiting toom.

These are given in Table 1.

A number of interesting facts emerge from Table 1. Consider, for example, the system with N = 0 and K = 1. In such a system a 2-customer has no waiting space and can enter service if and only if, upon arrival, he finds the system empty. Suppose we are interested in comparing this system to the one where no 2-customers are ever permitted to enter. For the latter M/PH/1 queue, it is elementary to verify that $\nu_1 = 2.976$, $\sigma_1 = 3.154$ and $e_{01} = 1 - \rho_1 = 0.2$. An elementary argument also yields that the proportion π_2 of services rendered to 2-customers in the system with N = 0, K = 1 is at most $0.2\,\frac{0.2}{0.2+0.8} = 0.04$; the actual value of π_2, as seen from Table 1, is 0.027. One could argue (naively) that since only about 3% of the services are exceptional, such a system cannot be very different from the M/PH/1 queue where no 2-customers are ever permitted. However, the effect of permitting 2-customers to enter the system (even in such a highly restricted manner) can have a considerable effect depending on the variability in $\tilde{H}_2$, and this is seen from the

Table 1. Selected Queue Characteristics For N = 0

$\tilde{H}_2$	Quantity	K = 1	K = 2	K = 4	K = 6
H_3	ν_1	4.885	6.480	6.989	7.095
	σ_1	5.384	5.780	6.344	6.451
	ν_2	-	0.869	2.729	4.673
	σ_2	-	0.337	0.680	0.852
	π_2	0.027	0.050	0.057	0.058
	e_{01}	0.135	0.081	0.064	0.060
	e_{02}	-	0.131	0.028	0.007
	f	-	0.869	0.834	0.826
E(2,.5)	ν_1	4.020	5.001	5.223	5.245
	σ_1	3.779	3.936	4.060	4.072
	ν_2	-	0.873	2.783	4.765
	σ_2	-	0.333	0.565	0.630
	π_2	0.027	0.052	0.058	0.059
	e_{01}	0.135	0.074	0.060	0.059
	e_{02}	-	0.127	0.013	0.0014
	f	-	0.873	0.844	0.841
E(8,2)	ν_1	3.760	4.553	4.677	4.683
	σ_1	3.470	3.552	3.593	3.595
	ν_2	-	0.873	2.812	4.806
	σ_2	-	0.332	0.492	0.519
	π_2	0.027	0.054	0.059	0.059
	e_{01}	0.135	0.070	0.059	0.059
	e_{02}	-	0.126	0.006	0.0003
	f	-	0.874	0.851	0.850
D	ν_1	3.674	4.399	4.492	4.495
	σ_1	3.384	3.436	3.458	3.459
	ν_2	-	0.874	2.825	4.822
	σ_2	-	0.332	0.461	0.475
	π_2	0.027	0.055	0.059	0.059
	e_{01}	0.135	0.068	0.059	0.059
	e_{02}	-	0.126	0.004	0.0001
	f	-	0.874	0.855	0.854

following table (obtained from Table 1) which yields the percentage increases that result in ν_1 and σ_1:

Table 2. Comparison of the System N = 0, K = 1 With The M/PH/1 Queue

Service time $\tilde{H}_2$	H_3	E(2,.5)	E(8,2)	D
% increase in ν_1	64.15	38.08	26.34	23.45
% increase in σ_1	70.70	19.82	10.02	7.29

It is interesting to note that the larger the variability in $\tilde{H}_2$, the larger is the effect of 2-customers. Comparison of the increases which occur in σ_1 especially show that in the presence of large variability in $\tilde{H}_2$, the 2-customers induce wide oscillations in the paths of the queue length process of 1-customers; indeed, the build ups which occur during the service of a 2-customer take a long time to dissipate.

The following table showing the percentage increases in ν_1 and σ_1 for K = 6 and K = 1 is also interesting.

Table 3. Comparison of The Model With K = 6 With The One With K = 1

Service time cdf	H_3	E(2,.5)	E(8,2)	D
% increase in ν_1	45.24	30.47	24.47	22.34
% increase in σ_1	19.81	7.75	3.61	2.22

From the above table it is clear that controlling the waiting room size is significantly helpful only in the case when $\tilde{H}_2$ exhibits large variability. While the increase in the waiting room size from 0 to 5 results only in a negligible 2.22% increase in σ_1 in the deterministic case, it results in nearly a 20% increase in the hyperexponential case. This again is indicative of the oscillatory effects induced by 2-customers in the case of higher variability in $\tilde{H}_2$. These considerations strengthen our belief that in the analysis of complex queueing models, restriction of attention to easily obtainable averages alone would be very inadequate and hazardous.

Table 1 also indicates that while an increase from 0 to 5 of the waiting room size roughly doubles the proportion of services rendered to 2-customers, such increase is effected at a significant inconvenience to 1-customers in the case of H_3. Such inconvenience turns out to be less for distributions $\tilde{H}_2$ with less variability.

A comparison of the case K = 6 with K = 4 indicates that further increase of K will not have a noticeable impact on 1-customers and will not result in any noticeable increase in the proportion of services rendered to 2-customers either. Thus at this point, any increase in π_2 can be effected only by increasing the value of the threshold N. This is also seen from the fact that e_{02}, the probability of emptiness for 2-customers, in nearly zero for K = 6.

The following table yields certain queue characteristics computed for the cases N = 1,2,6 and K = 2,6 for the cases where $\tilde{H}_2$ is the hyperexponential and where $\tilde{H}_2$ is deterministic. In table 4 the values given in parentheses pertain to the deterministic case.

Table 4 indicates that the value of N exerts little influence on π_2, the proportion of services to 2-customers. It also has little effect on ν_2 and σ_2. However, increase in N does have a significant adverse effect on the queue length of 1-customers.

A careful consideration of the dynamics of the queue explains these facts. Every time the queue length of 1-customers drops below N + 1, the priority switches to 2-customers; however, the 2-customer "on the average" stays in service so long that adequate 1-customers arrive during his service thereby letting the priority switch back to 1-customers. Thus although the choice of N may have some effect on the transient behavior of the system which starts with very few 1-customers, its long run effect on the proportion of 2-customers served is negligible. Note that $1 - \alpha_{0\cdot}^{(2)}$ is the probability of at least one 1-customer arriving during a service of a 2-customer. For models considered in these examples, $1 - \alpha_{0\cdot}^{(2)}$ is close to 1 (e.g., in the deterministic case, $1 - \alpha_{0\cdot}^{(2)} = 1 - e^{-\lambda_1 4} = 0.96$). Hence this phenomenon.

Table 4

	K = 2			K = 6		
Quantity	N = 1	N = 2	N = 6	N = 1	N = 2	N = 6
ν_1	7.350 (3.879)	8.220 (4.688)	11.805 (8.648)	8.087 (3.917)	9.080 (4.048)	13.057 (8.989)
σ_1	6.682 (5.273)	6.570 (5.086)	6.926 (5.460)	6.511 (5.257)	6.330 (5.181)	6.099 (5.139)
ν_2	.852 (.866)	.840 (.861)	.815 (.852)	4.668 (4.824)	4.666 (4.823)	4.658 (4.822)
σ_2	.355 (.341)	.367 (.346)	.389 (.355)	.865 (.469)	.875 (.471)	.893 (.476)
e_{01}	.034 (.012)	.026 (.008)	.009 (.002)	.002 (1.47×10^{-5})	.0014 ($8.95\ 10^{-6}$)	.0005 (1.84×10^{-6})
π_2	.052 (.056)	.054 (.057)	.057 (.058)	.058 (.059)	.059 (.059)	.059 (.059)

8. Concluding Remarks

As pointed out earlier, the queue characteristics in models of the kind discussed here usually exhibit a high amount of variability, and if Monte Carlo techniques were used, extensive simulation may be required to obtain reliable estimates of them. If one is interested in examining the system behavior over different choices of N and K, then one has a number of simulation runs to carry out. Thus the value and feasibility of simulation become doubtful for the analysis of the model discussed here. Our analysis also exhibits the highly involved dependence of even the simple measures of system performance on the parameters. Thus formal analytic methods of the classical type also do not appear to be of much help in obtaining useful insight into the behavior of the model. Considering these facts, the algorithmic approach becomes a matter of necessity and not merely one of choice.

In some follow-up papers we shall discuss the waiting times in the present model and also the generalization of our algorithms to certain cases of non-Poisson arrivals and to the case of an infinite waitimg room for 2-customers.

Among models with non-Poisson arrivals, of particular interest is the case where arrivals occur according to the Neuts' versatile Markovian point process, which permits through special cases the modelling of situations where the two streams are dependent such as e.g., situations where the arrival of 2-customers inhibits arrivals of 1-customers.

9. References

[1] Çinlar, E. (1969), "Markov Renewal Theory", Adv. in Appl. Probab., 1, 123-187.

[2] Conway, R. W., Maxwell, W. L. & Miller, L. W. (1967), Theory of Scheduling, Addison-Wesley, Reading, MA.

[3] Cox, D. R. (1955), "A Use of Complex Probabilities in the Theory of Stochastic Processes", Proc. Cambridge Phil. Soc., 51, 313-319.

[4] Heiman, D. & Neuts, M. F. (1973), "The Single Server Queue in Discrete Time - Numerical Analysis IV", Naval Res. Logist. Quart., 20, 753-766.

[5] Hunter, J. J. (1969), "On the Moments of Markov Renewal Processes", Adv. in Appl. Probab., 1, 188-210.

[6] Hunter, J. J. (1980), "Generalized Inverses and their Applications to Applied Probability Problems", Tech. Rep. No. VTR-8006, Virginia Polytechnic Institute and State University, Industrial Engineering and Operations Research, Blacksburg, VA 24061.

[7] Kesten, H. & Runnenburg, J. T. (1957), "Priority in Waiting Line Problems I and II", Koninkl. Ned. Akad. Wetenschap. Proc., Ser. A., 60, 312-336.

[8] Lucantoni, D. M. & Neuts, M. F. (1978), "Numerical Methods for a Class of Markov Chains Arising in Queueing Theory", Tech. Rep. No. 78/10, Applied Mathematics Institute, University of Delaware, Newark, Delaware.

[9] Matthews, D. E. (1979), "A Simple Method for Reducing Queueing Times in M/G/1", Oper. Res., 27, 318-323.

[10] Neuts, M. F. (1975), "Probability Distributions of Phase Type", Liber Amicorum Prof. Emeritus H. Florin, Department of Mathematics, University of Louvain, Belgium, 173-206.

[11] Neuts, M. F. (1976), "Moment Formulas for the Markov Renewal Branching Process", Adv. in Appl. Probab., 8, 690-711.

[12] Neuts, M. F. (1977), "The M/G/1 Queue with Several Types of Customers and Change-Over Times", Adv. in Appl. Probab., 9, 604-644.

[13] Neuts, M. F. (1978), "Renewal Processes of Phase Type", Naval Res. Logist. Quart., 25, 445-454.

[14] Neuts, M. F. (1979), "Queues Solvable without Rouché's Theorem", Oper. Res., 27, 767-781.

[15] Neuts, M. F. & Meier, K. S. (1981), "On the Use of Phase Type Distributions in Reliability Modelling of Systems with a Small Number of Components", O.R. Spectrum, 2, 227-234.

[16] Ramaswami, V. (1980), "The N/G/1 Queue and its Detailed Analysis", Adv. in Appl. Probab., 12, 222-261.

[17] Ramaswami, V. (1980), "The Busy Period of Queues which have a Matrix-Geometric Steady-State Probability Vector", Tech. Rep. No. 3-80, Dept. of Mathematics, Drexel University, Philadelphia, PA.

[18] Schrage, L. E. & Miller, L. W. (1966), "The Queue M/G/1 with the Shortest Remaining Processing Discipline", Oper. Res., 14, 670-684.

[1]Drexel University. [2]Bell Laboratories

[1]Research carried out during Summer 1980 at the Applied Mathematics Institute, University of Delaware under Contract No. ENG-7908351 of the National Science Foundation.

[2]Research supported by Contract No. AFOSR 77-3236B of the U.S. Air Force Office of Scientific Research.

Discussant's Report on
"Algorithmic Analysis of a Dynamic Priority Queue,"
by V. Ramaswami and D. M. Lucantoni

This paper presents the analysis of a two-queue non-preemptive priority queueing system, where the relative priority structure of the two queues is state-dependent. Specifically, type 1 customers are given priority service if the number of such customers present in the system exceeds a threshold level N; type 2 customers are given priority otherwise. In addition, it is assumed that the queue capacity for type 2 customers is finite. The analysis of this queueing system follows along the lines of the methodologies developed by M. F. Neuts and his colleagues in recent years, and is a prime example of the power of that general approach in tackling a large class of complex queueing problems. A particularly refreshing aspect of that approach is that the fairly involved mathematics justifies itself by resulting in efficient computational algorithms for the numerical solution of the problem.

While the technical aspects of this paper are extremely interesting, I will limit my discussion to some comments on the appropriateness of this queueing structure for real applications, and to a general comment on the "Neuts" methodologies.

In the Introduction it is implied that the type 1 customers are intrinsically the higher priority customers; this motivates allowing service of a type 2 customer to begin only when the number of type 1 customers is below threshold. Such a service discipline makes intuitive sense when the type 1 arrival and service rates are far in excess of the corresponding rates for the type 2 customers. Indeed, this service discipline is an obvious generalization of the ordinary non-preemptive static priority scheme, and one can see intuitively that raising the threshold level N will tend to increase the type 1 delays and reduce the type 2 delays; this might allow "tuning" the system to achieve an objective delay performance. However, this discipline would make little sense when the type 1 arrival rate is much less than that of the type 2 customers, for then it can be seen that the type 1 delay could be enormous. It is this sensitivity of the type 1 delay to the model parameters that causes one to question the suitability of this scheme, without modification, for a real application.

It seems that it might be easier to motivate a different version of this model: one where the type 2 customers have infinite queueing capacity, and the type 1 queueing capacity is finite, and close to N. If the type 1 customers can tolerate long delays, but the system cannot tolerate excessive losses of type 1 customers, it would make sense to use a dynamic priority service discipline as described above. Note, however, that now the type 1 customers are really the low priority customers! The basic analytic techniques used by the authors in this paper could be used to analyze the performance of this modified scheme.

Finally, while the matrix-based analyses and methodologies developed by Neuts, Ramaswami, Lucantoni, and others have definitely proven their worth in numerous examples including this one, their very success might be cause for a future analyst to forsake alternative solution techniques, e.g., those techniques based on Rouche's Theorem. For certain classes of problems (not necessarily for this problem), the latter technique might well yield a simpler analysis and just as efficient computational algorithms as the former.

Discussant: Dr. A. E. Eckberg, Bell Laboratories, Holmdel, New Jersey 07733.

STEADY-STATE ALGORITHMIC ANALYSIS OF M/M/c TWO-PRIORITY QUEUES WITH HETEROGENEOUS RATES

Douglas R. Miller

Abstract

An algorithm for steady-state analysis of M/M/c nonpreemptive two-priority queues with heterogeneous rates is presented. It is based on a computational analysis *à la* Neuts which exploits a partition of the full state space into blocks. Both M/G/1 and GI/M/1 paradigm block structures arise and are exploited in the analysis. The mean number of waiting customers and the mean delay for each priority class are calculated. This gives a partial solution to the "Probabilistic puzzler" posed by D. P. Heyman in the fall 1977 issue of *Applied Probability Newsletter*, and extends a result of A. Cobham [*Operations Research 2 (1954), 70-76*] to two-priority queues with unequal service rates. In addition, the probabilities of individual states are computed. The algorithm has been programmed and some examples computed for nonpreemptive systems with five servers.

0. Introduction

This paper presents an algorithm for computing steady-state probabilities and mean queue lengths and delay times for M/M/c queues with two priority classes. Cobham [2] computed mean delay times for multipriority M/M/c queues with homogeneous service rates. Heyman [3] pointed out the need to compute mean delay times where service rates for different priority classes are unequal. The algorithm presented below computes expectations for a system with two priority classes having heterogeneous rates; in addition, it can compute steady state probabilities for individual states. However it is doubtful whether this algorithm could be modified to a practical algorithm for systems involving more than two priority classes. The preemptive case is simpler than the nonpreemptive case; an algorithm for it can be developed using ideas presented in this paper.

The general approach to this analysis is a computational method developed by Neuts [8]. The simpler problem of single-server two priority Markovian queues was solved by Miller [5]. The approach consists of partitioning the full state space into blocks and discovering special structure of the invariant probability vector in relation to these blocks. It is assumed that the reader is familiar with [5,6,7,8].

In addition to solving the problem posed by Heyman, this algorithmic approach is of interest because it explores a new and complicated application of several ideas from the field of computational probability: quasi birth and death processes, M/G/1 paradigms, and matrix-geometric invariant vectors.

Using single precision arithmetic the algorithm appears to work well for low and moderate utilization factors, but there is some deterioration in the calculations for high utilization. This application demonstrates the need for developing good error analyses for this type of computation. This application may be a good vehicle for developing some such analyses.

1. State Space and Transition Matrices

Consider an M/M/c two priority nonpreemptive queueing system with arrival rates λ_1 and λ_2 and service rates μ_1 and μ_2. The state space can be described as follows. Let $x_{i,j,k}$ be the state with i first priority customers waiting, j second priority customers waiting, k first priority customers in service and c-k second priority customers in service, i, j $\geq$ 0, 0 $\leq$ k $\leq$ c. From them, define the blocks

$$H_{i,j} = \{x_{i,j,k} | 0 \leq k \leq c\}, \quad i,j \geq 0.$$

Let $x_{m,n}$ be the state with m customers in service of which n are first priority customers and no customers are waiting in queue, $0 \leq n \leq m \leq c$. Then define the blocks

$$H_m = \{x_{m,n} | 0 \leq n \leq m\}, \quad 0 \leq m \leq c.$$

(Note that $H_c = H_{0,0}$.) Thus, the state space is

$$S = \left(\bigcup_{m=o}^{c-1} H_m \right) \cup \left(\bigcup_{i \geq o} \bigcup_{j \geq o} H_{i,j} \right).$$

Now consider the probabilistic transition rates for the non-preemptive M/M/c two priority system on this state space. The matrices of transition rates between the blocks defined above can be denoted as

K_i: $H_i \longrightarrow H_{i+1}$, $\quad 0 \leq i \leq c-1$

J_i: $H_i \longrightarrow H_{i-1}$, $\quad 1 \leq i \leq c$

L_1: $H_{i,j} \longrightarrow H_{i+1,j}$, $\quad i,j \geq 0$

L_2: $H_{i,j} \longrightarrow H_{i,j+1}$, $\quad i,j \geq 0$

M_1: $H_{i,j} \longrightarrow H_{i-1,j}$, $\quad i \geq 1,\ j \geq 0$

M_2: $H_{0,j} \longrightarrow H_{0,j-1}$, $\quad j \geq 1.$

All transitions correspond to arrivals and departures thus there are no transitions within individual blocks. Therefore the submatrices of the transition rate matrix corresponding to transitions within a block are all diagonal:

$-D_i$: $H_i \longrightarrow H_i$, $\quad 0 \leq i \leq c$

$-D_c$: $H_{i,j} \longrightarrow H_{i,j}$, $\quad i,j \geq 0$

The transition structure is depicted in Figure 1.

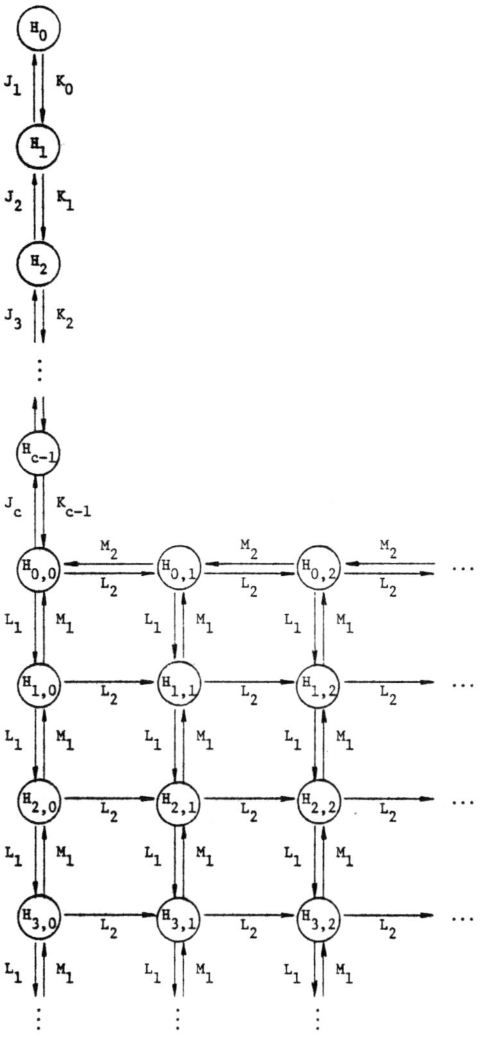

Figure 1. Partitioned state space (blocks) and transition scheme.

The matrix K_i is $(i+1) \times (i+2)$ dimensional; $(K_i)_{j,k}$ equals λ_1 for $k = j+1$, λ_2 for $k = j$, and 0 otherwise $(1 \leq j \leq i+1,\ 1 \leq k \leq i+2)$. The matrix J_i is $(i+1) \times i$ dimensional; $(J_i)_{j,k}$ equals $(i-j+1)\mu_2$ for $k = j$, $(j-1)\mu_1$ for $k = j-1$, and 0 otherwise $(1 \leq j \leq i+1,\ 1 \leq k \leq i)$. The matrices L_1, L_2, M_1, and M_2 are $(c+1) \times (c+1)$ dimensional; L_1 equals $\lambda_1 I$; L_2 equals $\lambda_2 I$; $(M_1)_{j,k}$ equals $(c-j+1)\mu_2$ for $k = j+1$, $(j-1)\mu_1$ for $k = j$, and 0 otherwise $(1 \leq j \leq c+1,\ 1 \leq k \leq c+1)$; $(M_2)_{j,k}$ equals $(c-j+1)\mu_2$ for $k = j$, $(j-1)\mu_1$ for $k = j-1$, and 0 otherwise $(1 \leq j \leq c+1,\ 1 \leq k \leq c+1)$. The matrix D_i is $(i+1) \times (i+1)$ dimensional; $(D_i)_{j,k}$ equals $(j-1)\mu_1 + (i-j+1)\mu_2 + \lambda$ for $k = j$ and 0 otherwise $(1 \leq j \leq i+1,\ 1 \leq k \leq i+1)$.

The blocks H_m, $0 \leq m \leq c-1$, and $H_{i,j}$, $i,j \geq 0$, can be combined into super blocks:

$$I_{-1} = \bigcup_{i=o}^{c-1} H_i, \quad I_i = \bigcup_{j=o}^{\infty} H_{i,j}, \quad i \geq 0.$$

The states in I_{-1} are exactly those corresponding to the existence of idle servers. The states in I_i, $i \geq 0$, are those with exactly i first priority customers awaiting service. The matrices of transition rates between and within these blocks can be denoted as

$B_{-1,-1}$: $I_{-1} \longrightarrow I_{-1}$

$B_{-1,0}$: $I_{-1} \longrightarrow I_0$

$B_{0,-1}$: $I_0 \longrightarrow I_{-1}$

$B_{0,0}$: $I_0 \longrightarrow I_0$

A_0: $I_i \longrightarrow I_{i+1}$, $i \geq 0$

A_1: $I_i \longrightarrow I_i$, $i \geq 1$

A_2: $I_i \longrightarrow I_{i-1}$, $i \geq 1$.

These transitions are depicted in Figure 2. The above matrices of

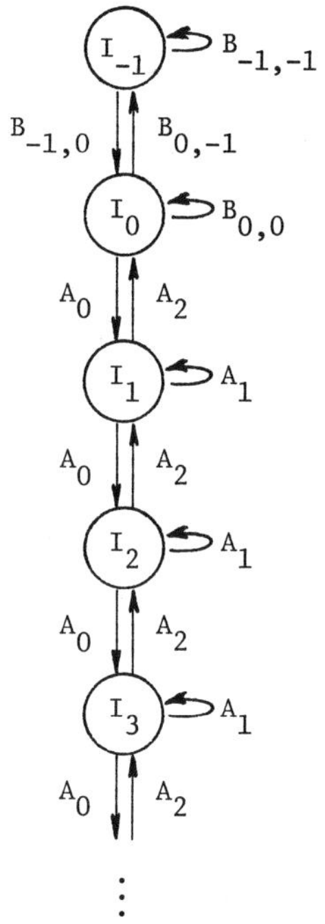

Figure 2. Partitioned state space (superblocks) and transition scheme.

transition rates can each be partitioned into submatrices corresponding to transitions between the subblocks of I_i, $i \geq -1$. These submatrices were defined earlier. The number of submatrices in one of these matrices varies according to the number of blocks in the corresponding superblocks. $(B_{-1,-1})_{j,k}$ equals K_{j-1} for $k = j+1$, $-D_{j-1}$ for $k = j$, J_{j-1} for $k = j-1$, and 0 otherwise ($1 \leq j \leq c$, $1 \leq k \leq c$). $(B_{-1,0})_{j,k}$ equals K_{c-1} for $(j,k) = (1,c)$ and 0 otherwise ($1 \leq j < \infty$, $1 \leq k \leq c$). $(B_{0,-1})_{j,k}$ equals J_c for $(j,k) = (c,1)$ and 0 otherwise ($1 \leq j \leq c$, $1 \leq k < \infty$). $(B_{0,0})_{j,k}$ equals L_2 for $k = j+1$, $-D_c$ for $k = j$, M_2 for $k = j-1$, and 0 otherwise ($1 \leq j < \infty$, $1 \leq k < \infty$). $(A_0)_{j,k}$ equals L_1 for $k = j$ and 0 otherwise ($1 \leq j < \infty$, $1 \leq k < \infty$). $(A_1)_{j,k}$ equals L_2 for $k = j+1$, $-D_c$ for $k = j$, and 0 otherwise ($1 \leq j < \infty$, $1 \leq k < \infty$).

$(A_2)_{j,k}$ equals M_1 for $k = j$ and 0 otherwise $(1 \leq j < \infty,\ 1 \leq k < \infty)$. Using this block structure and notation, the transition rate matrix for the M/M/c two-priority nonpreemptive queueing system is

$$P_S = \begin{pmatrix} B_{-1,-1} & B_{-1,0} & 0 & 0 & 0 & \cdots \\ B_{0,-1} & B_{0,0} & A_0 & 0 & 0 & \cdots \\ 0 & A_2 & A_1 & A_0 & 0 & \cdots \\ 0 & 0 & A_2 & A_1 & A_0 & \cdots \\ \vdots & \vdots & \vdots & \vdots & \vdots & \end{pmatrix} \qquad (1)$$

2. Invariant Measures

The process described above is a quasi birth and death process [4,8]. Consequently its steady-state probability vector is of matrix-geometric form [4,8]:

$$(\underline{\pi}_{-1}, \underline{\pi}_0, \underline{\pi}_0 R, \underline{\pi}_0 R^2, \underline{\pi}_0 R^3, \ldots).$$

3. The Matrix-Geometric Rate Matrix

The rate matrix, R, of the quasi birth and death process is the minimal solution of

$$A_0 + RA_1 + R^2 A_2 = 0. \qquad (2)$$

Furthermore, from the block structure of the process and the interpretation of the rate matrix [5,7,8], R must have the structure

$$R = \begin{pmatrix} R_0 & R_1 & R_2 & R_3 & R_4 & \cdots \\ 0 & R_0 & R_1 & R_2 & R_3 & \cdots \\ 0 & 0 & R_0 & R_1 & R_2 & \cdots \\ 0 & 0 & 0 & R_0 & R_1 & \cdots \\ 0 & 0 & 0 & 0 & R_0 & \cdots \\ \vdots & \vdots & \vdots & \vdots & \vdots & \end{pmatrix} \tag{3}$$

where each submatrix has dimension $(c+1) \times (c+1)$. Substituting (3) into (2) gives a system of equations:

$$L_1 - R_0 D_c + R_0^2 M_1 = 0 \tag{4}$$

$$R_{i-1} L_2 - R_i D_c + \sum_{j=0}^{i} R_j R_{i-j} M_1 = 0 \tag{5}$$

These can be solved numerically as follows. First consider (4). Let

$$\tilde{L}_1 = D_c^{-1} L_1, \quad \tilde{M}_1 = D_c^{-1} M_1, \quad \tilde{R}_0 = D_c^{-1} R_0 D_c;$$

then (4) becomes

$$\tilde{R}_0 = \tilde{L}_1 + \tilde{R}_0^2 \tilde{M}_1 \tag{6}$$

which can be solved using the usual iterative approach useful for matrix-geometric rate matrices [8]: Let

$$S_0 = 0, \quad S_1 = \tilde{L}_1 + S_0^2 \tilde{M}_1, \ \ldots, \ S_{i+1} = \tilde{L}_1 + S_i^2 \tilde{M}_1, \ \ldots$$

Then $S_i \uparrow \tilde{R}_0$, termwise. The usual procedure is to continue iterating until the maximum termwise difference between successive iterates is smaller than some ε, e.g., $\varepsilon = 10^{-7}$.

Now consider (5); it is equivalent to

$$\tilde{R}_1 = \tilde{R}_{i-1} \tilde{L}_2 + \sum_{j=1}^{i-1} \tilde{R}_j \tilde{R}_{i-j} \tilde{M}_1 + (\tilde{R}_0 \tilde{R}_i + \tilde{R}_i \tilde{R}_0)\, \tilde{M}_1, \quad i \geq 1 \tag{7}$$

For $i = 1$, $\tilde{R}_1$ can be found by using the above solution for $\tilde{R}_0$ and then using a similar iterative procedure which will converge monotonically to $\tilde{R}_1$. For $i \geq 1$, continue recursively, using the solutions for $\tilde{R}_j$, $j < i$, from previous steps and using the iterative procedure to get $\tilde{R}_i$. The desired R_i's are

$$R_i = D_c \tilde{R}_i D_c^{-1}, \quad i \geq 0.$$

Thus the rate matrix R can be computed up to any level of truncation. In this study the computation of the R_i's was truncated when the iterative procedure stopped in the first iteration, in which case the value was set to 0.

There is an internal accuracy check which can be used in the above numerical computation. Let

$$\tilde{R}^* = \sum_{i=0}^{\infty} \tilde{R}_1.$$

From (6) and (7) it follows that

$$\tilde{R}^* = \tilde{L}_1 + \tilde{R}^* \tilde{L}_2 + (\tilde{R}^*)^2 \tilde{M}_1. \tag{8}$$

The usual iterative method can be used to solve (8) for $\tilde{R}^*$. This matrix can then be compared to the sum of the individual solutions $\tilde{R}_i$, $i \geq 0$. This check was performed and virtually no error detected.

4. The M/G/1 Paradigm

According to the theory of matrix-geometric invariant vectors, if

$$\underline{0} = \underline{\pi} P_S \tag{9}$$

then

$$\underline{\pi} = (\underline{\pi}_{-1}, \underline{\pi}_0, \underline{\pi}_1, \underline{\pi}_2, \ldots)$$

where $\underline{\pi}_i$ corresponds to the vector on the super block I_i, $i \geq -1$, and

$$\underline{\pi}_{i+1} = \underline{\pi}_i R, \quad i \geq 0.$$

This can be used with (9) to temporarily reduce the problem of solving

(9) to consideration of

$$(\underline{0}\ \ \underline{0}) = (\underline{\pi}_{-1}\ \ \underline{\pi}_0)\, P_{I_{-1} \cup I_0}$$

where

$$P_{I_{-1} \cup I_0} = \begin{pmatrix} B_{-1.-1} & B_{-1,0} \\ B_{0,-1} & B_{0,0} + RA_2 \end{pmatrix}$$

is a matrix with negative diagonal entries, nonnegative off-diagonal entries, and row sums equal to 0; thus it can be thought of as a transition rate matrix for a Markov process on $I_{-1} \cup I_0$. Letting

$$C_0 = M_2, \quad C_1 = -D_c + R_0 M, \quad C_2 = L_2 + R_1 M_1, \quad C_i = R_{i-1} M_1, \quad i \geq 3,$$

gives

$$B_{0,0} + RA_2 = \begin{pmatrix} C_1 & C_2 & C_3 & C_4 & \cdots \\ C_0 & C_1 & C_2 & C_3 & \cdots \\ 0 & C_0 & C_1 & C_2 & \cdots \\ 0 & 0 & C_0 & C_1 & \cdots \\ \vdots & \vdots & \vdots & \vdots & \end{pmatrix}$$

Note that this has the M/G/1 paradigm structure discussed by Lucantoni and Neuts [4]. Following their approach, it will be easier to work with an embedded Markov chain. Define the submatrices of transition probabilities for this chain as

$$\tilde{K}_i = D_i^{-1} K_i, \quad \tilde{J}_i = D_i^{-1} J_i, \quad 0 \leq i \leq c,$$

$$\tilde{C}_1 = D_c^{-1} C_1 + I,$$

$$\tilde{C}_i = D_c^{-1} C_1, \quad i = 0, 2, 3, \ldots$$

This is a Markov chain on

$$I_{-1} \cup I_0 = \bigcup_{m=0}^{c-1} H_m \cup \bigcup_{j=0}^{\infty} H_{0,j} .$$

Each transition of this chain corresponds to a transition of the original process. Each transition of the original process out of a state in $I_{-1} \cup I_0$ is also a transition for this chain. If the process made a transition from I_0 to I_1, then the chain will make a transition from I_0 to I_0, the target state being the first return state in I_0. Thus the chain can make transitions from states of I_0 into themselves.

The following analysis uses ideas from Lucantoni and Neuts [4]. Consider the Markov chain $P_{I_{-1} \cup I_0}$ on this state space. Let G be the $(c+1) \times (c+1)$ matrix of hitting probabilities of states of $H_{0,j}$ starting at $H_{0,j+1}$, $j \geq 0$. G must satisfy

$$G = \sum_{\nu=0}^{\infty} \tilde{C}_\nu G^\nu,$$

and can be found iteratively starting with an initial 0 matrix; see [4]. Let V be the first passage probabilities of hitting states in $H_{0,j+1}$ starting from states in $H_{0,j}$ with $H_{0,j}$ a taboo set; then

$$V = \sum_{\nu=2}^{\infty} \tilde{C}_\nu G^{\nu-2}$$

Let W be the first passage probabilities of hitting states in $H_{0,j+1}$ with $H_{0,j}$ a taboo set starting from states in $H_{0,j+1}$; then

$$W = \sum_{\nu=1}^{\infty} \tilde{C}_\nu G^{\nu-1}$$

Finally, let $\tilde{S}_1$ be the matrix whose entries are the expected number of visits to states in $H_{0,j+1}$ from states in $H_{0,j}$ before returning to $H_{0,j}$; then

$$\tilde{S}_1 = \sum_{\nu=0}^{\infty} VW^\nu = V(I-W)^{-1}$$

Using the above relationships $\tilde{S}_1$ can be computed.

5. Probabilities of Idle Servers

Now consider an invariant vector for the Markov chain with state space $I_{-1} \cup I_0$ and the Markov transition matrix $\tilde{P}_{I_{-1} \cup I_0}$:

$$(\underline{z}_{-1}, \underline{z}_0) = (\underline{z}_{-1}, \underline{z}_0)\, \tilde{P}_{I_{-1} \cup I_0} \tag{10}$$

Partition the vector over the blocks H_m, $0 \leq m \leq c-1$, and $H_{0,j}$, $j \geq 0$:

$$(\underline{z}_{-1}, \underline{z}_0) = (\underline{z}_{-1,0}, \underline{z}_{-1,1}, \ldots, \underline{z}_{-1,c-1}, \underline{z}_{0,0}, \underline{z}_{0,1}, \underline{z}_{0,2}, \ldots)$$

From the structure of this process and an important property of taboo probabilities (Theorem 1 of [5], or see Chung [1], p. 53), it follows that

$$\underline{z}_{0,1} = \underline{z}_{0,0}\, \tilde{S}_1$$

This result applied to (10) gives

$$(\underline{z}_{-1}, \underline{z}_{0,0}) = (\underline{z}_{-1}, \underline{z}_{0,0}) \begin{pmatrix} 0 & \tilde{K}_0 & & & & 0 \\ \tilde{J}_1 & 0 & \tilde{K}_1 & & & \\ & \tilde{J}_2 & 0 & \tilde{K}_2 & & \\ & & & \cdots & & \\ & 0 & & \tilde{J}_{c-1} & 0 & \tilde{K}_{c-1} \\ & & & & \tilde{J}_c & \tilde{C}_1 + \tilde{S}_1 \tilde{C}_0 \end{pmatrix}$$

The dimension of the above square matrix is (c+1) (c+2)/2. For a moderate number of servers c, the invariant vector $(\underline{z}_{-1}, \underline{z}_{0,0})$ can be computed using existing numerical techniques. The corresponding invariant vector for the process has component vectors

$$\underline{y}_{-1,i} = \underline{z}_{-1,i}\, D_i^{-1}, \quad 0 \leq i \leq c-1$$

$$\underline{y}_{0,0} = \underline{z}_{0,0}\, D_c^{-1}.$$

In order to get a normalizing constant to convert this into the invariant probabilities, recall that the proportion of idle servers must equal

$$1 - \rho = 1 - \left(\frac{\lambda_1}{c\mu_1} + \frac{\lambda_2}{c\mu_2}\right)$$

where ρ is the utilization factor. Thus, let

$$\xi = \sum_{i=0}^{c-1} \frac{c-i}{c} \underline{y}_{-1,i} \ .$$

Then the invariant probabilities on $I_{-1} \cup H_{0,0}$ can be computed:

$$\underline{\pi}_{-1,i} = \underline{y}_{-1,i}\,(1-\rho)/\xi, \quad 0 \leq i \leq c-1,$$

$$\underline{\pi}_{0,0} = \underline{y}_{0,0}\,(1-\rho)/\xi.$$

6. Probabilities of States with Customers Waiting

Now it is possible to build the state space back up, computing the invariant probabilities of additional states. In order to compute the invariant probability vector $(\underline{\pi}_{0,1}, \underline{\pi}_{0,2}, \underline{\pi}_{0,3}, \ldots)$ over

$$\bigcup_{j=1}^{\infty} H_{0,j}$$

it is necessary to depart from the approach of Lucantoni and Neuts [4] because $\tilde{C}_0$ is singular. Instead, consider $\tilde{S}_i$, the matrix whose entries are the expected number of visits to states in $H_{0,j+i}$ from states in $H_{0,j}$ before hitting $H_{0,k}$, $k < j+i$;

$$\tilde{S}_i = \sum_{\nu=i+1}^{\infty} \tilde{C}_\nu \, G^{\nu-i-1} \, (I-W)^{-1}.$$

This follows by a simple sample path argument similar to the derivation of $\tilde{S}_1$ earlier. The invariant vector

$$(\underline{z}_{-1}, \underline{z}_{0,0}, \underline{z}_{0,1}, \underline{z}_{0,2}, \ldots)$$

for the Markov chain

$$P_{I_{-1} \cup I_0}$$

must satisfy

$$\underline{z}_{0,i+1} = \sum_{j=0}^{i} \underline{z}_{0,j} \tilde{S}_{i+1-j}, \quad i \geq 0.$$

This is a special case of the fundamental result for taboo probabilities (Theorem 1 of [5] or see Chung [1], p. 53). The invariant probabilities for the process must therefore satisfy

$$\underline{\pi}_{0,i+1} = \sum_{j=0}^{i} \underline{\pi}_{0,j} D_c \tilde{S}_{i+1-j} D_c^{-1}, \quad i \geq 0.$$

These can be computed recursively, starting from $\underline{\pi}_{0,0}$ which has already been computed.

Finally $\underline{\pi}_{i,j}$, $i \geq 1$, $j \geq 0$, can be computed using the matrix-geometric structure:

$$\underline{\pi}_{i+1,j} = \sum_{k=0}^{j} \underline{\pi}_{i,k} R_{j-k}.$$

Thus one can compute the invariant probability vector to any level of truncation $0 \leq i \leq I$, $0 \leq j \leq J$.

7. Sums and Means

By summing the above probabilities it is possible to get a separate calculation of the probability of no idle servers (this can be used as a consistency check on the numerical calculation) and the mean number of each type of customer awaiting service (then Little's formula can be applied to compute mean delay for each class).

$$\sum_{i,j=0}^{\infty} \underline{\pi}_{i,j} \underline{e}^t = \underline{\Pi} (I-R^*)^{-1} \underline{e}^t \tag{11}$$

$$\bar{q}_1 = \sum_{i=1}^{\infty} i \sum_{j=0}^{\infty} \underline{\pi}_{i,j} \underline{e}^t = \underline{\Pi} R^*(I-R^*)^{-2} \underline{e}^t \tag{12}$$

$$\bar{q}_2 = \sum_{j=1}^{\infty} j \sum_{i=0}^{\infty} \underline{\pi}_{i,j} \underline{e}^t = \lambda_2^{-1} \underline{\Pi} (S^{(1)} (I-S^{(0)})^{-1} - S^{(0)}) M_2 \underline{e}^t \tag{13}$$

where $\underline{e}^t$ is a (c+1) dimensional column vector consisting of all 1's, and

$$\underline{\Pi} = \underline{\pi}_{0,0}(I-S^{(0)})^{-1}$$

$$S^{(0)} = D_c \sum_{k=1}^{\infty} \tilde{S}_k D_c^{-1}$$

$$S^{(1)} = D_c \sum_{k=1}^{\infty} k \tilde{S}_k D_c^{-1}$$

8. Computational Experience

The algorithm described above has been programmed in single-precision Fortran and run on GWU's IBM370/3031. Cases with c=5 were run. Seventy five different cases were run corresponding to all combinations of ρ = .2, .5, .8, λ_1/λ_2 = .25, 15, 1,2,4, and μ_1/μ_2 = .25, .5, 1,2,4. Execution times for computing state probabilities and expectations varied from approximately 10 seconds per case with ρ = .2 to approximately 60 seconds per case with ρ = .8.

Various consistency checks were used in the computational procedure: i) independent calculations of R^* and $\sum_{i=0}^{\infty} R_i$ were compared; ii) the row sums of G were compared with unity; iii) the row sums of $\tilde{J}_c + \tilde{C}_1 + \tilde{S}_1 \tilde{C}_0$ were compared with unity; iv) the total probability computed was compared to unity; and v) moments were computed directly from the state probabilities and compared with values computed from equations (11) and (12). These consistency checks generally agreed to 5 or more digits.

The mean delay for cases with homogeneous service rates were calculated using Cobham's [2] approach. The values agreed with those computed by the above algorithm except in the case ρ = .8 where a discrepancy appeared in the fourth digit. This case illustrates the need for a more complete error analysis in this type of calculation.

9. Acknowledgements

The programming assistance of H. Arsham is gratefully acknowledged.

10. References

[1] K. L. Chung, Markov Chains with Stationary Transition Probabilities, Springer-Verlag, New York, 1967.

[2] A. Cobham, "Priority Assignment in Waiting Line Problems," Oper. Res., 2, (1954), 70-76.

[3] D. P. Heyman, "Problem: An M/M/c Queue with Priorities," Newsletter, Applied Probability Technical Section/College, O.R.S.A./T.I.M.S., Fall, 1977.

[4] D. M. Lucantoni and M. F. Neuts, "Numerical Methods for a Class of Markov Chains Arising in Queueing Theory," Technical Report No. 78/10, Department of Statistics and Computer Science, University of Delaware.

[5] D. R. Miller, "Computation of Steady-State Probabilities for M/M/1 Priority Queues," Oper. Res., 29, (1981), 945-958.

[6] M. F. Neuts, "Markov Chains with Applications in Queueing Theory, Which Have a Matrix-Geometric Invariant Probability Vector," Adv. in Appl. Probab., 10, (1978), 185-212.

[7] M. F. Neuts, "The Probabilistic Significance of the Rate Matrix in Matrix-Geometric Invariant Vectors," J. Appl. Probab., 17, (1980), 291-296.

[8] M. F. Neuts, Matrix-geometric Solutions in Stochastic Models, Johns Hopkins University Press, Baltimore, 1981.

This research was supported by the George Washington University Facilitating Fund and by Office of Naval Research Contract N00014-75-C-0729.

Department of Operations Research, School of Engineering and Applied Science, The George Washington University, Washington, D.C. 20052.

Discussant's Report on "Steady State Analysis of M/M/C Two-Priority Queues With Heterogeneous Rates," by Douglas Miller

This paper presents an interesting analysis of an M/M/C queueing system with two types of traffic. Type 1 traffic $(\lambda_1, \mu_1, \rho_1)$ has non-preemptive priority over type 2 $(\lambda_2, \mu_2, \rho_2)$. The paper presents a detailed equilibrium analysis based on the matrix geometric method of Neuts. The matrix geometric equilibrium form is achieved by blocking the state space based on the number of high priority customers waiting for service. The blocks are infinite dimensional, so the geometric parameter $\underset{\sim}{R}$ is also infinite dimensional. As such, the insight obtained from the matrix geometric form about the nature of the equilibrium distribution is somewhat limited. Nevertheless, this methodology leads to numerical algorithms for computing key system performance quantities such as waiting times and idleness probabilities. The algorithms seem to give highly accurate results when $\rho_1 + \rho_2$ is not near 1 and when μ_1/μ_2 or λ_1/λ_2 are not extreme. Consequently, this approach is especially attractive, and Professor Miller's analysis is quite elegant.

The author indicates that this work was inspired by a query of Heyman. Interestingly, this model is rather similar to models for voice-data communication systems, although there are several differences. Usually voice traffic operates as a loss system rather than a queueing system. Nevertheless, some aspects of the qualitative behavior will be preserved. Of particular importance is the case where μ_2/μ_1 is very large (order 10^4 is reasonable), but ρ_1 and ρ_2 are comparable. High priority is relatively infrequent but exhibits relatively long holding times. The phenomenon which occurs in this case can be seen most easily when $C = 1$. Here low priority traffic is blocked when high priority is using the server. It is convenient to focus on high priority busy periods which have mean length $\mu_1^{-1}(1-\rho_1)^{-1}$. During a time period T, $\lambda_2 T$ low priority arrivals are expected, thus on average $\frac{\mu_2}{\mu_1}\frac{\rho_2}{1-\rho_1}$ low priority arrivals will occur during the high priority busy period. It is important to notice that this number can be extremely large even for low values of ρ_1 and ρ_2. One only needs μ_2/μ_1 to be large. In this case, the mean low priority queue length can be misleading. The queue is either essentially empty or order μ_2/μ_1. This suggests that computational methods associated with this problem may be very unstable in this extreme range (as well as in the case $\rho_1+\rho_2 \approx 1$). This problem has been addressed by Feldman and Claybaugh (1980) using matrix geometric methods also.

The extreme cases cited earlier can be addressed in approximate ways including fluid flow approximations when μ_2/μ_1 is large and diffusion approximations when $\rho_1+\rho_2$ is near 1 (see Lahoczky and Gaver (1980) for a description). This family of methods should provide solutions over the full range of parameter values. It would also be useful to gain some insight into the eigenvalue structure of $\underset{\sim}{R}$, since the largest eigenvalue will control the tail behavior of the matrix geometric distribution.

The model studied by Professor Miller can be generalized in several interesting ways. When $C > 1$ and μ_2/μ_1 is large, it is useful to allocate some of the service for the exclusive use of low priority traffic. This can have the effect of a drastic reduction in low priority waiting times while creating only a moderate increase in the high priority waiting. The number so allocated must be optimized. One might also consider control strategies other than pure priority strategies.

References

[1] Feldman, R. and Claybaugh (1980), "A Computational Model for a Data/Voice Communication Queueing System," Texas A&M University Technical Report.

[2] Lehoczky, J. P. and Gaver, D. P. (1982), "Performance Evaluation of Voice-Data Queueing System," Proceedings of the Symposium on Applied Probability-Computer Science: The Interface, Birkhäuser-Boston, Boston, MA.

Discussant: Dr. John Lehoczky, Department of Statistics, Schenley Park, Carnegie-Mellon University, Pittsburgh, PA 15213.

PERFORMANCE MODELS OF COMPONENTS OF COMPUTER SYSTEMS

Bruce Clarke, Chairman

R. A. Geilleit & J. Wessels
G. B. Swartz
H. Mendelson
B. T. Doshi & E. H. Lipper

EXPLOITING SEEK OVERLAP

R. A. Geilleit*

and

J. Wessels**

1. Summary

In this paper it is demonstrated how with simple techniques it is possible to obtain insight into the effects of minor hardware alterations on the behaviour of a central computing facility. In the case described here the disks and disk control units formed the bottleneck. However, the only allowable alterations were in core size and central processor speed. The practical solution was to use such alterations in such a way that seek overlap was exploited as good as possible.

2. Introduction

Computer performance evaluation techniques may be used for different types of planning purposes. In this paper we will consider a case of medium term planning. The problem we were faced with was the question of how the CDC-Cyber 72-16 system of the National Aero Space Laboratory in the Netherlands would be able to cope with an increasing interactive workload for another two years by executing only minor hardware changes. The allowed hardware changes comprised core extension and extension of central processor capacity. For practical reasons no extensions with respect to the I/O-facilities were allowed and also software-based changes in facility use were not allowed.

The type of problem naturally required a solution in a relatively short time. A solution was obtained by modelling the system as a network of queues. This network of queues has been analyzed for several alternative hardware configurations applying decomposition and a home made iterative approach to account for the amount of overlap for seeks on disk units. The alternative configurations have been evaluated by comparing the numbers of active terminal users which can reasonably be processed. Following Denning and Buzen [2], this number has been defined as the saturation point (see section 2).

An important point in an analysis like this is the reliability of its results. It has to be conceded that the modelling process is smoothing reality in a way that models become nice and elegant but not really trustworthy in a quantitative sense. In fact, we also analyzed some alternative models for the same problem and this supported our belief that the saturation points are very robust. The utilization rates are less robust but give a good indication, the response times are rather sensitive to model changes.

In section 3 we will give a short problem description and the main features of the model. The analysis is treated in section 4 and in section 5 the results are presented and discussed.

3. The Problem and Its Model

The problem we are interested in makes it possible to consider only interactive jobs, since in the busy hours batch jobs don't get access. The system does not have a virtual memory, since in fact it lacks a really fast background memory device. The background memory for interactive jobs solely consists of disk units. Jobs have to be swapped out of central memory completely as soon as some interaction with terminals takes place. For reentrance into the core the job has to join the memory queue. Therefore figure 1 gives the basic set-up for the model.

In principle the consumption of a computation time portion may also lead to swap-out and renewed joining of the memory queue. However, this occurs rather seldomly and has hence been disregarded.

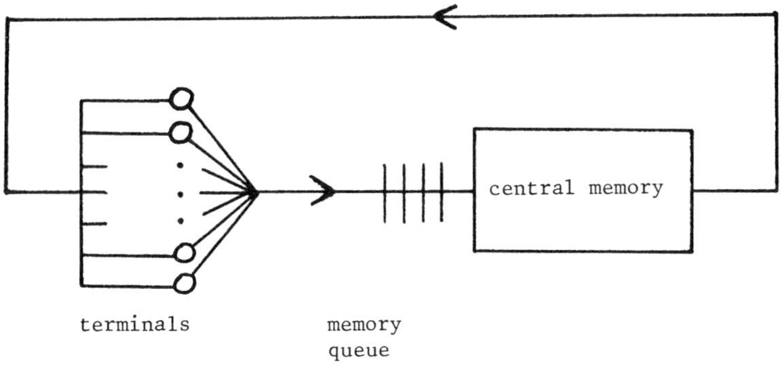

Figure 1. Basic Set-up for Job-flow Model

In principle the jobs from the memory queue are swapped-in on a first-come-first-served basis. However, the available and required amounts of memory also have an influence: if the job from the head of the line does not fit, then its successor gets a chance, etc. Measurements show that on the average 4 programs fit together in core (together with system programs and required library procedures).

Disk access is controlled by 2 disk control units which can work simultaneously on different disk units. Both disk control units can handle all disk units. Disks are used for storage of programs, but also for data. A program which gets access to central memory is swapped-in from disk to core and then the execution by the central processor starts. The central processor divides its capacity over all programs in core which are ready for execution (processor sharing). If a program requires some data handling, then its execution is stopped until the data handling has been completed. After the completion of the execution the program is swapped-out until a new request arrives. This gives the job flow of figure 2 for jobs in core.

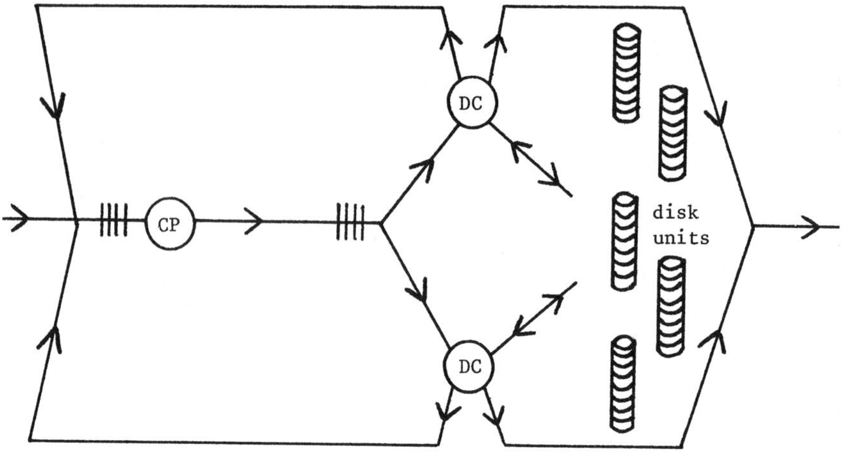

Figure 2. Job flow diagram for jobs in core. Jobs make several rounds through the central processor (CP) for execution and one of the disk control units (DC) for datahandling before leaving.

For this type of situation a natural and well established (e.g., [1]) approach would be to analyze core behaviour in a closed system and to use the results for the analysis of the traffic between the terminals and central memory (decomposition and aggregation). In this case the approach would lead to the model of figure 3 for the closed model for behaviour of jobs in core and to the model of figure 1 for the aggregated situation.

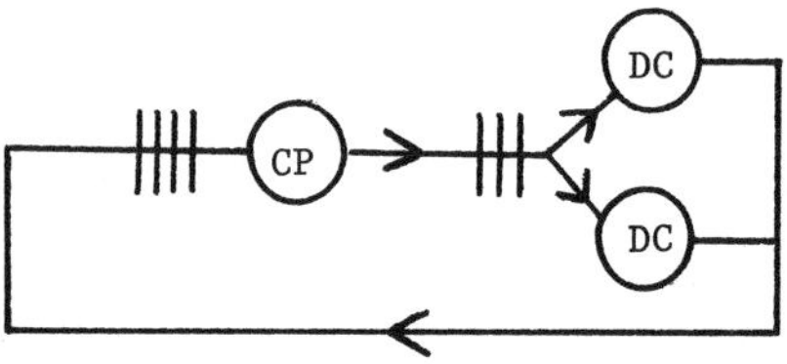

Figure 3. A Closed Model for the Behaviour of Jobs in Core

The alternative hardware configurations come down to several possibilities for the execution speed of the central processor and for the number of jobs which can be stored in central memory (core extension). So in principle, the aforementioned approach gives a way of analyzing the performance characteristics for several hardware configurations.

However, one difficulty which remains is due to the way the disk control units work. For simplicity we split-up the work a disk control unit has to do in 2 parts (for a more detailed description see Hunter [3]), viz. the seek and write/read (including latency) activities. During the seek the job has to wait, so - from the point of view of the job - the seek belongs to his service-time at the disk control unit. However, during this time the disk control unit can also help another job with its write/read activity or start seek activities for several other jobs as long as all the activities in which the two disk control units are involved regard different disk units. So, in fact, the work intensities of the disk control units depend heavily on the throughput (and conversely).

4. The Analysis

Because of the mutual interdependence of work intensities for the disk control units and the throughput, we have chosen an iterative approach. Before giving the approach, we will first give some data to show the relevance.

- On the standard central processor, the execution of a job between two data handling operations requires an average of 20 ms if the job would not have to share the processor with other jobs.
- The alternative central processors are modelled by giving them

a work-speed factor higher than the factor 1 for the standard processor; here we will consider the factors 1, 1.25, 2, 2.50.

≡ A seek requires on the average 30 ms, which is of the same order of magnitude as the average write/read operation which requires 33 ms.

≡ The time required for swapping is included in the average (a swap-in is a long read operation).

These data show that the system would be well balanced if the disk control units would not have to reserve any capacity for the seeks. However, if, on the other extreme, all seeks would have to be executed separately by the disk control units, then the disk control units would form a bottleneck for the system. So in practice the standard system will have a CP : DC workload proportion somewhere between $20 : \frac{33}{2}$ and $20 : \frac{63}{2}$.

In the standard system the seek overlap is relatively poor, since only an average of 4 jobs can be in core simultaneously. Extension of central memory (more jobs in core) would lead to better performance of the disk control units (more seek overlap) and the same holds for speeding up of the central processor. However, the question remains how much effect these changes will have.

A very simple analysis shows already that the system might be rather sensitive to the prospective hardware changes. Namely, it is worthwhile to perform an analysis of the closed system of figure 3 with a fixed number of jobs, with processor sharing and arbitrarily distributed CP service times (average workload w_{CP}) and with FCFS-queue discipline with exponentially distributed DC service times (average workload w_{DC}). For this analysis figure 4 gives the CP utilization and therefore the throughput as a function of the system parameters. In fact the only relevant parameters are:

$p = \frac{w_{DC}}{2w_{CP}}$, k = the number of jobs in the system.

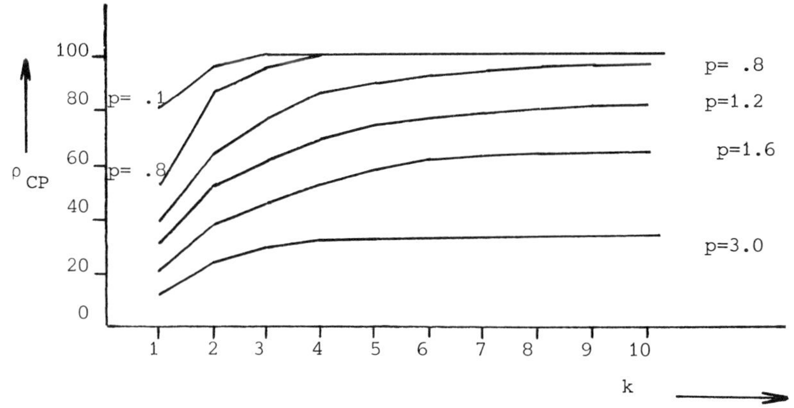

Figure 4. CP utilization; the figure gives CP utilization for fixed values of p as a function of k.

Figure 4 shows that for the relevant values of p (between .8 and 1.6 for ideal and no seek overlap respectively) and for k=4 the performance of the system is rather sensitive for variations in k and in p, which can be brought about by the relevant hardware changes. For example, core extension might bring k from 4 to 7, which would result in an essential increase of troughput for all relevant values of p, but it will moreover result in more seek overlap and therefore decrease of p, although it might increase cycletimes. Particularly in order to get grasp of the latter effects we execute an iterative analysis in which we analyze the closed system subsequently with varying values of w_{DC}. Namely, choose in the n-th iteration

$$w_{DC} = \alpha_n . 30 + 33$$

and start (e.g.) with $\alpha_1 = 1$. The analysis with $\alpha_1 = 1$ correspond to discarding seek overlap and blocking effects. Using the results of this analysis one can estimate the amount of seek overlap and blocking which corresponds to the resulting traffic intensities. This estimation procedure is relatively complicated. Nevertheless, it only provides relatively rough estimations. The basis of the estimation procedure is the fact that the seek time for a particular seek request is only attributed to the effective DC service time of this seek until the same DC starts a new service in the form of another seek or a write/read operation. The average attributed effective seek time is estimated by averaging estimates for this part of the seek time for different distributions of jobs over the servers in the closed network. So, for a given distribution of the jobs, one needs - for instance - the frequency density of the time of a new arrival at the DC and the probability that the seek for the new arrival will be blocked. Under some assumptions (for instance, for the form of the frequency density of the seek time), one can construct such estimates. This estimation leads to a new value for α: $\alpha_2 < 1$. After analyzing the system with $w_{DC} = \alpha_2 . 30 + 33$, again a new α is determined. Practically, this procedure appeared to converge. In this way we can analyze the closed system of figure 3 for fixed numbers of jobs k including the seek overlap effect.

For some results, see table 1. These results show that in particular increase of k from 4 to 7 and acceleration of the central processor with a factor 2 stimulate the seek overlap considerably without serious effects on the average cycle time.

For the analysis of the aggregated system a think time with average w_{TH} is introduced and the closed system of figure 1 is analyzed with the following features:

- ≡ maximally 4 jobs may enter central memory;
- ≡ each job requires an average of 20 CP-DC cycles before leaving central memory.

Table 1. Some results of the iterative analysis of figure 3 with w_{CP} = 20 ms in the standard configuration; an acceleration factor β for the central processor means that the effective value of w_{CP} becomes $20/_{\beta}$.

	acceleration factor central processor	α	average cycle time in ms
k=4	1	.72	117
	1.25	.60	106
	2	.43	93
	2.50	.37	89
k=7	1	.44	170
	1.25	.31	151
	2	.15	132
	2.50	.13	129

The preceding analyses for k = 0,1,2,3,4 are used to obtain load dependent output rates for the aggregated server. For the case of extended core, the same procedure is executed with a maximum of 7 jobs in core. This exercise is repeated for different values of w_{CP} corresponding to accelerated central processors (as in Table 1) and it is also repeated for different number of terminals (w_{TH} = 15 s).

The analyses are used to determine for each hardware alternative (value of w_{CP}, 4 or 7 jobs allowed in core) a number of terminals which can reasonably be handled by the configuration. Here we use the concept of saturation point as advocated by Denning and Buzen [2] and as explained in figure 5.

5. Results and Conclusions

Some of the results of the analyses of the aggregated model, using the results of the iterative analysis of the model of figure 3, are given in Table 2.

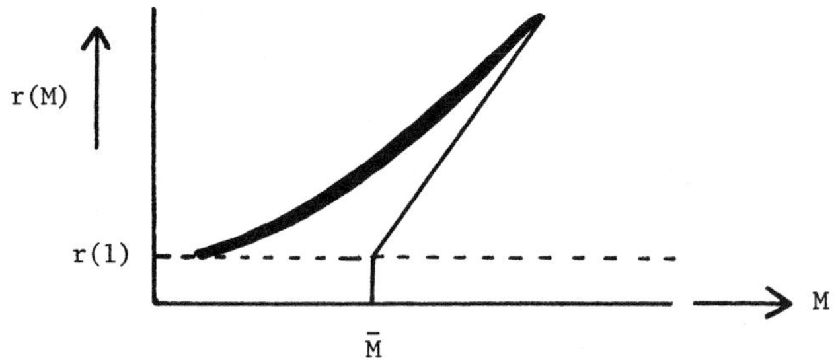

Figure 5. Saturation point for the number of terminals; for some hardware configuration the heavy line gives the average response time as function of M, the number of terminals; the straight line, which describes the asymptotic linear behaviour of the response time function, defines the saturation point $\bar{M}$. The response time function is computed by the method mentioned above.

It is again clear that k = 7 (extended central memory) and acceleration factor 2 for the central processor allow considerably more terminals to be active (43 against 28 for the standard configuration) without a loss in average response time. The extremely high DC utilizations are not so strange if one observes that they also contain seek activities, although a substantially smaller fraction than in the standard case. So, it appears sensible to adapt the system so that the DC's are heavily used since in that case they work more efficiently. In fact the adapted system (e.g., k = 7, acceleration factor 2) shows a fair increase in capacity.

The price to be paid for this capacity increase is a certain unbalance of the system, which will lead to instability and high variances.

Note that the quotient of $r(\bar{M})$ and r(1) in Table 2 is approximately constant over the alternative configurations.

Table 2. Saturation point analysis of the aggregated system; K indicates the maximally allowed number of jobs in core and therefore is related to the size of central memory; the first column gives the different alternatives for central processors and the second column gives the average response times if a job would have the system for itself; the third column gives the numbers of terminals for which the systems saturate; the other 3 columns give relevant performance data for the configurations with $\bar{M}$ terminals.

	accel. factor centr. proc.	response time for M=1 in seconds	saturation point $\bar{M}$	response time for $M=\bar{M}$ in seconds	CP-util. in %	DC-util. in %
k=4	1	1.66	28	3.57	60	89
	1.25	1.58	31	3.48	54	93
	2	1.46	35	3.16	39	96
	2.50	1.42	37	3.19	33	97
k=7	1	1.66	34	3.68	73	93
	1.50	1.58	38	3.48	66	96
	2	1.46	43	3.16	47	98
	2.50	1.42	44	3.08	39	99

6. Acknowledgement

The authors are grateful to the computing centre of the National Aero Space Laboratory of the Netherlands for the opportunity to obtain experience in analyzing a real life computer evaluation problem. In particular they thank Mr. G. Hameetman for the stimulating cooperation. Mr. H. Paquay of Eindhoven University of Technology is to be thanked for his contribution in the computations.

7. References

[1] K. M. Chandy, C. H. Sauer, Approximate methods for analyzing queueing network models of computer systems. Comput. Surveys, 10, (1978), 282-317.

[2] P. J. Denning, J. P. Buzen, The operational analysis of queueing network models. Comput. Surveys, 10, (1978), 225-261.

[3] D. Hunter, Modelling real DASD configurations. This volume.

*Currently at the Organization for Applied Scientific Research, The Hague; this research has been done while the author worked for the National Aero Space Laboratory of the Netherlands.

**Eindhoven University of Technology, Department of Mathematics and Computing Science.

ANALYSIS OF A SCAN SERVICE POLICY IN A GATED LOOP SYSTEM

G. B. Swartz

1. Introduction

A communication system of n buffered input terminals, numbered 1,2,3,...,n, connected to a computer by a single channel, is analyzed. The n terminals are serviced according to a SCAN policy: the terminals are polled in sequence 1,2,3,...,n,n,n-1,...,2,1. Data are removed one unit at a time from the terminals' buffer. Data arriving to the terminal during this time are stored in an auxilliary buffer and are served in the next poll to the terminal. When the buffer has been emptied, the channel is used for system overhead for a randomly determined length of time. The stationary distribution of buffer contents, assumed to exist if the total normalized arrival rate is less than one, is used to find the average buffer contents and the average intervisit time. General, independent arrival rates are assumed for the n terminals. The analysis follows that in Swartz [2].

The application of the model to a scanning disk is covered. A disk device used under a SCAN policy is analyzed by Coffman and Hofri [1]. Their paper gives additional references to analyses of disk service policies.

2. Mathematical Model

A system of n, $n \geq 2$, queues, numbered 1,2,...,n, of unlimited capacity are served in the order 1,2,...,n,n-1,...,2,1. Each terminal is served until its buffer is empty, and is followed by a walking time, or changeover time, as the server moves from one queue to the next. Only the data units that are buffered at the beginning of service are serviced during the poll of the terminal; arrivals during this time are kept in an auxilliary buffer.

The time interval $[0,\infty)$ is divided into contiguous slots, s_j: $[(j-1)\Delta, j\Delta)$, $j \geq 1$, each of which may accomodate the service of a single data unit. The entry of data into the system is described by the number of data units arriving in a slot s_j over a time period Δ. Let $X_j^{(i)}$ be the number of data units which arrive at the i-th terminal during the j-th slot. The processes

$$\chi^{(i)} = \{X_j^{(i)}: j \geq I\}, \quad 1 \leq i \leq n,$$

satisfy the following conditions:

(1) the $\chi^{(i)}$, $1 \leq i \leq n$, are independent; and

(2) for each i, $1 \leq i \leq n$, the random variables $\{X_j^{(i)}: 1 \leq j\}$ are nonnegative, integer-valued, independent, and identically distributed with mean μ_i, variance σ_i^2, and generating function $P_i(z)$.

The terminals are polled in the order

$$T^{(1)}, T^{(2)}, \ldots, T^{(n)}, T^{(n)}, T^{(n-1)}, \ldots, T^{(2)}, T^{(1)}.$$

Upon being polled, the terminal retains the use of the channel, which removes the data from its buffer at the rate of one data unit per slot. Data arriving at the terminal during this phase is placed in an auxilliary buffer. When the buffer contents reach zero, a special code word (EOM for end of message) is placed by the terminal in the next slot and the terminal relinquishes control of the channel. Thereafter, the channel, for an interval of random length, is not available to any terminal. We refer to this period as the reply interval, walking time, or changeover time. This time includes the EOM code. After the reply interval, the system continues with a poll of the next terminal. A poll of n terminals is called a SCAN of the system. A SCAN of $T^{(1)},\ldots,T^{(n)}$ followed by a SCAN of $T^{(n)},\ldots,T^{(1)}$ is called a cycle of the system. The order in which the n terminals are served is the polling order.

The terminals $T^{(1)},T^{(2)},\ldots,T^{(n)}$ are considered as being in positions $1,2,3,\ldots,n,n+1,n+2,\ldots,2n$. Terminal 1 is serviced in position 1 and 2n; terminal 2 is in position 2 and 2n-1, and terminal n is in position n and n+1.

The time duration of the reply interval, between completion of service to the queue in the k-th position and the start of service to the next terminal in the b-th cycle, is randomly determined by a random variable, $R_{b,k}$. The processes

$$\mathcal{R}^{(k)} = \{R_{b,k}: 1 \leq b\}, \quad 1 \leq k \leq 2n,$$

satisfy the following conditions:

(1) the $\mathcal{R}^{(k)}$, $1 \leq k \leq 2n$, are independent; and

(2) for each k, $1 \leq k \leq 2n$, the random variables $\{R_{b,k}: 1 \leq b\}$ are positive, integer-valued, independent, and identically distributed with mean r_k, variance ν_k^2, and generating function $R_k(z)$.

The contents of each queue at time Δj is $W_j^{(i)}$, $1 \leq i \leq n$. The contents of the auxilliary buffer at time Δj is $A_j^{(i)}$. The terminals have initial contents $W_0^{(i)}$ with joint generating function $F_0(z_1,z_2,\ldots,z_n)$. Service starts at $T^{(1)}$ in position 1, and the equations of evolution are:

$$W_j^{(1)} = W_{j-1}^{(1)} - 1,$$

$$W_j^{(i)} = W_{j-1}^{(i)} + X_j^{(i)},\ 2 \le i \le n,$$

$$A_j^{(1)} = \sum_{\nu=1}^{j} X_\nu^{(1)},$$

until the instant of completion at the terminal in position 1, denoted by $\bar{\tau}_{1,1}$, when $W_{\bar{\tau}_{1,1}}^{(1)} = 0$. In the next slot terminal $T^{(1)}$ puts an EOM code on the channel; the server requires $R_{1,1}$ time slots before the server starts at the next position. During the walking time, arrivals continue and the equations of evolution are

$$W_j^{(i)} = W_{j-1}^{(i)} + X_j^{(i)},\ 1 \le i \le n$$

and at $j = \bar{\tau}_{1,1+1}$, $W_j^{(1)} = A_{\bar{\tau}_{1,1}}^{(1)}$. Service starts at time $\Delta\tau_{1,2}$ where $\tau_{1,2} = \bar{\tau}_{1,1} + R_{1,1}$. The system proceeds in this way for each cycle. The constant Δ is assumed to be 1 in the remaining sections. Thus, the time of completion of service at position k in the b-th cycle is denoted as $\bar{\tau}_{b,k}$, $1 \le k \le 2n$. The time that service starts at the k-th position in the b-th cycle is $\tau_{b,k}$. The queue, in position k, is given by the mapping π as specified in Table 1 below.

Table 1. MAPPING π

Position k	1	2	3	···	n	n+1	n+2	···	2n-1	2n
Queue $\pi(k)$	1	2	3	···	n	n	n-1	···	2	1

In general, for $1 \le k \le 2n$,

$$\tau_{b,k} = \begin{cases} \bar{\tau}_{b,k-1} + R_{b,k-1}, & 2 \le k \le 2n \\ \bar{\tau}_{b,2n} + R_{b,2n} & \end{cases}$$

and

$$\bar{\tau}_{b,k} = \tau_{b,k} + W_{\tau_{b,k}}^{\pi(k)}.$$

The evolution equations are, with $\ell = \pi(k)$,

$$W_j^{(i)} = \begin{cases} W_{j-1}^{(i)} + X_j^{(i)}, & i \neq \ell \\ W_{j-1}^{(\ell)} - 1, & i = \ell \end{cases},$$

and

$$A_j^{(\ell)} = \sum_{\nu=1}^{j} X_\nu^{(\ell)} \quad \text{for } \tau_{b,k} \leq j \leq \bar{\tau}_{b,k}$$

and

$$W_j^{(i)} = W_{j-1}^{(i)} + X_j^{(i)} \quad \text{for } \bar{\tau}_{b,k} \leq j \leq \tau_{b,k+1},$$

or ($\tau_{b+1,1}$ if $k = 2n$).

We assume in Section 3 that under the condition $\sum_{i=1}^{n} \mu_i < 1$, that the limiting distribution of the vector

$$(W_{\tau_{b,k}}^{(1)}, W_{\tau_{b,k}}^{(2)}, \ldots, W_{\tau_{b,k}}^{(n)})$$

as b increases, exists. In Section 4 we use the limiting distribution to calculate queue lengths.

3. The Invariant Distribution

The contents of the n queues when service starts at position k in the b-th cycle is the vector

$$\underline{W}_{\tau_{b,k}} = (W_{\tau_{b,k}}^{(1)}, \ldots, W_{\tau_{b,k}}^{(n)})$$

for $1 \leq k \leq 2n$ and $b \geq 1$. The generating function of $\underline{W}_{\tau_{b,k}}$ is

$$F_{b,k}(z_1, z_2, \ldots, z_n) = E\left(\prod_{r=1}^{n} z_r^{W_{\tau_{b,k}}^{(r)}}\right).$$

The operation of serving the queue at position k and proceeding to the next queue, takes $\underline{W}_{\tau_{b,k}}$ to $\underline{W}_{\tau_{b,k+1}}$ for $1 \leq k \leq 2n-1$, and $\underline{W}_{\tau_{b,2k}}$ to $\underline{W}_{\tau_{b+1,1}}$. We represent this by the action of an operator, $\mathcal{E}_k$, acting on $F_{b,k}(z_1,z_2,\ldots,z_n)$. C_n is the set of complex n-tuples.

Definition 3.1. $\underline{L}_k: C_n \to C_n$ is the mapping

$$\underline{L}_k(z_1,\ldots,z_n) = \left(L_k^{(1)}(z_1,\ldots,z_n),\ldots,L_k^{(n)}(z_1,\ldots,z_n)\right)$$

in which

$$L_k^{(i)}(z_1,\ldots,z_n) = \begin{cases} z_i & i \neq \pi(k) \\ \prod_{r=1}^{n} P_r(z_r), & i = \pi(k) \end{cases},$$

for $1 \leq i \leq n$ and $1 \leq k \leq 2n$.

One can show the following as in [2].

Theorem 3.1. <u>The operator</u> $\mathcal{E}_k$ <u>defined by</u>

$$\mathcal{E}_k\left(F(z_1,\ldots,z_n)\right) = R_k\left(\prod_{r=1}^{n} P_r(z_r)\right)F\left(\underline{L}_k(z_1,\ldots,z_n)\right)$$

<u>satisfies</u>

$$\mathcal{E}_k(F_{b,k}) = \begin{cases} F_{b,k+1}, & 1 \leq k \leq 2n-1 \\ F_{b+1,1}, & k = 2n \end{cases}.$$

Definition 3.2. Let $B_j: C_n \to C$ be

$$B_j(z_1,\ldots,z_n) = R_j\left(\prod_{r=1}^{n} P_r(z_r)\right), \quad 1 \leq j \leq 2n.$$

We express the operator as

$$\mathcal{E}_k(F(z)) = B_k(z)F(\underline{L}_k(z))$$

in terms of the vector $z = (z_1,z_2,\ldots,z_n)$.

For a string $\underline{k} = (k_1,k_2,\ldots,k_n)$, we have

Definition 3.3. Let $\underline{L}_{\underline{k}}: C_n \to C$ have components

$$\underline{L}_{\underline{k}}(z) = (L_{\underline{k}}^{(1)}(z),\ldots,L_{\underline{k}}^{(n)}(z))$$

with

$$L_{\underline{k}}^{(j)}(z) = \begin{cases} z_j & , \ \underline{k} = \varphi \\ L_{k_1}^{(j)}(z) & , \ M = 1 \\ L_{k_1,\ldots,k_{M-1}}(\underline{L}_{k_M}(z)), & 1 < M \end{cases}$$

for $1 \leq j \leq n$.

We can then obtain

Lemma 3.2. For $1 \leq j \leq n$,

$$L_{\underline{k}}^{(j)}(z) = L_{k_1,\ldots,k_r}^{(j)}\left(\underline{L}_{k_{r+1},\ldots,k_M}(z)\right)$$

and

$$L_{t,\ldots,2n,(1,\ldots,2n)_b}^{(\pi(t))}(z) = \prod_{r=1}^{n} P_r\left(L_{t+1,\ldots,2n,(1,\ldots,2n)_b}^{(r)}(z)\right).$$

The first and second SCANS, respectively, are represented by the operators

$$S_1 = \mathcal{E}_n \circ \mathcal{E}_{n-1} \circ \cdots \circ \mathcal{E}_1 \text{ and } S_2 = \mathcal{E}_{2n} \circ \mathcal{E}_{2n-1} \circ \cdots \circ \mathcal{E}_{n+1},$$

and a cycle of service by the operator $S = S_2 \circ S_1$. With $S^M = S(S^{M-1})$, we obtain

$$(S^M F)(z) = \prod_{b=0}^{M-1} \prod_{t=1}^{2n} B_t\left(\underline{L}_{t+1,\ldots,2n,(1,\ldots,2n)_b}(z)\right) \cdot F\left(\underline{L}_{(1,\ldots,2n)_M}(z)\right).$$

We have, on taking the limit:

Theorem 3.2. If $(S^M F)(z)$ converges to a generating function $F^*(z)$, then $F^*(z)$ is independent of F and is the unique solution to $SF^* = F^*$, and

$$F^*(z) = \prod_{b=0}^{\infty} \prod_{t=1}^{2n} B_t\left(\underline{L}_{t+1,\ldots,2n,(1,\ldots,2n)_b}(z)\right).$$

Applying the definition of B_t and Lemma 3.2, we obtain:

Corollary 3.1.

$$F^*(z) = \prod_{b=0}^{\infty} \prod_{t=1}^{2n} R_t\left(L_{t,\ldots,2n,(1,\ldots,2n)_b}^{(\pi(t))}(z)\right).$$

As shown in Swartz [2], S^M converges if $E(F^*)$ exists. The expected value is found in the next section.

4. The Expected Value

For this presentation the results will be given for the particular example of three queues. The generating function F^* was for service starting at position 1. We will denote by F_k^* the generating function when service starts at position k, $1 \le k \le b$. Denote the derivative of F_k^* with respect to z_j, $1 \le j \le z$, by $f_{j,k}$ with each $z_i = 1$. Using the symbolism $E_j\ F^*$ for this derivative and the fact that $E(Q(P)) = EQ \cdot EP$, we obtain

$$F^*(z) = R\left(L_6^{(1)}(z)\right) R_5\left(L_{56}^{(2)}(z)\right) R_4\left(L_{456}^{(3)}(z)\right)$$

$$\cdot\ R_3\left(L_{3456}^{(3)}(z)\right) R_2\left(L_{23456}^{(2)}(z)\right) R_1\left(L_{12\cdots 6}^{(1)}(z)\right)$$

$$\cdot \prod_{b=1}^{\infty} \prod_{t=1}^{6} R_t\left(L_{t,\ldots,b,(1,\ldots,6)_b}^{(\pi(t))}(z)\right).$$

We obtain, on taking the expectation

$$f_{1,1} = r_6\lambda_{6,0} + r_5\lambda_{5,0} + r_4\lambda_{4,0} + r_3\lambda_{3,0} + r_2\lambda_{2,0} + r_1\lambda_{1,0}$$

$$+ \sum_{b=1}^{\infty} \sum_{t=1}^{6} r_t\lambda_{t,b}\,,$$

where

$$\lambda_{t,b} = E\left(L_{t,\ldots,t,(1,\ldots,6)_b}^{(\pi(t))}(z)\right).$$

Thus,

$$f_{1,1} = \sum_{t=1}^{6} r_t\ A_t$$

where

$$A_t = \sum_{b=0}^{6} \lambda_{t,b}, \quad 1 \leq t \leq 6.$$

From the recurrence in Lemma 3.2, we obtain a system of six equations:

$$\begin{pmatrix} 1 & -\mu_2 & -\mu_3 & 0 & 0 & -\mu_1 \\ 0 & 1 & -\mu_3 & 0 & -\mu_2 & -\mu_1 \\ 0 & 0 & 1 & -\mu_3 & -\mu_2 & -\mu_1 \\ 0 & 0 & -\mu_3 & 1 & -\mu_2 & -\mu_1 \\ 0 & -\mu_2 & -\mu_3 & 0 & 1 & -\mu_1 \\ -\mu_1 & -\mu_2 & -\mu_3 & 0 & 0 & 1 \end{pmatrix} \begin{pmatrix} A_1 \\ A_2 \\ A_3 \\ A_4 \\ A_5 \\ A_6 \end{pmatrix} = \begin{pmatrix} 0 \\ 0 \\ 0 \\ 0 \\ 0 \\ \mu_1 \end{pmatrix}$$

with the solution

$$A_t = \begin{cases} \dfrac{\mu_1^2}{(1-\mu)(1+\mu_1)} & , \quad 1 \leq t \leq 5 \\[2ex] \dfrac{\mu_1}{1+\mu_1} + \dfrac{\mu_1^2}{(1-\mu)(1+\mu_1)} , & t = 6. \end{cases}$$

where $\mu = \mu_1 + \mu_2 + \mu_3$ is the total arrival rate. Substituting into the equation for $f_{1,1}$, one obtains

$$f_{1,1} = \frac{r\mu_1^2}{(1-\mu)(1+\mu_1)} + \frac{r_6\mu_1}{1+\mu_1} = \frac{\mu_1}{1+\mu_1}\left[r_6 + \frac{r\mu_1}{1-\mu}\right]$$

where

$$r = \sum_{t=1}^{6} r_t .$$

Since $F_1^* = \mathscr{E}_6 \circ F_6^*$, we have

$$E_1F_1^* = E_1\left[R_6\left(\prod_{r=1}^{3} P_r(z_r)\right) F_6^*\left(\underline{L}_6(z)\right)\right] ,$$

$$E_1F^* = \mu_1 r_6 + E_1(F_6(P_1(z_1)P_2(z_2)P_3(z_3), z_2, z_3)),$$

$$f_{1,1} = \mu_1 r_6 + f_{1,6}\mu_1.$$

Thus,

$$f_{1,6} = \frac{\mu_1}{1+\mu_1}\left[\frac{r}{1-\mu} - r_6\right] .$$

Notice that, in general, for n queues

$$f_{1,1} + f_{1,2n} = \frac{r\mu_1}{1-\mu}$$

which is the mean content of terminal one, without scanning, as shown in Swartz [3] but with larger total walking time r.

Similarly, the heuristic statement of stationarity, that the average contents at the start of service equals the average number of arrivals since the terminal last started, can be shown to be true. The average time to service a queue at any position is equal to the average contents when service starts. Thus, the intervisit time, V_k, from position k, the time to return to terminal $\pi(k)$, is for k = 1 and 6

$$V_1 = \frac{r(1-\mu_1)}{1-\mu} - r_6 \quad \text{and} \quad V_6 = r_6$$

and the average cycle time

$$T_C = \frac{r}{1-\mu} \quad .$$

Comparing the intervisit time for gated polling (non-SCAN) V_G and SCAN polling with equal total overhead time r, we have

$$V_G = \frac{r(1-\mu_1)}{1-\mu}$$

versus, in the case of n terminals,

$$V_1 = \frac{r(1-\mu_1)}{1-\mu} - r_{2n} \quad \text{or} \quad V_{2n} = r_{2n}.$$

Thus, SCAN intervisit time is always less for equal overhead time. The average intervisit time $AVG(V_1) = (V_1 + V_{2n})/2$ is:

$$AVG(V_1) = \frac{r(1-\mu_1)}{2(1-\mu)} \quad .$$

For any terminal $T^{(i)}$, the stationarity condition requires

$$\mu_i T_c = f_{i,k} + f_{i,k'}$$

where k and k' are the two positions of $T^{(i)}$. Since

$$T_c = V_k + V_{k'} + f_{i,k} + f_{i,k'}$$

we have

$$AVG(V_k) = \frac{T_c(1 - \mu_i)}{2}$$

5. Application to Disk Service Policy

Let the i-th cylinder of a disk have requests for data stored in a list called Terminal i. Let the requests for data in cylinder i be modeled by the discrete random variable $X_j^{(i)}$, representing the number of data items requested at time j. Assume each data item requires Δ time units to retrieve. Let r_i be the average time to go from cylinder i to $i+1$. All requests for a cylinder are obtained before the controller moves the disk head to the next cylinder. After all cylinders are serviced, the controller rescans, starting with the last cylinder rather than returning to the first cylinder. Since the return is approximately the same as the sum of the time to go from cylinder to cylinder, the analysis shows that the average time between servicing any terminal is always less by scanning, than by polling. Let the actual return time be b and the sum of the individual overhead times be $(n-1)a$, then polling has a percentage increase over SCAN of

$$\frac{V_G - AVG(V_k)}{AVG(V_k)} = \frac{b}{(n-1)a}$$

which is 1.25 percent, using numbers given in Hofri [4].

6. References

[1] Coffman, E. G., Jr. and Hofri, M., "On the Expected Performance of Scanning Disks," SIAM J. Comput., 1981.

[2] Swartz, G. B., "Polling a Loop System," J. Assoc. Comput. Mach., January 1980, pp. 42-59.

[3] Swartz, G. B., "Polling in a Gated Loop System," submitted for publication, June 1980.

[4] Hofri, M., "Disk Scheduling: FCFS vs. SSTF Revisited," Comm. ACM, November 1980, pp. 645-653.

Monmouth College, West Long Branch, New Jersey 07764

LINEAR PROBING AND RELATED PROBLEMS

Haim Mendelson

Abstract

Consider a random-access file which consists of a given number of buckets. Each bucket contains a fixed number of slots. Storage and retrieval of records are performed using linear probing: when a record is hashed into a full bucket, the file is scanned sequentially (in a cyclic manner) until a non-full bucket is found. The problem is related to some versions of an applied probability model which arises in economics (analysis of a decentralized market). The probabilities underlying the behavior of the addressing system are determined, and the relevant performance measures are derived.

1. Introduction

Consider a direct-access file which is used to store a collection of data records. A record is identified by a unique key, ω. The file consists of N buckets (or blocks) numbered by $1,2,3,\ldots,N$, where each bucket can contain up to b records.

Let Ω be the set of all possible keys ω. A fixed function, called the hash-function

$$h : \Omega \to \{1,2,3,\ldots,N\}$$

is used to assign keys $\omega\varepsilon\Omega$ into buckets. When a record with key ω is to be added to the file, it is assigned to bucket number $h(\omega)$, and stored there as long as the bucket is not full. If bucket $h(\omega)$ is full, then there is a collision that has to be resolved. One of the simplest collision resolution methods[1] is the linear probing scheme: If bucket $h(\omega)$ is full, then the next consecutive bucket, number $h(\omega)+1$, is tried, and so on until a non-full bucket is found. If bucket number N is reached and found full, then the scan goes on cyclically to bucket number 1. Thus, the file is considered to be cyclic, with bucket number $N+j$ equivalent to bucket j. It is assumed that keys are sampled from Ω independently, and that $h(\cdot)$ is constructed so that $h(\omega)$ has a uniform distribution over $\{1,2,3,\ldots,N\}$.

Knuth (1974) has described this setting in terms of a "musical chairs" game: Consider a set of N empty seats, arranged in a circle. Each seat can accommodate up to b persons. A person appears at a random spot just outside the circle and dashes (in a fixed direction) to the first seat which contains room for him (i.e., is not full). This is repeated $N\cdot b$ times, until all seats are full. The problem is to find the probability distribution of the number of seats the $(k+1)^{st}$ person ($k = 0,1,2,\ldots,Nb-1$) has to pass before he can find a seat. The same problem can also be formulated as a "parking problem" (see Blake and Konheim (1977, pp. 604-605), and Knuth (1973, p. 545, problems 29-31)).

Linear probing may be used either for management of tables in main storage, or for files on a secondary storage device. In the former case, it may be assumed that each bucket contains a single slot, and a special case of the general problem is obtained. This case has been analyzed by Konheim and Weiss (1966), Knuth (1973), and Mendelson and Yechiali (1980). The approach of Konheim and Weiss (1966) has been extended to the general case by Blake and Konheim (1977), where the asymptotic behavior as the number of records and buckets tend together to infinity so that their ratio possesses a finite limit was studied.

In this paper we approach the analysis of linear probing by applying some ramifications of the ballot theorem. The method of analysis is presented in section 2. In section 3 we relate linear probing to a different applied probability model which arises in economics (analysis of a decentralized market). Section 4 considers the modification of the problem to the case of "finite populations". We conclude the paper in section 5 by considering performance measures for linear probing. Some of the results are given without proof; the proofs may be found in Mendelson (1980). The reader may also consult Mendelson (1982) for additional applications and generalizations.

2. Analysis of Linear Probing

Consider a random-access file containing k records with keys $\omega_1,\omega_2,\ldots,\omega_k (k<Nb)$. We make the standard assumption that $\omega_1,\omega_2,\ldots,\omega_k$ are selected from the key space Ω so that the bucket numbers $h(\omega_1)$, $h(\omega_2),\ldots,h(\omega_k)$ are uniform i.i.d. random variables. Let X_j be a random-variable counting the number of records assigned (tentatively) by $h(\cdot)$ to bucket j $(j=1,2,3,\ldots,N)$. It should be emphasized that $X_1,X_2,\ldots,X_N$ correspond to the <u>tentative</u> assignment of buckets generated by the hash function $h(\cdot)$, and <u>not</u> to the actual placement of records after the collision resolution method has been applied. The joint distribution of $(X_1,X_2,X_3,\ldots,X_N)$ is multinomial symmetric with parameters k (= total number of trials) and $(1/N,1/N,\ldots,1/N)$ (= hitting probabilities). The random variables $X_1,X_2,\ldots,X_N$ are <u>interchangeable</u>: For each permutation π of $(1,2,3,\ldots,N)$, the random vectors $(X_1,X_2,\ldots .X_N)$ and $(X_{\pi(1)},X_{\pi(2)},\ldots,X_{\pi(N)})$ have the same joint distribution. In other words, the operation of renumbering buckets in the file is a probability-preserving transformation. A special permutation that will be used in the sequel is $\pi(j) = N + 1 - j$ for

$j=1,2,3,\ldots,N$, which amounts to renumbering the buckets in a reverse order. This creates a dual random vector $(X_1^*,X_2^*,\ldots,X_N^*)$, where $X_j^* = X_{n-j}$, and $(X_1^*,X_2^*,\ldots,X_N^*) \sim (X_1,X_2,\ldots,X_N)$.

We also define $S_n = \sum_{j=1}^{n} X_j$ $(S_0 \equiv 0)$ to be the total number of records tentatively hashed (by $h(\cdot)$) to buckets $1,2,3,\ldots,n$ (note again that S_1, $S_2,\ldots,S_N$ correspond to the tentative assignment by $h(\cdot)$, rather than to actual placement). Clearly, the random vectors $(S_1,S_2,\ldots,S_N)$ and $(S_1^*,S_2^*,\ldots,S_N^*)$ have the same joint distribution, where

$$S_n^* = \sum_{j=1}^{n} X_j^* = S_N - S_{N-n} \qquad \text{for } n=1,2,3,\ldots,N.$$

Consider the family of probabilities

$$P_{n,r}^{d} = P\{S_1<b-d, S_2<2b-d,\ldots,S_n<nb-d \mid S_n=r\} \tag{1}$$

$$\text{for } n=1,2,3,\ldots,N;\ r=0,1,2,\ldots,;\ d=0,1,2,\ldots,b-1.$$

$P_{n,r}^{d}$ is the probability that in a file with n buckets and r records, a given bucket (bucket 1, say) contains more than d empty slots. We shall show how the problem of finding the underlying probabilities for the system can be reduced to the problem of finding the probabilities (1). Our treatment here generalizes the approach introduced by Mendelson and Yechiali (1980).

Inspection of a typical configuration of the system reveals that the records fill some portions of the file, which we call strings. We say that a string with length of m buckets and residual q $(m=0,1,2,\ldots,[k/b];\ q=0,1,2,\ldots,b-1)$ starts at bucket j if:

(i) bucket $(j-1)$ is not full;

(ii) buckets $j,j+1,j+2,\ldots,j+m-1$ are all full;

(iii) bucket $j+m$ contains exactly q records.

(For convenience of exposition, we identify bucket number j with bucket numbers $j\pm N$ for all j.) When $h(\omega)\varepsilon\{j,j+1,\ldots,j+m\}$, we say that the above string is the string of ω, or that key ω belongs to the string starting at bucket j.

Define $a(m,q|k)$ as the probability that a string with length m (buckets) and residual q (records) starts at bucket j when there are k records in the file. Due to the cyclic symmetry of the system, this probability is independent of j, and we shall henceforth assume that j=1. It will be shown in the following sections how the probabilities $a(m,q|k)$ can be used to derive the relevant performance measures. We now set forth to compute $a(m,q|k)$. A string with length m and residual q will start at bucket 1 if and only if the following three events occur simultaneously (note that the S_j correspond to the tentative assignment by the hash function, not to the actual placement):

(E_1) $\quad S_{m+1} = mb + q$

(E_2) $\quad S_j \geq jb \quad$ for all $j=1,2,3,\ldots,m$

(E_3) $\quad S_N - S_{N-j} < jb \quad$ for all $j=1,2,3,\ldots,N-m-1$.

Furthermore, given E_1, the events E_2 and E_3 are conditionally independent. Hence,

$$a(m,q|k) = P(E_3|E_1) \cdot P(E_2|E_1) \cdot P(E_1), \tag{2}$$

where

$$P(E_1) = \binom{k}{mb+q} \cdot \left(\frac{m+1}{N}\right)^{mb+q} \cdot \left(1 - \frac{m+1}{N}\right)^{k-(mb+q)}, \tag{3}$$

since $S_{m+1} \sim B(k,(m+1)/N)$.

Using the interchangeability of the X_j's, we renumber buckets $(m+2,m+3,\ldots,N-1,N)$ as $(1,2,3,\ldots,N-m-2,N-m-1)$, and obtain

$$P(E_3|E_1) = P\{S_j<jb \text{ for } j=1,2,3,\ldots,N-m-1 | S_{N-m-1} = k-(mb+q)\} \tag{4}$$
$$= P^0_{N-m-1,k-(mb+q)}$$

in light of definition (1).

To compute $P(E_2|E_1)$, note that by duality (i.e., by renumbering buckets $1,2,\ldots,m$ in a reverse order), S_j^* may be substituted for S_j $(j=1,2,3,\ldots,m+1)$ in the computation of $P(E_2|E_1)$:

$$P(E_2|E_1) = P\{S_j^* \geq jb \text{ for } j=1,2,3,\ldots,m|S_{m+1}^*=mb+q\}.$$

Now, given $S_{m+1} = S_{m+1}^* = mb+q$, we have $S_j^* = S_{m+1} - S_{m+1-j}$, hence

$$P(E_2|E_1) = P\{S_{m+1}^* - S_{m+1-j} \geq jb \text{ for } j=1,2,3,\ldots,m|S_{m+1}=mb+q\}$$

$$= P\{S_{m+1-j} < (m+1-j)b - (b-q-1) \text{ for } j=1,\ldots,m|S_{m+1}=mb+q\}.$$

Substituting now $i=m+1-j$ and noting that the $(m+1)^{st}$ inequality holds trivially, we obtain

$$P(E_2|E_1) = P\{S_i < ib - (b-q-1) \text{ for all } i=1,2,3,\ldots,m+1|S_{m+1}=mb+q\} =$$

$$= P_{m+1,mb+q}^{b-(q+1)} . \tag{5}$$

Combining equations (2)-(5), we obtain

$$a(m,q|k) = \frac{1}{N^k}\binom{k}{mb+q}(m+1)^{mb+q}(N-m-1)^{k-(mb+q)} \cdot P_{m+1,mb+q}^{b-(q+1)}$$

$$\cdot P_{N-m-1,k-(mb+q)}^{0} \tag{6}$$

for all $k=0,1,2,\ldots,Nb-1$; $q=0,1,2,\ldots,b-1$;

$m=0,1,2,\ldots,[k/b]-1, \min\{[k/b],N-2\}$.

The above results do not apply to the case where $m = N-1$, since the probabilities $P_{0,k-(N-1)b-q}^{0}$ are not defined. It can be shown (see Mendelson (1980)) that the results apply to this case as well if we define

$$P_{0,r}^{0} = \begin{cases} 1 & r = 0 \\ 0 & r \neq 0. \end{cases} \tag{7}$$

In the next section we relate our analysis to an applied-probability model from a different area.

3. A Decentralized-Market Model

Gould (1978) has suggested the following model for the operation of a decentralized market, where there is no clearing house to match sellers with buyers. There are N equal sellers numbered by $1,2,3,\ldots,N$, each having b units of the traded good in stock (b is the expected-profit-maximizing quantity for each of the sellers, and its determination is one of the problems at hand). At the prevailing market price, each customer wants to purchase exactly one unit of the product. The consumer selects an arbitrary seller at random (i.e., the identities of selected sellers form a sequence of uniform i.i.d. random variables). If the seller so selected has sufficient inventory, the consumer buys one unit; otherwise, the seller "stocks out", and the consumer does not get the product.

Adding several features to this basic model, Gould (1978) examined the nature of equilibrium and analyzed the economic implications of this market mechanism. Gould (1978, p. 39, Appendix A) also provided an approximate analysis for a generalization that allows multiple searching, where an unsatisfied buyer will select an additional seller at random.

Consider now the following modification. When seller i stocks out, his customers are referred to seller $\pi(i)$, where $\pi(\cdot)$ is a permutation of $(1,2,3,\ldots,N)$. Thus, a consumer who started with seller i may be advised to go to seller $\pi(i)$ if seller i stocks out, then to seller $\pi^2(i) = \pi(\pi(i))$ if seller $\pi(i)$ also stocks out, and so on. The probabilities of interest include the distribution of the number of units sold by a seller and the distribution of the length of search.

These probabilities may be combined with the price and cost parameters to yield an economic analysis of this market.

Note that without loss of generality, we may assume[2] that $\pi(i) = i+1$ for $1 \leq i < N$, and $\pi(N) = 1$. Now, the current model is isomorphic to the model of linear probing, and we can directly apply the expressions derived there. Given that k units of product have already been sold, $a(m,q|k)$ is the probability that the route to be followed by an incoming consumer includes m unsucdessful trials with stocked-out sellers, and a success with the $(m+1)^{st}$ seller, who still has (b-q) units available for sale.

Turning to the probabilities of interest, here $P^d_{N,k}$ (as defined by (1)) is the probability that a seller has more than d units available for sale. It follows that the probability of having sold s units is $P^{b-(s+1)}_{N,k} - P^{b-s}_{N,k}$ for $s < b$, and $1 - P^0_{N,k}$ for $s = b$. Finally, the probability that the search route of a consumer consists of m sellers that have stocked out, and only the $(m+1)^{st}$ has the product available for sale, is

$$\sum_{q=0}^{b-1} \sum_{j \geq m} a(j,q|k).$$

4. Finite Key Space

Assume now that (in our original linear-probing problem) the key space Ω consists of a finite number of keys, $N \cdot T$, where the cardinality of $h^{-1}(j)$ is T for all $j=1,2,3,\ldots,N$. Once a record with key $\omega \epsilon \Omega$ has been added to the file, it cannot be selected again for insertion. Note that under these new assumptions, the analysis of section 2 remains valid with only two modifications: the binomial probability in equation (3) is replaced by the hypergeometric probability

$$P(E_1) = \frac{\binom{(m+1)T}{mb+q}\binom{(N-m-1)T}{k-(mb+q)}}{\binom{NT}{k}} \quad ,$$

and the probabilities $P^d_{N,r}$ are the ones corresponding to the multi-hypergeometric distribution,

$$P\{X_1=x_1, X_2=x_2, \ldots, X_N=x_N\} = \binom{NT}{k}^{-1} \cdot \prod_{j=1}^{N} \binom{T}{x_j} \cdot \Phi_{\{\sum_{j=1}^{N} x_j = k\}} \tag{8}$$

Thus, expression (6) for $a(m,q|k)$ is replaced by

$$a(m,q|k) = \binom{NT}{k}^{-1} \cdot \binom{(m+1)T}{mb+q} \cdot \binom{(N-m-1)T}{k-(mb+q)} \cdot P^{b-(q+1)}_{m+1,mb+q} \tag{9}$$

$$\cdot\ P^0_{N-m-1,k-(mb+q)} \quad .$$

This clearly demonstrates the advantage of using symmetry arguments for the analysis: the results are independent of the specific details as long as the symmetry prevails. It can readily be seen that the infinite-population treatment of section 2 amounts to replacing hypergeometric probabilities by the corresponding binomial limits. Clearly, the same modification applies to the finite-population version of the decentralized market model of section 3.

It is possible to generalize and consider applications and properties of probabilities of the form (1) in the more general setting where $(X_1, X_2, \ldots, X_N)$ is a vector of interchangeable random variables. Such applications and properties are given in Mendelson (1982) for a generalization of the classical ballot problem (Bertrand (1887), Takács (1967)). One of these properties, which we shall apply here, relates to the computation of the probabilities $P^d_{n,r}$. We have the following theorem.

Theorem 4.1. $\{P^d_{n,r}: n=0,1,2,\ldots;\ r=0,1,2,\ldots\}$ uniquely satisfy the recursive difference equations

$$P^d_{n,r} = \begin{cases} \binom{nT}{r}^{-1} \sum_{j=0}^{r} \binom{(n-1)T}{j} \binom{T}{r-j} P^d_{n-1,j} & 0 \le r < nb-d \\ 0 & r \ge nb-d, \end{cases} \tag{10}$$

for n=1,2,3,..., with the boundary conditions

$$P^d_{0,r} = \begin{cases} 1 & r = 0 \\ 0 & r \ne 0. \end{cases} \tag{11}$$

In the infinite-population case, equation (10) is replaced by

$$P^d_{n,r} = \begin{cases} \frac{1}{n^r} \sum_{j=0}^{r} \binom{r}{j} (n-1)^j P^d_{n-1,j} & 0 \le r < nb-d \\ 0 & r \ge nb-d. \end{cases} \tag{12}$$

Note that formally, the transition from the "finite population" to the "infinite population" case is accomplished (as usual) by replacing the hypergeometric terms in (9) and (10) by the corresponding limiting (as $T \to \infty$) binomial probabilities.

In the next section we return to the original "infinite population" linear probing problem, and consider the performance measures of the system.

5. Operating Costs

The costs of operating a file system consist of Input/Output (I/O) costs, Central Processing Unit (CPU) costs, and storage costs. The storage costs are easily calculated once the number of buckets, N, is given. Clearly, the CPU costs are dominated by the I/O costs, which are determined by the number of buckets that have to be read from or written onto the file. The relevant I/O performance measures for a file with N buckets and k records are (see Knuth (1973)):

(i) $D_N(k)$, the expected number of extra probes needed for an unsuccessful search,

(ii) $C_N(k)$, the expected number of extra probes needed for a successful retrieval,

where a probe is an I/0 operation (reading one bucket). In both cases, we do not count the initial probe at bucket $h(\omega)$.

It is easily seen that when each of the records in the file is equally likely to be processed, the operating costs may be derived from $\{D(k)\}_{k=0}^{Nb-1}$. First, under linear probing the probe sequence examined upon addition of a record is the same sequence followed in an unsuccessful search for the same key value. Thus, the operation of record addition is in fact an unsuccessful search which is followed by storing the new record. Next, an update in-place is in fact composed of a successful search and a store operation. Finally, under linear probing the sequence of buckets tested in a successful search is the same probe sequence that was followed when the record was inserted. Thus (see Knuth (1973, p. 528)),

$$C_N(k) = \frac{1}{k} \cdot \sum_{j=0}^{k-1} D_N(j). \tag{13}$$

Let $U_N(k)$ be a random variable counting the number of extra probes needed to find a non-full bucket for key ω in a file with N buckets and k records. Then,

$$D_N(k) = E[U_N(k)].$$

For $q=0,1,2,\ldots,b-1$, we define the partial expectations

$$D_{N,q}(k) = E[U_N(k) \cdot I_{\left\{\substack{\text{the string of key} \\ \omega \text{ has residual } q}\right\}}],$$

where I_E is the indicator of event E. Then,

$$D_N(k) = \sum_{q=0}^{b-1} D_{N,q}(k). \tag{14}$$

Expression (14) splits the expected cost $D_N(k)$ into b terms, where the qth term is the contribution of strings with residual q. Let $P_N(m,q|k)$ be the probability that the string of key ω has length m (i.e., $U_N(k) = m$) and residual q. Clearly,

$$D_{N,q}(k) = \sum_{m=0}^{\infty} m \cdot P_N(m,q|k). \tag{15}$$

But

$$P_N(m,q|k) = \sum_{j \geq m} a(j,q|k), \tag{16}$$

since the event defining $P_N(m,q|k)$ occurs if and only if for some $j \geq m$, bucket h(ω) is the $(j-m+1)^{st}$ bucket of a string with total length j and residual q.

$D_{N,q}(k)$ may be computed from the recursive scheme given by the following theorem (for proof, see Mendelson (1980)).

Theorem 5.1. Let X be a binomial random variable with parameter k and 1-1/N: $X \sim B(k,1-1/N)$. Then,

$$D_{N,q}(k) = \begin{cases} E\, D_{N-1,q}(X) & k < (N-1)b+q \\ \frac{1}{2} N(N-1) P^{b-(q+1)}_{N,(N-1)b+q} & k = (N-1)b+q \\ 0 & k > (N-1)b+q \end{cases} \tag{17}$$

Consider now the implication of theorem 5.1 on the special case where b=1. We shall need the following lemma.

Lemma 5.2. For all N = 1,2,3,...,

$$\sum_{j \geq 0} \binom{N-1}{j} \frac{(j+1)!}{N^{j+1}} = 1.$$

Proof. We have

$$\binom{N-1}{j}\frac{(j+1)!}{N^{j+1}} = \frac{j+1}{N}\cdot\prod_{i=1}^{j}(1-i/N)$$

(where the empty product $\prod_{i=0}^{0}$ is defined to be 1). Letting $a_i \equiv 1 - i/N$, we obtain

$$\sum_{j\geq 0}\binom{N-1}{j}\frac{(j+1)!}{N^{j+1}} = \sum_{j=0}^{N-1}(1-a_{j+1})\prod_{i=1}^{j}a_i =$$

$$= \sum_{j=0}^{N-1}\left(\prod_{i=1}^{j}a_i - \prod_{i=1}^{j+1}a_i\right) = 1 - \prod_{i=1}^{N}a_i = 1.$$

Q.E.D.

The case where b=1 is distinguished by the fact that here, $P_{n,r}^{0}$ are readily available from the ballot theorem:

Theorem 5.3. When b=1,

$$P_{n,r}^{0} = (1-r/n)^{+}.$$

This is actually a restatement of the ballot theorem for the case at hand (see e.g., Takács (1967, p. 10)). A direct proof may also be obtained by induction, using theorem 4.1. Since only d=0 is relevant when b=1 (since then, a non-full bucket must be empty), we can immediately find $D_{N,0}(k)$.

Theorem 5.4. When b=1,

$$D_{N,0}(k) = \begin{cases} 1/2\sum_{j=0}^{k}\binom{k}{j}\frac{(j+1)!}{N^{j}} - 1/2 & k=0,1,2,\ldots,N-1 \\ 0 & k=N,N+1,N+2,\ldots \end{cases} \quad (18)$$

for N = 2,3,4,...

Proof. By induction on N. The case N=2 is obvious. Assume that $D_{N-1,0}(k)$ are given by (18). By theorem 5.1, there exists a random variable $X \sim B(k,1-1/N)$ such that for all k = 0,1,2,...,N-2

$$D_{N,0}(k) = E\, D_{N-1,0}(X) = \sum_{i=0}^{k} \binom{k}{i} (1-1/N)^{i}(1/N)^{k-i} \cdot$$

$$\cdot \left(1/2 \sum_{j\geq 0} \binom{i}{j} \frac{(j+1)!}{(N-1)^{j}} - 1/2\right) = 1/2 \sum_{j\geq 0} \binom{k}{j} \frac{(j+1)!}{N^{k}} \cdot$$

$$\cdot \sum_{i\geq 0} \binom{k-j}{i-j} (N-1)^{i-j} - 1/2 = 1/2 \sum_{j\geq 0} \binom{k}{j} \frac{(j+1)!}{N^{j}} - 1/2$$

while for k=N-1,

$$D_{N-1,0}(N-1) = 1/2\; N(N-1)P^{0}_{N,N-1} = 1/2(N-1)$$

where the last equality follows from theorem 5.3 with n=N, r=N-1. But by Lemma 5.2,

$$1/2 \sum_{j\geq 0} \binom{N-1}{j} \frac{(j+1)!}{N^{j}} - 1/2 = 1/2(N-1).$$

Q.E.D.

Our proof of theorem 5.4 provides one more way of proving theorem K of Knuth (1973, p. 530).

We consider next the evaluation of the Central Processing Unit (CPU) costs. The relevant CPU cost is the cost of search <u>within</u> buckets. Assuming that the same sequence of storage locations is followed in both successful and unsuccessful searches for a given key ω, we obtain an analog of (13) with CPU costs, and therefore it is sufficient to treat unsuccessful searches. Clearly, the expected cost of search within a bucket depends on the search technique employed. We assume that records are stored (and retrieved) within a bucket sequentially (of course there is a variety of other possibilities). Then, the expected total cost of CPU probes within buckets is given by $b \cdot D_N(k) + q$, where q is the residual of the string of the required key, ω. The term $b \cdot D_N(k)$ is proportional to the expected I/O cost, so it is sufficient to investigate the distribution of the string residual, q.

Let $P_N(\cdot,q|k)$ be the probability that the residual of a string corresponding to the sampled key is equal to q. Obviously,

$$P_N(\cdot,q|k) = \sum_{m=0}^{[k/b]} P_N(m,q|k) = \sum_{m=0}^{[k/b]} \sum_{j\geq m} a(j,q|k) =$$

$$= \sum_{j\geq 0} (j+1) \cdot a(j,q|k).$$

Substituting $a(j,q|k)$ from equation (6), we obtain

$$P_N(\cdot,q|k) = \frac{1}{N^k} \sum_{j\geq 0} (j+1)\binom{k}{jb+q} \cdot (j+1)^{jb+q} \cdot (N-j-1)^{k-(jb+q)} \cdot$$

$$\cdot P_{j+1,jb+q}^{b-(q+1)} \cdot P_{N-j-1,k-(jb+q)}^{0}$$

and we have

Theorem 5.5. Let $X \sim B(k,1-1/N)$. Then,

$$P_N(\cdot,q|k) = \begin{cases} E\, P_{N-1}(\cdot,q|X) & k < (N-1)b+q \\ N\, P_{N,(N-1)b+q}^{b-(q+1)} & k = (N-1)b+q \\ 0 & k > (N-1)b+q. \end{cases}$$

6. References

[1] Bertrand, J., "Solution d'un Problème," C. R. Acad. Sci. Paris Ser. A-B, 105, (1887), 369.

[2] Blake, I. F., and Konheim, A. G., "Big Buckets Are (Are Not) Better!," J. Assoc. Comput. Mach., 24, (1977), 591-606.

[3] Gould, John P., "Inventories and Stochastic Demand: Equilibrium Models of the Firm and Industry," J. Business, 51, (1978), 1-42.

[4] Knuth, D. E., The Art of Computer Programming, Vol. 3, Section 6.4, Addison Wesley (1973), 506-518.

[5] Knuth, D. E., "Computer Science and its Relation to Mathematics," Amer. Math. Monthly, 81, (1974), 323-343.

[6] Konheim, A. G., and Weiss, B., "An Occupancy Discipline and Applications," SIAM J. Appl. Math., 14 (November 1966), 1266-1274.

[7] Lum, V. Y., Yuen, P. S. T., and Dodd, M., "Key to Address Transform Techniques: A Fundamental Performance Study on Large Existing Formatted Files," Comm. ACM, 14, (1971), 228-239.

[8] Maurer, W. D., and Lewis, T. G., "Hash Table Methods," Comput. Surveys, 7, (1975), 5-19.

[9] Mendelson, H., and Yechiali, U., "Performance Measures for Ordered Lists in Random Access Files," J. Assoc. Comput. Mach., 26, (1979), 654-667.

[10] Mendelson, H., and Yechiali, U., "A New Approach to the Analysis of Linear Probing Schemes," J. Assoc. Comput. Mach., (July 1980), 474-483.

[11] Mendelson, H., "Analysis of Linear Probing with Buckets," Working Paper, Graduate School of Management, University of Rochester (1980).

[12] Mendelson, H., "A Batch-Ballot Problem and Applications," J. Appl. Probab., (forthcoming, 1982).

[13] Morris, R., "Scatter Storage Techniques," Comm. ACM, 11, (1968), 38-44.

[14] Peterson, W. W., "Addressing for Random-Access Storage," IBM J. Res. Develop., 1, (1957), 130-146.

[15] Scheuermann, P., "Overflow Handling in Hashing Tables: A Hybird Approach," Inform. Systems, 4, (1979), 183-194.

[16] Takács, L., Combinatorial Methods in the Theory of Stochastic Processes, Wiley and Sons, New York (1967).

[17] Van Der Pool, J. A., "Optimum Storage Allocation for Initial Loading of a File," IBM J. Res. Develop., 16, (1972), 579-586.

7. Endnotes

[1] For discussions of hashing and different collision resolution methods, see Knuth (1973), Peterson (1957), Morris (1968), Lum et al. (1971), Van Der Pool (1972), Maurer and Lewis (1975), Scheuermann (1979), Mendelson and Yechiali (1979).

[2] If $\pi^j(i) = i$ for some $j < N$, it is possible to consider each closed loop $(i, \pi(i), \pi^2(i), \ldots, \pi^{j-1}(i))$ separately.

Graduate School of Management, University of Rochester, Rochester, New York 14627

COMPARISONS OF SERVICE DISCIPLINES IN A QUEUEING SYSTEM WITH DELAY DEPENDENT CUSTOMER BEHAVIOUR

Bharat T. Doshi
and
Edward H. Lipper

Abstract

In a variety of data processing and call processing systems the customer, unknown to the system, turns 'bad' at a random time after its arrival. That is, serving a customer with waiting time in excess of this random time results in a 'bad' (unsuccessful) service. If the performance of such systems is measured by the rate at which it serves good customers ('goodput'), then this performance is determined by the delay distribution and the distribution of the time at which the customer turns bad. Given the offered load, the service demand of the customers and the customer behavior, the performance is determined by the service discipline. It is then of interest to identify the best discipline and compare its performance with that of other disciplines. These problems are studied here for an M/G/1 queue. Optimal disciplines are derived for two special cases of the customer behavior. For a more complex customer behavior, the 'goodputs' under three different disciplines are numerically compared and a desired discipline is identified.

1. Introduction and Summary

In a variety of data processing and call processing systems the performance is significantly affected by the behavior of the customers when they are waiting in the queue for the service to start. To illustrate this point consider a call processing system in which the customers are the calls and the service is giving dial tone to the customers. If a customer starts dialing after the dial tone is given, then the system receives all the digits properly and the call is properly handled. If, on the other hand, the customer starts dialing before the dial tone is given, then the system will not receive all the dialed digits and the call will be unsuccessful. If we measure the performance of the system by the rate at which it serves good calls, then for given service rate and load this performance depends on both the distribution of the delay in giving dial tone and the (random) time at which a customer starts dialing after picking up the phone. Similar situations may arise in time shared data processing systems and in the emergency rooms of hospitals. Suppose we do not have any control over the service rate, offered load and the customer behavior but we know their relevant parameters. Then the performance is determined by the delay distribution. Since the service rate and the offered load are fixed, the delay distribution mainly depends on the service discipline. Thus, we are interested in finding the best service discipline for the given customer behavior and in comparing the performance under this discipline with the other disciplines. To our knowledge there is no published literature dealing with this problem. However, substantial work has been done in Bell Laboratories comparing the first in first out discipline with variations of the last in first out discipline for local switching systems. This will be available in a forthcoming paper [3] and also in [4]. In this paper we study the problem discussed above for a simple single server $(M|G|1)$ queueing model. The customer arrival process is Poisson with rate $\lambda > 0$. The service time distribution is $F(\cdot)$ with mean $\frac{1}{\mu}$. When the server completes a service it receives a reward $P(t)$ where t is the waiting time in the queue of the customer who completed the service. If $0 \leq P(t) \leq 1$, then $P(t)$ can be considered as the probability that the customer gets good service if the service begins at t time units after its arrival. But we need not restrict to this case. At each service completion epoch the server can observe the number of customers in the queue and their current waiting times. Based on this information the server selects one of the waiting customers for the next service. The strategy used to select the customer for the next service based on the available information is called the discipline. We impose the following restrictions on the permissible disciplines:

(1) A customer in service cannot be preempted.

(2) The server cannot remain idle if at least one customer is waiting in the queue.

Let π be a service discipline. If the steady state waiting time distribution $W_\pi(\cdot)$ exists under π, then the reward rate is given by

$$V_\pi = \lambda \int_{0-}^{\infty} P(t)\, dW_\pi(t). \tag{1.1}$$

We use V_π as a measure of performance of the discipline π. If $0 \leq P(t) \leq 1$, then V_π can also be considered the throughput of good customers ('goodput'). A discipline π^* is optimal if

$$V_{\pi^*} = \max_\pi V_\pi . \tag{1.2}$$

In Section 2 we formalize our model and introduce the necessary notations. In Section 3 we consider two special cases of the function $P(\cdot)$: (1) $P(\cdot)$ is concave and (2) $P(\cdot)$ is convex decreasing.

Under additional assumptions we show that the first-in-first-out (FIFO) discipline is optimal in case (1) and the last-in-first-out (LIFO) discipline is optimal in case (2).

In Section 4 we consider a special case of case (2) above. We assume that

$$P(t) = e^{-\alpha t} \qquad (t \geq 0) \tag{1.3}$$

for some $\alpha > 0$. Also, the service times are assumed to be exponentially distributed with mean $\frac{1}{\mu}$. For given α, μ and $\zeta = \lambda/\mu$ we obtain expressions for the 'goodput' (reward rate) under the optimal (LIFO) discipline and under the FIFO discipline. The numerical results are presented in Figures 1 and 2. Clearly, the FIFO discipline provides poor performance under heavy load (ζ close to 1) and serves only bad customers for $\zeta \geq 1$. The LIFO discipline provides much better performance and, in the limit as $\zeta \uparrow \infty$, serves only good customers. However, if α/μ is large enough, the performance under the LIFO discipline degrades as $\zeta \uparrow 1$ (see Figure 2) and then becomes better as $\zeta \uparrow \infty$. This peculiar behavior can be attributed to the two restrictions we have imposed on the permissible disciplines, namely, that a service cannot be preempted and that the server cannot remain idle if customers are waiting in the queue. For the reasons explained below in Remark 1, we do not want to consider preemption. However, the performance near $\zeta = 1$ can be improved by relaxing restriction 2. Hence we consider the LIFO with time out discipline. Here, the server remains idle if the current waiting times of all the customers in the queue exceed T. Otherwise the customers are served last in first out. The 'goodput' under the LIFO with 'time out' discipline is obtained in Appendix 1. This 'goodput', for various values of T, is compared with that from LIFO in Figure 3 and 4. With an appropriate value of T the LIFO with 'time out' discipline provides significant improvement over the LIFO discipline.

In Section 5 we consider a more complex customer behavior. Namely,

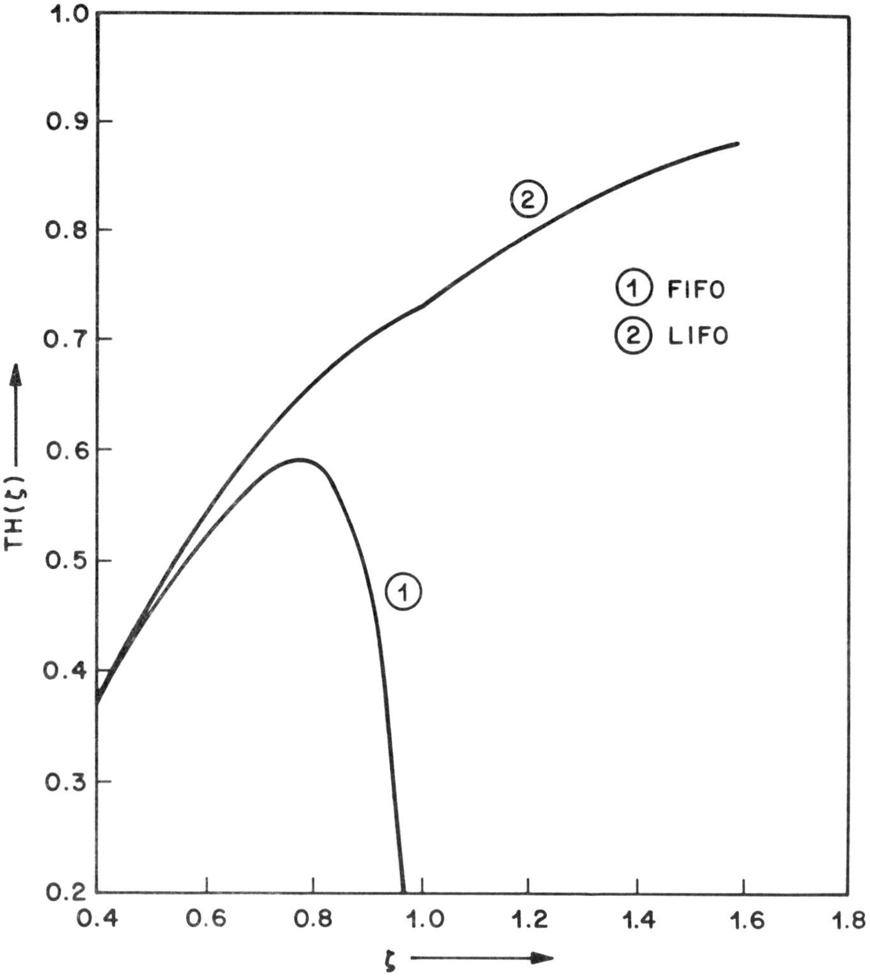

FIGURE 1: THROUGHPUT OF GOOD CUSTOMERS, TH(ζ), FOR FIFO AND LIFO

$\mu = 1.5, \quad \alpha = 0.15, \quad P(t) = e^{-\alpha t}$

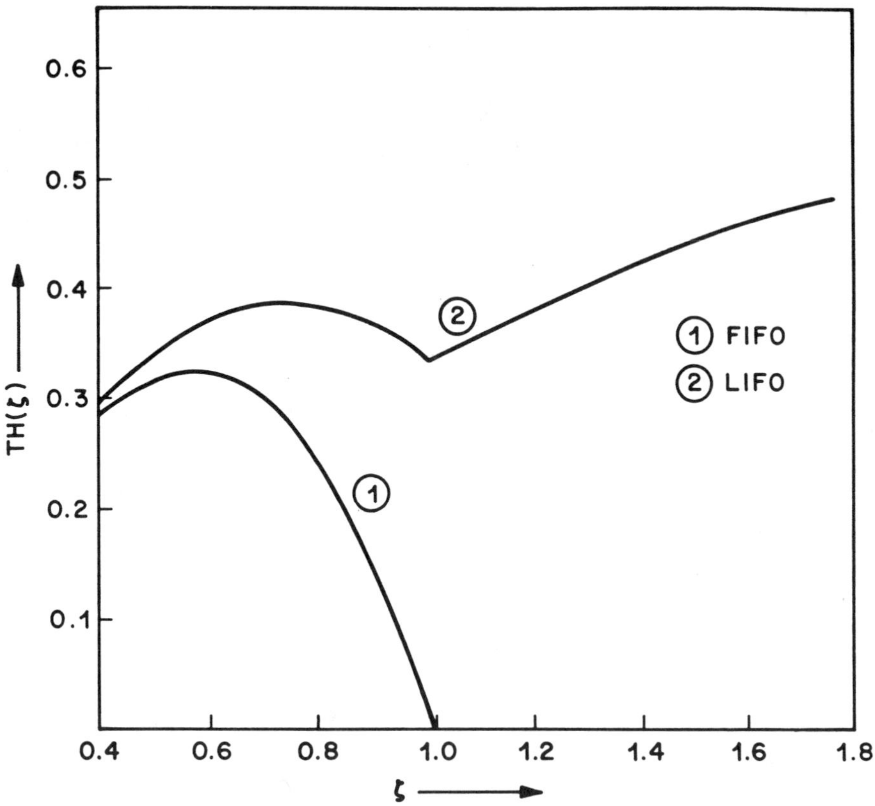

FIGURE 2: TH(ζ) FOR FIFO AND LIFO

$\mu = 1.5, \quad \alpha = 2.0, \quad P(t) = e^{-\alpha t}$

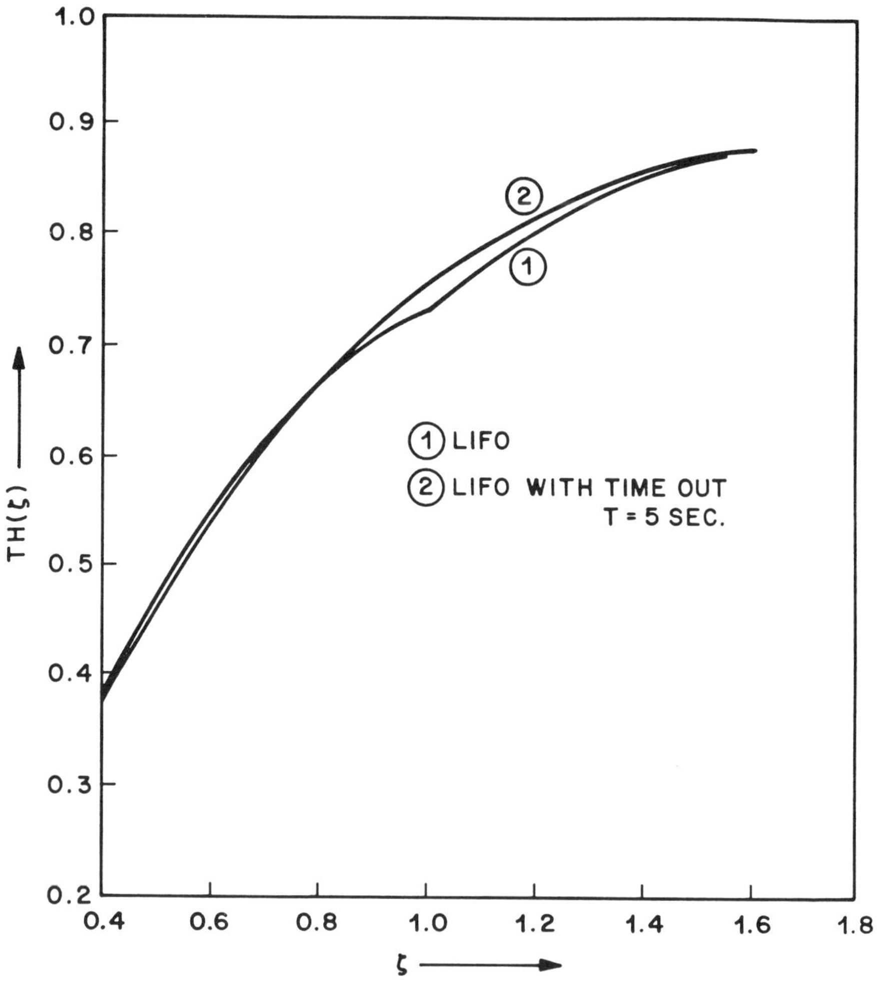

FIGURE 3: TH(ζ) FOR LIFO AND LIFO WITH TIME OUT

$\mu = 1.5$, $\alpha = 0.15$, $P(t) = e^{-\alpha t}$

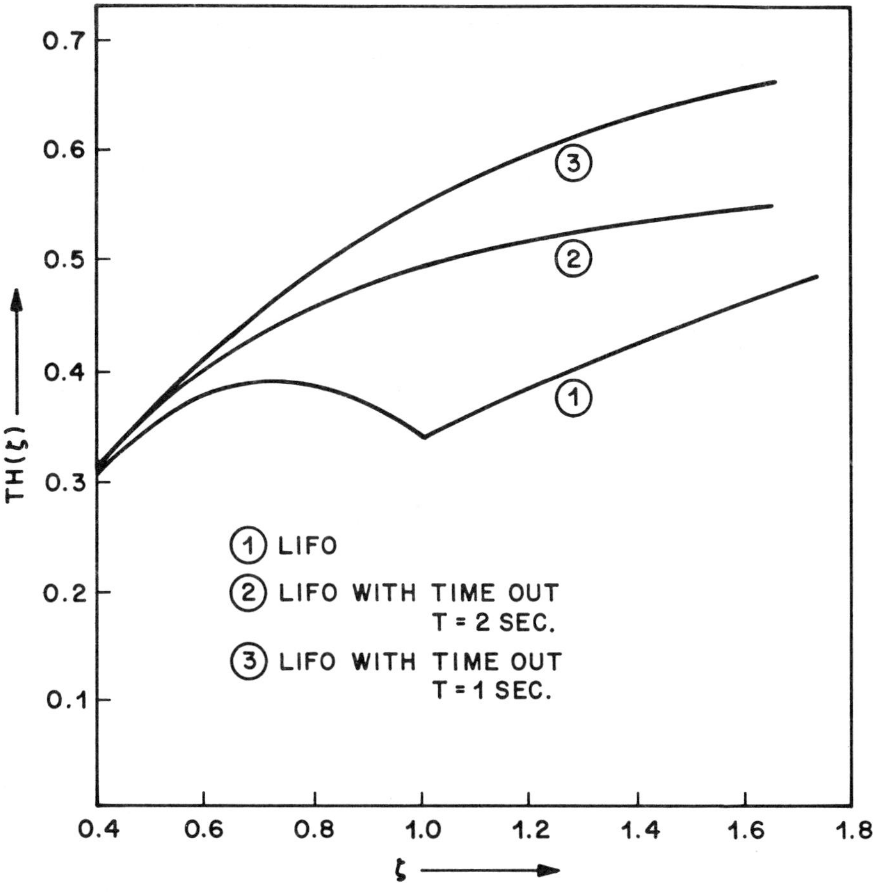

FIGURE 4: TH(ζ) FOR LIFO AND LIFO WITH TIME OUT.

$\mu = 1.5, \quad \alpha = 2.0, \quad P(t) = e^{-\alpha t}$

$$P(t) = \begin{cases} 1 & \text{if } t \leq S \\ e^{-(t-S)} & \text{if } t > S. \end{cases} \tag{1.4}$$

It is argued that in this case, as in other cases of more complex customer behavior, no simple discipline based only on the order of arrivals could be optimal. Therefore, we select three 'reasonable' disciplines and compare their performances. The disciplines considered are FIFO, LIFO and a mixture of FIFO and LIFO. The 'Mixture" discipline works as follows:

We set up two queues, Q1 and Q2. Q1 has nonpreemptive priority over Q2. On arrival the customer is put at the back of Q1. Thus Q1 is served first in first out. If a customer in Q1 does not begin service at U time units after its arrival, then it is transferred to the front of Q2. Thus Q2 is served last in first out. For this Mixture discipline π^U we obtain the 'goodput' in Appendix 2. The results obtained in Appendix 1 and 2 may be of interest in their own right.

The 'goodputs' from the FIFO, LIFO and the Mixture disciplines are compared in Figures 5 and 6. Once again, if we work near $\zeta = 1$, FIFO is not desirable. The Mixture discipline with U = S always performs better than the LIFO discipline. If we select other values of U, then the performance improves for some values of ζ and degrades for other values of ζ. Depending on the expected range of ζ we can select the value of U which achieves desirable performance in this range.

Since we have the waiting time distribution under the FIFO, LIFO and the Mixture disciplines we can compare their performance for any customer behavior.

Remark 1. In most applications where the customer behavior affects the 'goodput' the customer may turn bad even during its service time. In such cases the customer behavior function P(·) should have the total system time as its argument. However, if no preemption is permitted, then there is no loss of generality in using the waiting time in the queue as the argument of $P(\cdot)$. Suppose, for example, that $P(t)$ is the reward obtained by serving a customer with the total system time t. Let

$$\bar{P}(t) = \int_0^\infty P(t+y)dF(y). \quad (1.5)$$

Then, equivalently, we can use $\bar{P}(t)$ as the reward obtained by serving a customer with waiting time t. The relationship between $P(\cdot)$ and $\bar{P}(\cdot)$ is much more complicated if preemption is permitted. Also, the time required in preempting a customer and restarting its service later on may be a significant fraction of the service time. For these reasons we do not permit preemption in our model.

2. Model and Notation

We consider an $M|G|1$ queue. The arrival process is Poisson with rate $\lambda > 0$. The service time distribution is $F(\cdot)$ with finite mean $b = \frac{1}{\mu}$ and variance σ^2. Let $\zeta = \frac{\lambda}{\mu} = \lambda b$.

The system obtains a reward $P(t)$ when it completes the service of a customer who waited t time units in the queue before starting the service. $P(\cdot)$ is called the customer behavior function. We assume that $P(\cdot)$ is bounded on every compact interval. At each service completion epoch the system can observe the number of customers in the queue and also their current waiting times. Based on this information the system selects one of the waiting customers, if any, for the next service. If the queue is empty at a service completion epoch, then the server remains idle until the next arrival epoch and then servers the new arrival immediately. We assume that the service of any customer cannot be preempted and that the server cannot remain idle if a customer is waiting in the queue. Thus the decisions as to which customer is to be served next are made only at service completion epochs.

The state of the system seen by the server at a service completion epoch can be represented by $Z = (N, Y^N)$ where N is the number of customers in the queue and $Y^N = (Y_1^N, Y_2^N, \ldots Y_N^N)$ with Y_i^N being the current waiting time of the ith oldest customer in the queue. We will write $Z = o$ if the queue is empty. Let $Z(n) = (N(n), Y^N(n))$ be the state of the system at the nth service completion epoch. Then $\{Z(n): n \geq 1\}$ is a Markov chain. The state space of this Markov chain is

$$Z = \{0\} \cup \{(M, Y^M);\ M=1,2\ldots, Y_M^M < Y_{M-1}^M \cdots < Y_1^M < \infty\}. \tag{2.1}$$

Let $I_M = \{0,1,2\ldots M\}$ and $I = \{0,1,\ldots\}$. A service discipline is a function

$$\pi : Z \to I \tag{2.2}$$

with

$$\pi(0) = 0, \ \pi((M, Y^M)) \in I_M - \{0\} \ (M \geq 1).$$

$\pi((M, Y_M)) = k$ implies that the kth oldest customer will be selected for the next service when the state $Z = (M, Y^M)$ is observed at a service completion. $\pi(0) = 0$ implies that the server will remain idle until the next arrival if the queue is empty at a service completion epoch.

Suppose at time 0 the queue is empty. Let $V_\pi(s)$ be the expected reward obtained under service discipline π until time s. Consider all service disciplines under which

$$V_\pi = \lim_{s\to\infty} \frac{V_\pi(s)}{s} \tag{2.3}$$

exists and equals $\lambda \int P(t)\,dW_\pi(t)$ where $W_\pi(\cdot)$ is the steady state waiting time distribution under π. We compare service disciplines with V_π as a measure of performance. A service discipline π^* is optimal if

$$V_{\pi^*} = V = \max_\pi V_\pi \,. \tag{2.4}$$

3. Optimal Service Disciplines in Special Cases

In this section we obtain optimal service disciplines for two special cases of the customer behavior function $P(\cdot)$: (1) $P(\cdot)$ is concave, and (2) $P(\cdot)$ is convex decreasing. Under appropriate assumptions we prove that the first-in-first-out (FIFO) discipline is optimal in case (1) and the last-in-first-out (LIFO) discipline is optimal in case (2). In our notation the FIFO discipline π^* is defined by

$$\pi^*((M,Y^M)) = 1 \text{ for all } Z = (M,Y^M),$$

and for the LIFO discipline π^{**}

$$\pi^{**}((M,Y^M)) = M \text{ for all } Z = (M,Y^M).$$

Assume that $\zeta < 1$. We have already assumed that the system is empty at time 0. Let BC denote the first time at which the system becomes empty again. Then BC is the busy cycle. The distribution of BC is the same for all service disciplines π. Also, since $\zeta < 1$,

$$E_\pi[BC] = E[BC] < \infty \; . \tag{3.1}$$

We will only consider those disciplines π for which

$$E[V_\pi(BC)] < \infty \; . \tag{3.2}$$

For this class of disciplines (see Ross [5])

$$V_\pi = \frac{E[V_\pi(BC)]}{E[BC]} \; . \tag{3.3}$$

Since the denominator of the right hand side in Equation (3.3) is the same for all π, we may use $E[V_\pi(BC)]$ as a measure of performance. In particular, a discipline π' is optimal if

$$E[V_{\pi'}(BC)] = \underset{\pi}{\text{Max}}\; E[V_\pi(BC)]. \tag{3.4}$$

We now show that FIFO and LIFO disciplines satisfy (3.4) in cases 1 and

2, respectively.

Theorem 1. If $P(\cdot)$ is concave, then the discipline π^* (FIFO) satisfies Equation (3.4) and hence is optimal.

Proof. We will prove that the total reward in a busy cycle is maximized by the FIFO discipline for each realization of the arrival and the service processes. Equation (3.4) will then follow immediately on taking expectations. By a realization of the service process we mean a sequence $\{S_i\}_{i=1}^{\infty}$ where S_i is the service time of the ith customer served. This is assumed to be independent of the discipline. By a realization of the arrival process we mean a sequence $\{a_i\}_{i=1}^{\infty}$ where a_i is the time between i^{th} and $(i+1)^{st}$ arrival. Suppose we have been given sequences $\{S_i\}$ and $\{a_i\}$. These sequences completely determine the lengths of the busy cycles irrespective of the discipline used. Consider the first busy cycle. Suppose at the nth service completion in this cycle the state of the system is (M, Y^M). Consider any discipline π. Since we know the realizations $\{S_i\}$ and $\{a_i\}$ we know the times (counting from this service completion epoch) at which the currently waiting M customers will begin service under π. Suppose the ith oldest customer begins service at time t_i under π. Note that one of the t_i's, say t_k, is zero. That is, k^{th} oldest customer will be selected under π for the next service. Consider a discipline π' which selects the oldest customer at time 0 and begins the service of the current k^{th} oldest customer at time t_1. The services of all other customers start at the same times under the disciplines π and π'. Hence the rewards obtained in the remaining part of the current busy cycle from disciplines π' and π differ by

$$D(\pi',\pi) = P(Y_1^M) + P(Y_k^M + t_1) - P(Y_k^M) - P(Y_1^M + t_1). \qquad (3.5)$$

Since P is concave and $Y_k^M < Y_1^M$, $D(\pi',\pi) \geq 0$. Thus π' is better than π.

In particular, the best current strategy is to select the oldest customer for the next service. The above argument can be easily extended to show that at each service completion the remaining reward is maximized by serving the oldest customer next. This establishes the optimality of FIFO when $P(\cdot)$ is concave.

Theorem 2. If $P(\cdot)$ is convex, then the policy π^{**} (LIFO) is optimal.

Proof. The proof is basically the same as that of Theorem 1. Here, instead of π', we define a discipline π'' which selects the youngest customer for the next service and begins the service of the Kth oldest customer at time t_M. Then

$$D(\pi'',\pi) = P(Y_M^M) + P(Y_K^M + t_M) - P(Y_K^M) - P(Y_M^M + t_M) \qquad (3.6)$$

Since $P(\cdot)$ is convex and $Y_M^M \leq Y_K^M$, $D(\pi'',\pi) \geq 0$. The rest of the argument is now obvious.

Remark 2. From the proofs of Theorems 1 and 2 we can also assert the following:

Suppose at a service completion epoch the waiting time of the youngest customer in the queue exceeds t. If $P(\cdot)$ is concave on (t,∞), then the optimal discipline will select the oldest customer for the next service. On the other hand, if $P(\cdot)$ is convex on (t,∞), then the optimal discipline will select the youngest customer for the next service.

Remark 3. Since Theorems 1 and 2 are proved for each realization, they are valid for a much more general system than the one considered here. In fact, as long as $E[BC] < \infty$ and $E[V_\pi(BC)] < \infty$, Theorems 1 and 2 hold for a $G|G|C$ queue. For a $G|GI|1$ queue we can get the following stronger result:

Let F denote the distribution of the service times and $F^{(n)}$ be its n-fold convolution with itself. If

$$\int_{y=0}^{\infty} [P(t+y) - P(t)]dF^{(n)}(y)$$

is nonincreasing (nondecreasing) for all $t \geq 0$, then FIFO (LIFO) is optimal.

This result follows because t_1 in Equation (3.5) is now necessarily the sum of n independent service times, for some $n \geq 1$.

Remark 4. Suppose that $P(\cdot)$ is convex decreasing and that $\lim_{t\to\infty} P(t) = 0$. Also assume that the system earns zero reward from a customer who does not get served. Then the LIFO discipline is optimal even when $\zeta \geq 1$. For $\zeta \geq 1$, $E_\pi[BC] = \infty$ for all π. Hence the arguments of Theorem 2 do not apply. Therefore, we will use a different approach in this case.

Consider any discipline π. Suppose we observe the state $Z = (M, Y^M)$ at a service completion epoch. Suppose $\{S_i\}_{i=1}^{\infty}$ and $\{a_i\}_{i=1}^{\infty}$ are the sequences of the service times and the interarrival times beginning at the current service completion epoch. Suppose, as in Theorem 2, that the M waiting customers begin their services at times t_1, $t_2, \ldots t_M$ under π. Here $t_i = \infty$ if the ith oldest customer never gets served under π, Suppose $t_k = 0$. That is, the next customer to be served under π is the kth oldest. Let π'' be the discipline which serves the youngest customer next and serves the kth oldest customer at time t_M. The other customers get served at the same times under π and π''. Let $D(\pi'', \pi, t)$ denote the difference between the reward obtained in the next t time units under π'' and that under π. Then

$$\lim_{t\to\infty} D(\pi'', \pi, t) = P(Y_M^M) + P(Y_K^M + t_M) - P(Y_K^M) - P(Y_M^M + t_M), \tag{3.7}$$

where $P(t) = 0$ if $t = \infty$. Once again, since $P(\cdot)$ is convex and decreasing, we have

$$\lim_{t\to\infty} \frac{D(\pi'', \pi, t)}{t} \geq 0. \tag{3.8}$$

It now follows that the optimal action, at each service completion epoch, is to serve the youngest customer next.

The above argument can also be used to show that if P(t) is concave increasing with $\lim_{t\to\infty} P(t) = 0$ and if the server is not allowed to remain idle when a customer is waiting, then FIFO is optimal even when $\zeta \geq 1$. If we allow the server to remain idle when a customer is waiting, then it would be optimal to serve no customer. Thus the situation where $\zeta \geq 1$ and P(t) is concave increasing with $\lim_{t\to\infty} P(t) = 0$ is not of practical relevance.

Remark 5. After the first version of this paper was written, the authors became aware of [6] in which Vasiček proved Theorem 2 for a G|G|C queue in a different context. He does not prove the stronger results of Remarks 3 and 4.

4. Exponential Customer Behavior Function

In this section we compare the FIFO discipline and the LIFO discipline for a special case of our general model. We assume that

$$P(t) = e^{-\alpha t} \quad (t \geq 0) \tag{4.1}$$

for some $\alpha > 0$. We also assume that the service time distribution is exponential with mean $\frac{1}{\mu}$. Let $\zeta = \lambda/\mu$. For this M|M|1 queue we derive expressions for the average reward V_π under FIFO (π^*) and LIFO (π^{**}) disciplines. As shown in Section 3, the LIFO discipline is optimal for the customer behavior function given by (4.1). Therefore, the comparison of V_{π^*} and $V_{\pi^{**}}$ will indicate how much worse is the FIFO discipline compared to the optimal discipline.

Since $0 \leq P(t) \leq 1$ for all t, we can consider P(t) as the probability that a cusotmer is 'good' if his service begins at t time units after his arrival. V_π then represents the steady state rate of serving good customers. We will call it 'goodput' under discipline π.

4.1 'Goodput' under the FIFO Discipline

Suppose $\zeta < 1$. Then under the FIFO discipline the steady state distribution of the waiting time W is given by (See Cooper [2], p. 76)

$$P[W>t] = \zeta e^{-\mu(1-\zeta)t} \quad \text{for} \quad t > 0, \tag{4.2}$$

and

$$P[W=0] = 1 - \zeta . \tag{4.3}$$

Hence

$$V_{\pi*} = \lambda\left[(1-\zeta) + \mu\zeta(1-\zeta)\int_0^\infty e^{-\alpha t}\, e^{-\mu(1-\zeta)t} dt\right] \tag{4.4}$$

$$= \lambda\left[(1-\zeta) + \frac{\mu\zeta(1-\zeta)}{\alpha + \mu(1-\zeta)}\right]$$

$$= \lambda\left[\frac{(\alpha+\mu)(1-\zeta)}{\alpha + \mu(1-\zeta)}\right] .$$

for $\zeta \geq 1$, $V_{\pi*} = 0$. Let

$$TH_{\pi*} = \frac{V_{\pi*}}{\mu} = \frac{\zeta(1-\zeta)(\alpha+\mu)}{\alpha + \mu(1-\zeta)} . \tag{4.5}$$

$Th_{\pi*}$ is called the normalized 'goodput'. It represents the fraction of time the server spends in serving good customers.

4.2 'Goodput' under the LIFO Discipline

Let $0 < \zeta < \infty$. Let $B(\cdot)$ denote the distribution of the busy period in the $M|M|1$ queue. Finally, let $\hat{B}(\cdot)$ be the Laplace Stieltjes transform of $B(\cdot)$. Then

$$P[W=0] = \begin{cases} 1 - \zeta \text{ if } \zeta < 1 \\ 0 \quad \text{if } \zeta \geq 1 , \end{cases} \tag{4.6}$$

and for $t > 0$

$$P[W>t] = \begin{cases} \zeta[1-B(t)] \text{ if } \zeta < 1 \\ [1-B(t)] \text{ if } \zeta \geq 1. \end{cases} \tag{4.7}$$

Hence, for $\zeta < 1$,

$$V_{\pi**} = \lambda[1-\zeta+\zeta\int_0^\infty e^{-\alpha t}dB(t)] \tag{4.8}$$

$$= \lambda[1-\zeta+\zeta\hat{B}(\alpha)].$$

For $\zeta \geq 1$

$$V_{\pi**} = \lambda\hat{B}(\alpha). \tag{4.9}$$

Also

$$TH_{\pi**} = \frac{V_{\pi**}}{\mu} \tag{4.10}$$

$$= \begin{cases} \zeta[1-\zeta+\zeta\hat{B}(\alpha)] & \text{if } \zeta < 1 \\ \zeta\hat{B}(\alpha) & \text{if } \zeta \geq 1, \end{cases}$$

where (see Cooper [2], p. 198)

$$\hat{B}(\alpha) = \frac{\mu(1+\zeta) + \alpha - \sqrt{(\mu(1+\zeta)+\alpha)^2 - 4\zeta\mu^2}}{2\mu\zeta} . \tag{4.11}$$

4.3 Numerical results

We consider two cases:

(1) $\mu = 1.5$, $\alpha = 0.15$

(2) $\mu = 1.5$, $\alpha = 2$

For these cases the values of $TH_{\pi*}$ and $TH_{\pi**}$ as functions of ζ are plotted in Figures 1 and 2. $TH_{\pi*}(\zeta)$ increases with ζ, reaches a maximum and then drastically reduces as ζ approaches 1. It can be shown that the maximum occurs at

$$\zeta = \frac{(\alpha+\mu) - \sqrt{\alpha(\alpha+\mu)}}{\mu} . \tag{4.12}$$

Because of this drastic degradation in performance at heavy loads the FIFO discipline is not desirable for the assumed customer behavior. Under the LIFO discipline, for case (1), $TH_{\pi**}(\zeta)$ monotonically increases with ζ and approaches 1 as $\zeta \uparrow \infty$. For case 2, $TH_{\pi**}(\zeta)$

increases with ζ, reaches a local maximum, then decreases as ζ approaches 1 and once again increases to 1 as $\zeta \uparrow \infty$. It can be shown that this decrease near $\zeta = 1$ occurs if $\alpha > (\sqrt{5}-2)\mu$. Since the LIFO discipline is optimal the performance cannot be improved if we insist on the conditions of our model. However, significant improvements can be achieved if we let the server be idle even if customers are waiting in the queue. To understand this consider the following situation:

Suppose at a service completion the current waiting times of all the customers in the queue exceed t and P(t) is very small. That means that the next customer to be served is, with a large probability, a bad customer. But it will keep the server busy for the duration of its service and the customers arriving during this service duration will have to be delayed thus increasing the probabilities of their turning bad. One way of avoiding this problem is to allow preemption. As explained in Remark 1, however, we then have to include the overhead involved in preemption and define the customer behavior as a complicated function of the initial and the subsequent waiting times of the customer. If preemption is not permitted, then the performance can still be improved by not serving any customer whose waiting time exceeds some number T. This suggests the following discipline.

4.3 LIFO with 'Time Out' Discipline

We select a number $T > 0$ and decide not to serve any customer who has waited more than T time units. This is equivalent to 'timing out' and removing the customer from the queue at T time units after arrival. The other customers are served according to the LIFO discipline. We will denote this LIFO with 'time out' discipline by π_T.

It is shown in Appendix 1 that

$$TH_{\pi_T}(\zeta) = \frac{V_{\pi_T}(\zeta)}{\mu} \tag{4.13}$$

$$= \frac{\zeta[1-\zeta B(T) + \zeta\int_0^T e^{-\alpha t} dB(t)]}{1 + \zeta(1-B(T))} ,$$

Here, the busy period distribution $B(\cdot)$ is given by (see Cooper [2], p. 222)

$$B(t) = \int_0^{\mu t} \frac{e^{-(1+\zeta)x}}{\sqrt{\zeta}} I_1(2x\sqrt{\zeta}) \frac{dx}{x} , \tag{4.14}$$

where I_1 is the Bessel function of the first kind. It is shown in Appendix 1 that, for a fixed ζ, there exists a unique $T = T(\zeta)$ which maximizes $TH_{\pi_T}(\zeta)$ over all T and that $T(\zeta)$ is the unique solution of

$$e^{-\alpha T} - \zeta + \zeta e^{-\alpha T}(1-B(T)) + \zeta\int_0^T e^{-\alpha t} dB(t) = 0. \tag{4.14}$$

Figures 3 and 4 give the values of $TH_{\pi_T}(\zeta)$ as a function of ζ for various values of T for cases 1) and 2) respectively. Also, we reproduce $TH_{\pi**}(\zeta)$ in Figures 3 and 4. Clearly, π_T provides a significant improvement over $\pi**$. The improvement is more significant for case 2 where α is much larger than in case 1.

5. Other Customer Behavior Functions

In Section 3 we showed that the FIFO (LIFO) discipline is optimal when $P(\cdot)$ is concave (convex decreasing). These disciplines are easy to implement because they select customers for service on the basis of the order of their arrival epochs and not on the actual current waiting times. With more complicated customer behavior functions no such optimal discipline may exist. For example, suppose $P(\cdot)$ is given by

$$P(t) = \begin{cases} 1 & \text{if } t \le S. \\ e^{-\alpha(t-S)} & \text{if } t > S. \end{cases} \tag{5.1}$$

Suppose the service times are deterministic with value b. Suppose, at a service completion, there are two customers waiting in the queue with waiting times t_1 and t_2, $t_1 < t_2$. Finally, suppose that ζ is extremely small. Thus there is a very small probability of an arrival during the next service time. Consider three situations:

(i) $t_2 > t_1 > S$. In this case Remark 2 following Theorem 2 indicates that the optimal action is to select customer 1 for the next service.

(ii) $t_1 + b < S < t_2 + b$. In this case customer 2 should be selected for the next service.

(iii) $t_1 < S < t_1 + b < t_2 + b$ with

$$P(t_2) - P(t_2+b) < P(t_1) - P(t_1+b) = 1 - P(t_1+b). \tag{5.2}$$

In this case customer 1 should be selected for the next service.

Thus, no simple discipline based on the order of arrival of the cusotmers can be optimal. For larger values of ζ the problem of finding the best action is even more complicated. In such cases we can only compare 'reasonable' disciplines and select the one with desirable 'goodput' performance.

In this section we consider the customer behavior function defined by (5.1). We again assume that the service times are exponentially distributed with mean $\frac{1}{\mu}$. We look at three disciplines: FIFO (π^*), LIFO(π^{**}) and Mixture (π^U). The Mixture disciplines π^U works as follows:

We set up two queues, Q1 and Q2. Q1 has a non-preemptive priority over Q2. On arrival the customer is put at the back of Q1. Thus Q1 is served according to FIFO discipline. If a customer is in Q1 at time U after its arrival epoch (that is, its service has not begun yet), then it is transferred to the front of Q2. Thus Q2 is served according to the LIFO discipline.

5.1 'Goodput' under the FIFO Discipline

We have for $\zeta < 1$,

$$TH_{\pi*}(\zeta) = \zeta\left[(1-\zeta) + \zeta\mu(1-\zeta)\int_0^\infty e^{-\mu(1-\zeta)t}P(t)dt\right] \tag{5.3}$$

$$= \zeta\left[1-\zeta + \zeta(1-e^{-\mu(1-\zeta)S}) + \frac{\zeta\mu(1-\zeta)}{\mu(1-\zeta)+\alpha}e^{-S\mu(1-\zeta)}\right].$$

For $\zeta \geq 1$, $TH_{\pi*} = 0$.

5.2 'Goodput' under the LIFO Discipline

For $\zeta < 1$,

$$TH_{\pi**}(\zeta) = \zeta\left[1 - \zeta + \zeta\int_0^\infty P(t)dB(t)\right] \tag{5.4}$$

$$= \zeta\left[1 - \zeta + \zeta B(S) + \zeta\int_S^\infty e^{-\alpha(t-S)}dB(t)\right]$$

$$= \zeta\left[1 - \zeta + \zeta e^{\alpha S}\hat{B}(\alpha) + \zeta\int_0^S (1-e^{-\alpha(t-S)})dB(t)\right].$$

For $\zeta \geq 1$

$$TH_{\pi**}(\zeta) = \zeta\left[e^{\alpha S}\hat{B}(\alpha) + \int_0^S (1-e^{-\alpha(t-S)})dB(t)\right]. \tag{5.5}$$

5.3 'Goodput' Under the Mixture Discipline

We first specify the notations required to express the 'goodput' under the Mixture discipline π^U. Let N denote the number of customers arriving in time U. Thus

$$P[N=k] = \frac{e^{-\lambda U}(\lambda U)^k}{k!}. \tag{5.6}$$

Let $B_U(\cdot)$ denote the distribution of the busy period generated by N + 1 customers. Finally, let $W(\cdot)$ denote the distribution of the waiting time of an arbitrary customer. It is shown in Appendix 2 that

$$W(0) = \begin{cases} 1-\zeta & \text{if } \zeta < 1 \\ 0 & \text{if } \zeta \geq 1. \end{cases} \tag{5.7}$$

and

$$\frac{dW(t)}{dt} = \begin{cases} A_o(\zeta)e^{\mu(\zeta-1)t} & \text{if } t \leq U \\ \dfrac{A_o(\zeta)e^{\mu(\zeta-1)U}}{\mu} \dfrac{dB_U(t-U)}{dt} & \text{if } t > U, \end{cases} \tag{5.8}$$

where

$$\frac{1}{A_0(\zeta)} = \begin{cases} \dfrac{1}{\mu\zeta(1-\zeta)}(1-e^{-\mu(1-\zeta)U}) + \dfrac{1}{\mu\zeta}e^{-\mu(1-\zeta)U} & \zeta < 1. \\ \dfrac{1}{\mu(\zeta-1)}(e^{\mu(\zeta-1)U}-1) + \dfrac{1}{\mu}e^{\mu(\zeta-1)U} & \zeta \geq 1. \end{cases} \tag{5.9}$$

If $U \leq S$, then the normalized 'goodput' is given by

$$TH_{\pi^U}(\zeta) = \zeta[1 - \zeta + \int_0^\infty W'(t)P(t)dt] \tag{5.10}$$

$$= \zeta\Big[1 - \zeta + \frac{Ao(\zeta)}{\mu(1-\zeta)}(1 - e^{-\mu(1-\zeta)U})$$

$$+ \frac{Ao(\zeta)e^{-\mu(1-\zeta)U}}{\mu}(B_U(S-U) + \int_{S-U}^\infty e^{-\alpha(t-S+U)}dB_U(t))\Big]$$

for $\zeta < 1$, and

$$TH_{\pi^U}(\zeta) = \zeta\int_0^\infty W'(t)P(t)dt \tag{5.11}$$

$$= \zeta\Big[\frac{Ao(\zeta)}{\mu(\zeta-1)}(e^{\mu(\zeta-1)U}-1)$$

$$+ \frac{Ao(\zeta)e^{\mu(\zeta-1)U}}{\mu}(B_U(S-U) + \int_{S-U}^\infty e^{-\alpha(t-S+U)}dB_U(t))\Big]$$

for $\zeta \geq 1$. On the other hand if $U > S$, then

$$TH_{\pi^U}(\zeta) = \Big[1 - \zeta + \frac{Ao(\zeta)}{\mu(1-\zeta)}(1 - e^{-\mu(1-\zeta)S}) \tag{5.12}$$

$$+ \frac{e^{\alpha S}Ao(\zeta)}{\mu(1-\zeta+\alpha/\mu)}(e^{-\mu(1-\zeta+\alpha/\mu)S} - e^{-\mu(1-\zeta+\alpha/\mu)U})$$

$$+ e^{-\alpha(U-S)}\hat{B}(\alpha)e^{-\lambda U(1-B(\alpha))}\, \underline{e}^{-\mu(1-\zeta)U}\, Ao(\zeta)\Big]$$

for $\zeta < 1$, and

$$
\begin{aligned}
TH_{\pi^U}(\zeta) = \zeta A_0(\zeta)\Bigg[&\frac{1}{\mu(\zeta-1)}\left(e^{\lambda(\zeta-1)S}-1\right) \\
&+ \frac{e^{\alpha S}}{\mu(\zeta-1-\alpha/\mu)}\left(e^{\mu(\zeta-1-\alpha/\mu)U}-e^{\mu(\zeta-1-\alpha/\mu)S}\right) \\
&+ e^{-\alpha(U-S)}\hat{B}(\alpha)e^{-\lambda U(1-\hat{B}(\alpha)}\frac{e^{-\mu(\zeta-1)U}}{\mu}\Bigg]
\end{aligned}
\tag{5.13}
$$

for $\zeta \geq 1$.

5.4 Numerical Results

Once again we calculate $TH_{\pi^U}(\zeta)$ for various values of ζ and U and for

(i) $\mu = 1.5$, $\alpha = 0.15$, $S = 1.0$

(ii) $\mu = 1.5$, $\alpha = 2.0$, $S = 1.0$.

We also calculate $TH_{\pi^*}(\zeta)$ and $TH_{\pi^{**}}(\zeta)$ for these cases. The results are plotted in Figures 5, 6 and 7. Figures 5 and 6 give the 'goodput' for the FIFO, LIFO and the mixture with U = S. Once again it is clear that the FIFO discipline is not desirable at heavy loads. The mixture discipline π^S provides uniformly better performance over the LIFO discipline for all values of ζ. The improvement is more significant for $\alpha = 2.0$ than for $\alpha = 0.15$. The 'goodputs' under the mixture discipline for various values of U are given in Figure 7. Note that the mixture discipline π^U with U = 0 is the same as the LIFO discipline π^{**}. For each value of ζ the 'goodput' increases as U increases, reaches a maximum and then decreases as $U \uparrow \infty$. The maximum 'goodput' occurs at some $U = U(\zeta) > 1$. $U(\zeta)$ decreases with ζ and approaches 1 as $\zeta \uparrow \infty$. Thus π^U for $U \leq S$ provides uniformly better performance than π^{**} over the entire range of ζ. For $U > S$, π^U is better than π^{**} for some values of ζ and worse for the other values of ζ. Ideally, therefore, we would like a dynamic mixture discipline with U depending on the load. Failing that we should select the value of U at or near S. Since some uncertainty always exists in our knowledge of the customer behavior, the

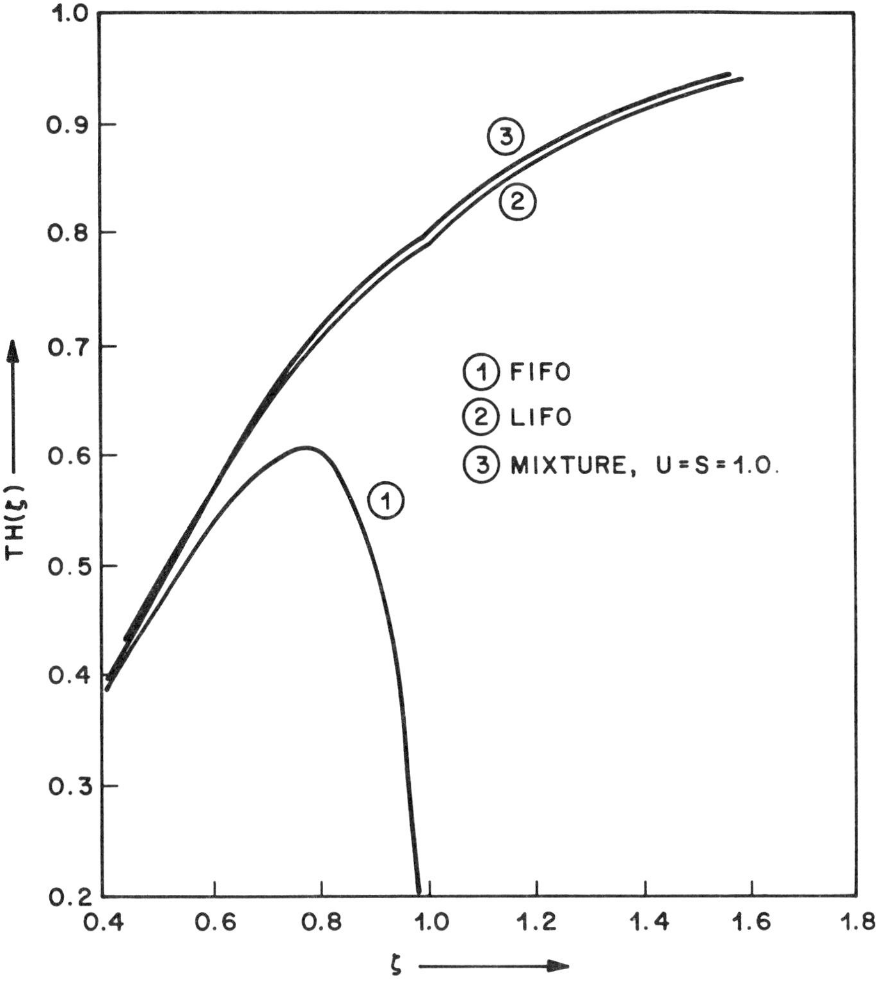

FIGURE 5: TH(ζ) FOR FIFO, LIFO AND MIXTURE.

$$\mu = 1.5,\ \alpha = 0.15,\ U = S = 1.0,\ P(t) = \begin{cases} 1 & t \leq S = 1.0 \\ e^{-\alpha(t-S)} & t > S. \end{cases}$$

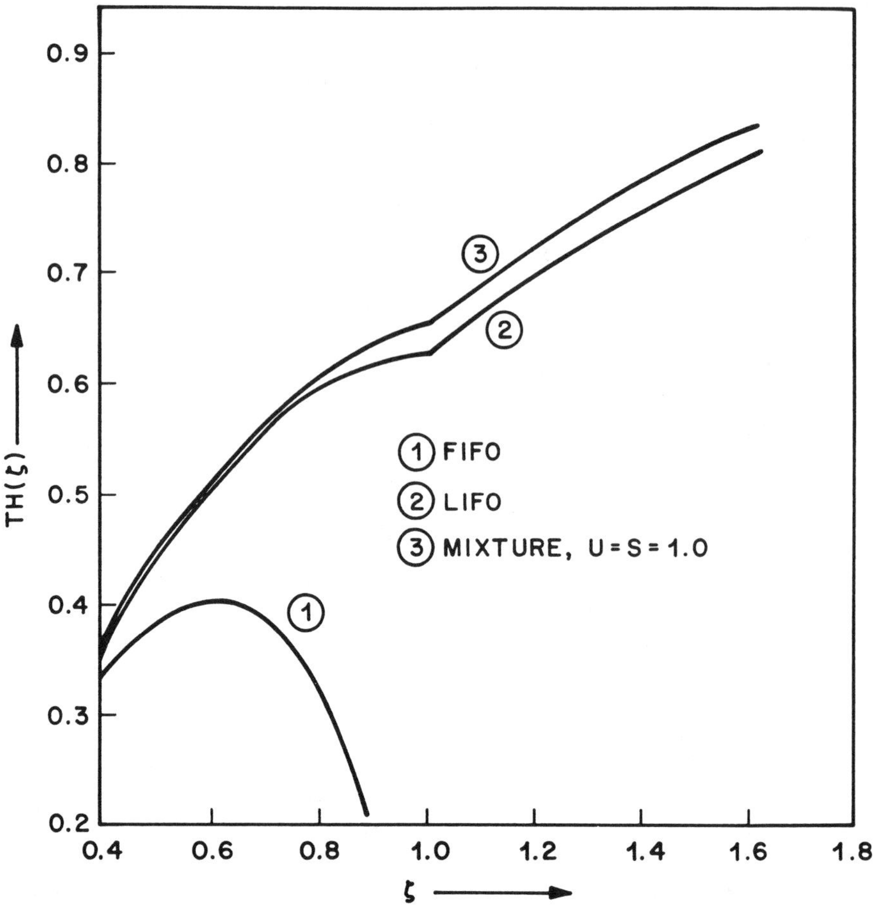

FIGURE 6: TH(ζ) FOR FIFO, LIFO AND MIXTURE

$$\mu = 1.5, \quad \alpha = 2.0, \quad U = S = 1.0, \quad P(t) = \begin{cases} 1 & t \leq S = 1.0 \\ e^{-\alpha(t-S)} & t > S. \end{cases}$$

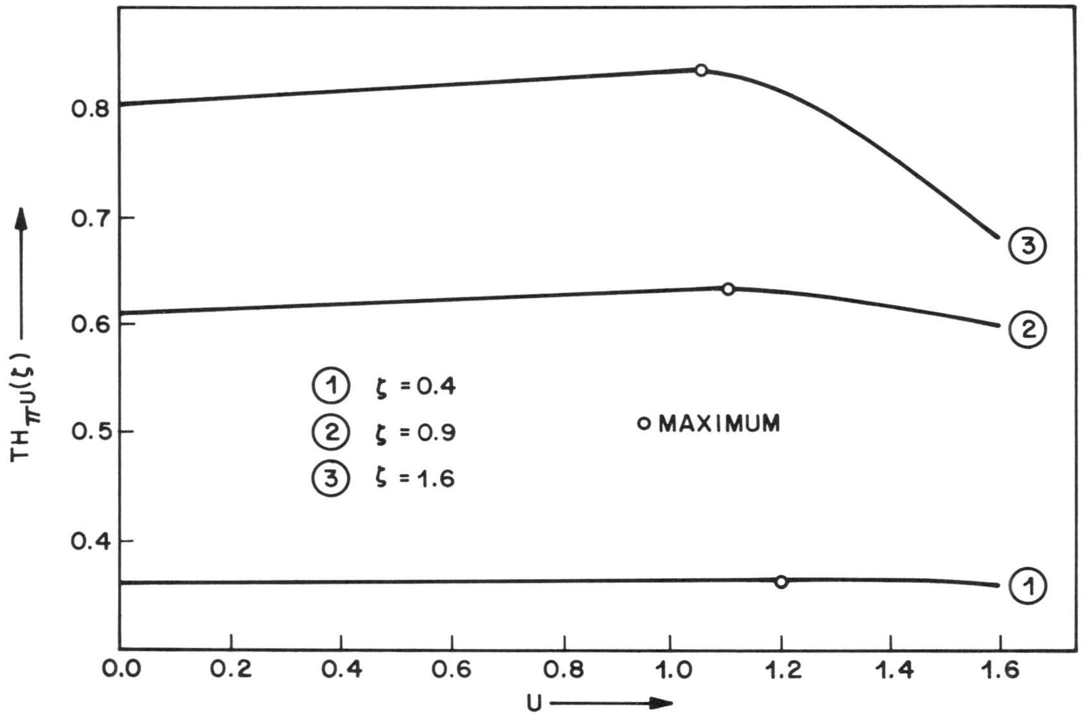

FIGURE 7: $TH_{\pi U}(\zeta)$ FOR VARIOUS VALUES OF U AND ζ.

$$\mu = 1.5, \quad \alpha = 2.0, \quad P(t) = \begin{cases} 1 & t \le S = 1.0 \\ e^{-\alpha(t-S)} & t > S. \end{cases}$$

results in Figure 7 suggest that we should use U equal to or slightly below S.

6. Acknowledgement

The authors would like to thank S. Stidham and A. Lemoine for bringing reference [6] to their attention. Thanks are also due to the referee for a couple of useful suggestions.

7. Appendix 1

7.1 'Goodput' for the LIFO with Time Out Discipline

In this Appendix we obtain an expression for the 'goodput' from the LIFO with 'time out' discipline π_T. We also characterize the value of T which maximizes the 'goodput' over all possible values of T.

We first obtain the waiting time distribution of an arbitrary customer. Let WT denote the waiting time of an arbitrary customer and

$$P[R] = P[\text{Time out}] = P[WT > T] = 1 - W(T), \tag{7.1}$$

$$P[NR] = P[\text{getting served}] = P[WT \leq T] = W(T), \text{ and for } t \leq t \tag{7.2}$$

$$W(t) = P[WT \leq t]. \tag{7.3}$$

Let P_o and P denote the steady state probabilities of the server being idle and busy, respectively. Then

$$W(t) = P_o + PB(t), \tag{7.4}$$

$$P[R] = P[1 - B(T)], \tag{7.5}$$

and

$$P[NR] = P_o + PB(T). \tag{7.6}$$

Note that $\lambda P[NR]$ is the rate at which the customers are served. But this rate also equals μP. Thus

$$\lambda(P_o + PB(T)) = \mu P, \tag{7.7}$$

and

$$P_o + P = 1. \tag{7.8}$$

Solving (7.7) and (7.8) we get

$$P = \frac{\zeta}{1 + \zeta(1-B(T))}, \tag{7.9}$$

$$P_o = \frac{1 + \zeta B(T)}{1 + \zeta(1-B(T))}, \tag{7.10}$$

$$P[R] = \frac{\zeta(1-B(T))}{1 + \zeta(1-B(T))}, \tag{7.11}$$

$$P[NR] = \frac{1}{1 + \zeta(1-B(T))}, \tag{7.12}$$

and for $0 < t \leq T$

$$W(t) = \frac{1 - \zeta(B(T)-B(t))}{1 + \zeta(1-B(T))}. \tag{7.13}$$

Note that

$$\begin{aligned} TH_{\pi_T}(\zeta) &= TH(T,\zeta) \\ &= \zeta[W(o) + \int_0^T W'(t)P(t)dt] \\ &= \frac{\zeta[1- B(T)+\zeta\int_0^T e^{-\alpha t}dB(t)]}{1 + \zeta(1-B(T))}. \end{aligned} \tag{7.14}$$

Let $\bar{B}(t) = 1 - B(t)$. Then

$$\begin{aligned} \frac{\partial TH(T,\zeta)}{\partial T} &= \frac{\zeta}{(1+\zeta\bar{B}(T))^2} \\ &\quad \times [(1+\zeta\bar{B}(T))(-\zeta B'(T)+\zeta e^{-\alpha T}B'(T)) \\ &\quad + \zeta B'(T)(1-\zeta B(T) + \zeta\int_0^T e^{-\alpha t}dB(t))] \end{aligned} \tag{7.15}$$

Let $N(T,\zeta)$ denote the quantity within the squared brackets in (7.15). Then

$$\frac{N(T,\zeta)}{\zeta B'(T)} = e^{-\alpha T} - \zeta + \zeta e^{-\alpha T}\bar{B}(T) + \zeta\int_0^T e^{-\alpha t}dB(t). \tag{7.16}$$

The right hand side of (7.16) is decreasing in T. It equals 1 at T = 0 and it approaches

$$-\zeta + \zeta\int_0^\infty e^{-\alpha t}dB(t) < 0$$

as $T \to \infty$. So there is a unique value of T for which $\frac{\partial TH(T,\zeta)}{\partial T} = 0$. This value of T is the unique solution of

$$e^{-\alpha T} - \zeta + \zeta e^{-\alpha T}\bar{B}(T) + \zeta\int_0^T e^{-\alpha t}dB(t) = 0. \tag{7.17}$$

8. Appendix 2

8.1 Waiting Time Distribution for the Mixture Discipline π

In this appendix we derive the distribution of the waiting time of an arbitrary customer for the Mixture discipline π^U. Let

$$P[NR] = P[\text{customer gets served from Q1}], \tag{8.1}$$

$$P[R] = P[\text{customer gets transferred to Q2}].$$

Let $W(\cdot)$ denote the distribution of the waiting time of an arbitrary customer. Then, for $t > U$

$$W'(t) = P[R]B_U'(t-U). \tag{8.2}$$

In order to specify the waiting time distribution completely we now need P[R], P[NR] and the conditional distribution of the waiting time given that the customer is served from Q1.

The behavior of the customers in Q1 is equivalent to the one in the following queueing system:

Let VT(t) denote the load (virtual waiting time) in the system consisting of the server and Q1. If, at the arrival epoch of a customer, we have $VT \leq U$, then this customer joins Q1. If $VT > U$, then the arriving customer joins Q2. Thus Q1 behaves like an $M|M|1$ queue with customers reneging at time U after arrival. However, there is an important difference. In the M|M|1 queue the load VT becomes zero when the queue is empty at a service completion and remain zero until the next arrival epoch. In our queueing system, when Q1 is empty at a service completion epoch and at least one customer is present in Q2, a customer from Q2 enters the service and VT jumps up by a service time. If Q2 is empty when VT reaches zero, then VT remains zero until the next arrival epoch.

Let $V(\cdot)$ denote the steady state distribution of VT. Then

$$P[R] = 1 - V(U), \tag{8.3}$$

$$P[NR] = V(U), \tag{8.4}$$

and for $t \leq U$

$$W(t) = V(t). \tag{8.5}$$

So our problem reduces to that of finding $V(\cdot)$. We first derive an integral equation for $f(x) = \frac{dV(x)}{dx}$, $\mu > 0$. Let $D(x)$ denote the average number of downcrossings of level x per unit time by the process $VT(\cdot)$ in the steady state. Similarly, let $U(x)$ denote the number of upcrossings of state x per unit time by the process $VT(\cdot)$ in the steady state. In the steady state we have

$$D(x) = U(x) \text{ for } \mu > 0. \tag{8.6}$$

Brill and Posner [1] have showed that

$$f(x) = D(x) \text{ for } x > 0. \tag{8.7}$$

Let J denote the number of customers served from Q2 per unit time in the steady state. Then

$$J = \begin{cases} \lambda P[R] \text{ for } \zeta \leq 1 \\ f(o^+) \text{ for } \zeta > 1. \end{cases} \tag{8.8}$$

Also, U(x) is given by

$$U(x) = \lambda\int_0^{x\wedge U} f(y)e^{-\mu(x-y)}dy + \lambda V(0)e^{-\mu x} + Je^{-\mu x} \tag{8.9}$$

Equations (8.6), (8.7) and (8.9) give the desired integral equation

$$f(x) = \lambda\int_0^{x\wedge U} f(y)e^{-\mu(x-y)}dy + \lambda V(0)e^{-\mu x} + Je^{-\mu x} \tag{8.10}$$

On differentiating (8.10) we get

$$f'(x) = \begin{cases} (\lambda-\mu)f(x) & 0 < x < U \\ -\mu f(x) & x > U. \end{cases} \tag{8.11}$$

Thus for some constant A_o

$$f(x) = \begin{cases} A_o e^{(\lambda-\mu)x} & 0 < x < U \\ A_o e^{\lambda U-\mu x} & x \geq U. \end{cases} \tag{8.12}$$

We also have

$$V(0) + \int_0^{\infty} f(x)dx = 1. \tag{8.13}$$

For $\zeta \geq 1$, $V(0) = 0$. Hence the condition (8.13) gives

$$\frac{1}{A_o} = \frac{1}{\lambda-\mu}(e^{(\lambda-\mu)U}-1) + \frac{1}{\mu}e^{(\lambda-\mu)U} \tag{8.14}$$

$$= \frac{1}{\mu(\zeta-1)}(e^{\mu(\zeta-1)U}-1) + \frac{1}{\mu}e^{\mu(\zeta-1)U}.$$

For $\zeta < 1$, $V(0) = 1 - \zeta$ and

$$\frac{1}{A_o} = \frac{1}{\mu\zeta(1-\zeta)}(1-e^{-\mu(1-\zeta)U}) + \frac{1}{\mu\zeta}e^{-\mu(1-\zeta)U}. \tag{8.15}$$

From (8.5) and (8.12) we have

$$W(0) = \begin{cases} 1-\zeta & \text{if } \zeta < 1 \\ 0 & \text{if } \zeta \geq 1, \end{cases} \tag{8.16}$$

and for $0 < t < U$

$$W'(t) = A_o e^{(\lambda-\mu)t} = A_o e^{\mu(\zeta-1)t}. \tag{8.17}$$

Also,

$$P[R] = 1 - V(U)$$

$$= \frac{A_o}{\mu} e^{(\lambda-\mu)U} = \frac{A_o}{\mu} e^{\mu(\zeta-1)U}.$$

Thus, for $t > U$

$$W'(t) = \frac{A_o}{\mu} e^{\mu(\zeta-1)U} B_U'(t-U). \tag{8.18}$$

9. References

[1] Brill, P. H. and Posner, M. J. M. (1977), "Level Crossings in Point Processes Applied to Queues: Single Server Case," Oper. Res., 25, 662-674.

[2] Cooper, R. B. (1972), Introduction to Queueing Theory, MacMillan, New York.

[3] Forys, L., Fredericks, A. A. and Kersten, P., "Improving the Overload Performance of Local Switching System," to be submitted to Bell System Tech. J.

[4] Forys, L., Fredericks, A. A., Hejny, G. and Borchering, J., "Coping with Overloads," Bell Labs Record, July-August, 1981.

[5] Ross, S. (1970), "Average Cost semi-Markov Decision Processes," J. Appl. Probab., 7, 649-656.

[6] Vasiček, O. A. (1977), "An Inequality for the Variance of Waiting Time Under a General Queueing Discipline," Oper. Res., 25, 879-884.

Bell Laboratories, Holmdel, New Jersey 07733.

Discussant's Report on "Comparisons of Service Disciplines in a Queueing System with Delay Dependent Customer Behaviour," by Bharat T. Doshi and Edward H. Lipper

The problem considered by the authors, in which customers may turn "bad" after a certain random length of time in the queue, is interesting and new, at least to me. The authors have modelled this problem in the context of an M/G/1 queue, in which a customer whose waiting time in the queue is t has a probability P(t) of being "good". The problem is thus seen to be a special case of a more general problem in which a customer who spends t time units in the queue earns a reward P(t). The objective is to choose a service discipline that minimizes the steady-state expected reward per unit time.

One of the main results in the paper is a proof under mild regularity conditions that the FIFO discipline is optimal (among non-preemptive, work-conserving, service-time independent disciplines based on the elapsed times in queue of the customers present at a service completion) if P(t) is concave, and that the LIFO discipline is optimal if P(t) is convex decreasing. The proofs are based on (1) using the renewal-reward theorem to express the steady-state mean reward rate as the ratio of the expected reward during a busy cycle to the expected length of a busy cycle, and (2) using a sample-path argument to show that the actual total reward during a busy cycle is maximized by the appropriate discipline (FIFO, if P(t) is concave; LIFO, if P(t) is convex). Since the expected busy-cycle length is discipline-invariant, the desired result then follows from (1) and (2). The sample-path argument involves pairwise interchange of the order of service and is reminiscent of familiar proofs in queueing theory, for example, the proof of the optimality of the shortest-remaining-processing-time discipline (e.g., by Conway, Maxwell, and Miller).

The authors observe, correctly, that by its very nature their sample-path argument is valid for much more general systems than M/G/1, since it does not depend on distributional assumptions. In fact, if we reformulate the problem in the setting of cost minimization rather than profit maximization, we have the following result, which is, I think, part of the folklore:

Let C(t) be the waiting cost incurred by a customer whose waiting time in the queue is t time units. If C(t) is convex (concave), then the customer average waiting cost is minimized by the FIFO (LIFO) discipline, for any realization of the input (interarrival and service times).

I have not seen a formal proof of this result in the literature, but I would not be surprised if there is one somewhere. (Vasicek [6] has a steady-state probabilistic version for G/G/c systems.) But it is "obvious" that a sample-path, interchange argument like that used by the authors would be the preferred method of proof, for both its robustness and its intuitive appeal.

For the special case of negative exponential (and therefore convex decreasing) reward function P(t) and exponential service-time distributions, the authors derive expressions for the average reward under FIFO and LIFO and make numerical comparisons. A variant of LIFO, in which the server remains idle if all customers in the queue have been waiting longer than some prescribed length of time T, is also studied. This "time-out" discipline avoids inefficiencies that arise in pure LIFO under heavy traffic. Finally, a simple reward function that is neither convex nor concave is studied. In this case no discipline based just on the order of arrival is optimal.

A final observation:
For the original problem where P(t) is the probability that a customer

with waiting time t is good, P(t) cannot be decreasing if it is concave. But in practice the goodness probability normally would be decreasing. Thus in this case P(t) cannot be concave and thus the FIFO discipline would probably not perform well in general.

Discussant: Dr. Shaler Stidham, Jr., Department of Industrial Engineering, Box 5511, North Carolina State University, Raleigh, North Carolina 27650.

PROBABILISTIC SCHEDULING
John Bruno, Chairman

M. Pinedo
R. R. Weber
H. G. Badr, I. Mitrani,
& J. R. Spirn

ON THE OPTIMAL ORDER OF STATIONS IN TANDEM QUEUES

Michael Pinedo

Abstract

Consider m nonidentical service stations which have to be set up in a sequence $j_1, j_2, \ldots, j_m$. At time t = 0 the system is empty and there is an infinite number of customers waiting in front of station j_1. Customers have to be served first at station j_1, after finishing service there they have to go through station j_2, etc. Both the cases of infinite waiting room and of zero waiting room between the stations are considered; in the case of zero waiting room, blocking may occur. We study how the output process depends on the sequence in which the stations are set up.

1. Introduction and Summary.

Consider m non-identical stations and an infinite number of customers. The time it takes station h to serve customer i is a random variable S_i^h with distribution F_h. These m stations have to be set up in a sequence $j_1, \ldots, j_m$. A customer has to go first through station j_1, after finishing its service there it has to go through station j_2, etc. At time t = 0 the system is empty, an infinite number of customers are present in front of station j_1 and the process starts. Both the cases of infinite waitimg room and of zero waiting room between the stations are considered. In case of infinite waiting room customer i when finished on machine j_k will, if station j_{k+1} is busy serving a preceding customer, wait in the room in front of station j_{k+1}. The case of zero waiting room can cause what is called blocking: When customer i is being served at station j_k while customer i+1 has finished its service at station j_{k-1}, customer i+1 may not leave station j_{k-1}, preventing subsequent customers (i+2,i+3,...) from moving on.

We are interested in how the output process is affected by the sequence $j_1,\ldots,j_m$ in which the stations are set up, Especially the output rate in steady state will be considered, i.e., the reciprocal of the mean interdeparture time. This will be called the *capacity*. In the case of infinite waiting room between the stations it is clear that the capacity does *not* depend on the sequence of the stations; the capacity of the system is then simply determined by the station with the largest expected service time, i.e., the slowest station. Nevertheless, the station sequence still has some effect on the output process and the problem of interest now is to find the sequence which minimizes the expected time customer i, i = 1,2,..., leaves the system. An overview of the available literature on tandem queues with and without blocking is given in Section 2.

In Section 3 the model with infinite waiting room between the stations is considered. In this section it is shown that optimal sequences and sequences which are near optimal should have so-called "bowl" shapes. This means that, if the stations can be ordered according to their expected service times, a sequence which orders the stations initially in increasing order of their expected service times and afterwards in decreasing order should be optimal or near optimal. Such a sequence suggests the shape of an upside down bowl. When the expected service times of the stations are identical and the stations can be ordered according to their variance, a sequence which orders the stations initially in decreasing order of their variance and afterwards in increasing order of their variance should be optimal or near optimal; again the sequence has a bowl shape, but now based on the variance.

In Section 4 we consider the case where there is no waiting room between the stations. Three conditions are discussed under which the capacity does *not* depend on the sequence.

In Section 5 we again consider the case where there is no waiting room between the stations. We first consider a case where all stations have the same expected service time, but only two of them have a positive variance i.e., are nondeterministic. It is shown, under some special conditions on the distributions, that the capacity of the system is maximized when either one of these two is put in front of the sequence and the remaining one at the end of the sequence. A second case is considered where m-2 stations have identical service time distributions and the two remaining stations have larger expected service times. It is shown, again under some special conditions on the distributions that either one of the two stations with larger expected service times has to be put at the beginning of the sequence and the other one at the end of the sequence. These two special results in Section 5 do not give enough indication of what shapes optimal sequences would have in a more general problem.

In the last section we discuss the results obtained and state some conjectures. Although we have not much evidence we believe that in the zero waiting room case optimal sequences have "saw-tooth" shapes. This means that if the stations can be ordered according to their expected service times one should *alternate* between stations with large expected service times and small expected service times, while putting at the beginning as well as at the end of the sequence stations with large expected service times. Such a sequence suggests the shape of a "saw-tooth". When the expected service times of the stations are identical and the stations can be ordered according to their variance, a sequence which alternates between stations with large variances and stations with small variances, while putting at the beginning as well as at the end of the sequence stations with large variances, should be optimal or near optimal.

We believe that the results obtained in this paper give an indication of how these tandem queues behave under heavy traffic, as under heavy traffic the average number of customers waiting in front of the first station tend to be quite large.

The following terminology will be used throughout: Random variable A is said to be stochastically larger than random variable B, when $P(A \geq t) \geq P(B \geq t)$ for all t.

2. The Literature

In the literature these models have been considered repeatedly. Usually these models were indeed described as tandem queues but once in awhile similar models came up under the name of Flow Shops where stations were called machines or processors and customers were called jobs or tasks.

Avi-Itzhak [1] considered the zero waiting room case where the customers are not all available at time $t = 0$, but come in according to an arbitrary arrival process. He showed that when station $h, h=1,\dots,m$, has a deterministic service time D_h, the departure epochs of the customers do not depend on the sequence in which the stations are set up. Tembe and Wolff [6] treated the infinite waiting room case. Again the customers were assumed to come in according to an arbitrary arrival process. They showed that with two stations, one with a deterministic and the other with a stochastic service time distribution, the departure epoch of customer $i, i=1,2,\dots,$ is stochastically minimized when the deterministic station is set up in front. They also considered the case of m stations with *nonoverlapping* service time distributions, i.e., for any two stations h and k $P(S_i^h \geq S_i^k)$ is either zero or one. They showed that the departure epoch of customer i,i=1,2,..., is stochastically minimized, if the customers have to go first through the station with the largest expected service time, then through the station with the second largest expected service time, etc. Dattatreya [2] and Muth [4] both considered the zero waiting room case where all customers are present at time t = 0. They both showed, independently, that the capacity of the system when setting up the stations in sequence $j_1, j_2, \ldots, j_m$ is the same as when setting up the stations in sequence $j_m, j_{m-1}, \ldots, j_1$. In fact, they proved a much stronger result: The output process of a particular sequence is probabilistically equivalent to the output process of the reverse sequence. Dattatreya [2] moreover showed that, when there are two stations, one with a deterministic and one with a stochastic service time distribution, no waiting room in between and an arbitrary arrival process, the departure epoch of customer i,i=1,2,..., is stochastically smaller when the deterministic station is set up in front. Also, for m stations with nonoverlapping service time distributions he showed that the departure epoch of customer i,i=1,2,..., is stochastically minimized if the customers have to go first through the station with the largest expected service time, then through the station with the second largest expected service time, etc.

Pinedo [5] looked at a Flow Shop model with m identical machines and n nonidentical tasks, all tasks being available at time t = 0. The processing times of task i on the different machines were taken as independent draws from distribution G_i. Both the cases of no waiting room and infinite waiting room between the machines were considered.

The problem of interest was to find the sequence in which the tasks have to go through the machines in order to minimize the completion time of the last task stochastically. For the case of infinite waiting room it was shown that when two tasks have stochastic processing time distributions, not necessarily identical distributions, and the remaining $n-2$ tasks have deterministic distributions, again not necessarily identical, then the completion time of all tasks (the so-called makespan) is stochastically minimized by any sequence which puts either one of the stochastic tasks first in the sequence and the other stochastic task last. In [5] a sequence $j_1,\ldots,j_n$ was called a SEPT-LEPT sequence if there exists an i such that $E(X_{j_1}) \leq E(X_{j_2}) \leq \cdots \leq E(X_{j_i})$ and $E(X_{j_i}) \geq E(X_{j_{i+1}}) \geq \cdots \geq E(X_{j_n})$. It was shown that when the processing time distributions of the tasks are nonoverlappingly distributed, a sequence is optimal if it is a SEPT-LEPT sequence, i.e., if it has a bowl form. The case of no waiting room was also considered in this paper. For this case it was shown that if the processing times are nonoverlapping, a sequence is optimal *if and only if* it is SEPT-LEPT. The conclusion in [5] (based on the results just mentioned and results for more special cases) was that both for the zero and infinite waiting room case the following rule of thumb holds: Tasks with a shorter expected processing time have to be scheduled more toward the beginning and toward the end, while tasks with a larger expected processing time have to be scheduled more toward the middle of the sequence. The variance in the processing times of the tasks has the following influence: Tasks with a larger variance have to be scheduled more toward the beginning and toward the end, while tasks with a smaller variance have to be scheduled more toward the middle of the sequence. So in [5] sequences with a bowl form appeared to be very important. In [5] was also observed that optimal sequences in the case of no waiting room are similar to optimal sequences in the case of infinite waiting room.

3. The Infinite Waiting Room Case

In this section we consider the infinite waiting room case and will look for sequences of stations which stochastically minimize the departure epoch of customer $i, i=1,2,\dots$. In [5] a duality relationship was pointed out between the case of m different tasks on n identical machines and n identical customers at m different stations. This was done by a simple interchange between machines and tasks (see Lemma 1 in [5]). Using this duality relationship and the results for the model discussed in [5] we can analyze easily the following cases. Consider a tandem queue where two stations have stochastic service time distributions, not necessarily identical, and $m-2$ stations have deterministic service time distributions, again not necessarily identical.

Theorem 1. The departure epoch of customer $i, i=1,2,\dots$, is stochastically minimized when the customers go first through one of the stochastic stations, after that through the $m-2$ deterministic stations, which may be set up in any order, and finally through the second stochastic station.

Proof. Similar to the proof of Theorem 1 in [5]. □

Note that Theorem 1 does *not* state that the sequences with the stochastic stations at the beginning and at the end are the only sequences that stochastically minimize the customers departure epochs.

Consider now the case where the m stations have service time distributions which are nonoverlapping.

Theorem 2. The departure epoch of customer $i, i=1,2,\dots$, is stochastically minimized if the station sequence is a SEPT-LEPT sequence.

Proof. Similar to the proof of Theorem 2 in [5]. □

Again, note that Theorem 2 does *not* state that SEPT-LEPT sequences are the only sequences that minimize the customers departure epochs.

So although the capacity of the system is not affected by the sequence of the stations, it is clear that certain sequences are more advantageous than others. In general, we believe that in case of infinite intermediate storage, in order to minimize the departure epoch of each customer stochastically, stations with longer expected service times and smaller variances in the service times should be set up more toward the middle of the sequence and stations with shorter expected service times and larger variances in the service times should be set up more toward the beginning and toward the end of the sequence.

4. Zero Waiting Room Cases Where the Capacity Does not Depend on the Sequence

Avi-Itzhak [1] showed that if all stations have deterministic service time distributions, not necessarily identical, the sequence in which the stations are set up does not affect the departure epochs of the customers. Of course, the capacity does not depend then on the sequence either. In this section we consider two more cases where the capacity of the system remains the same when the sequence of the stations is altered.

Consider the case of $m-1$ identical stations with deterministic service times and one station with arbitrary stochastic service time distributions. From Dattatreya's results [2] follows that for an arbitrary arrival process the station sequence which has the stochastic station at the end stochastically minimizes the departure time of customer i. From the Reversibility Property follows that in the case where all customers are present at time $t=0$, there are (at least) two sequences which minimize the departure time of customer i stochastically: The one which puts the stochastic station last and the one which puts the stochastic station first. Now consider the capacity of this system.

Theorem 3. When $m-1$ stations are identical and have deterministic service times and one station has an arbitrary stochastic service time distribution the capacity does not depend on the sequence.

Proof. As there are no waiting rooms, the capacity can be determined by examining the arrivals or departures from any station in the sequence. Consider the arrivals at the stochastic station with $m \geq 2$. Assume that customer i, $i \geq 2$, has just entered the stochastic station. Independent of the sequence, the time until the next arrival at the stochastic station is $\max\{1, S_i\}$ where S_i is the service time at the stochastic station. Hence, the capacity does not depend on the sequence. □

Now consider the case where we have m stations with nonoverlapping service time distributions. Dattatreya showed that when the customers come in according to an arbitrary arrival process, the stations should be ordered in decreasing order of their expected service times in order to minimize stochastically the departure time of customer $i, i=1,2,\ldots$. When all customers are available at time $t=0$, it can be shown easily that any SEPT-LEPT sequence minimizes stochastically the departure epoch of customer $i, i=1,2,\ldots$. This can be generalized as follows: Let the service time of one particular station be longer than the service time of any other station with probability one. This implies that the service time distribution of this station does not overlap with the service time distribution of any other station.

Theorem 4. The capacity of a system where the service time of one particular station is larger than the service time of any other station with probability one does not depend on the sequence. The capacity of the system is the reciprocal of the expected service time of the slowest station.

Proof. Let station j_ℓ be the slowest in the sequence $j_1,\ldots,j_m$. When customer i is being served at station j_ℓ customer i cannot be blocked by preceding customers in subsequent stations; this implies that customers i-1,i-2,..., which are being served respectively at station $j_{\ell+1},j_{\ell+2},\ldots$, will finish their service before customer i finishes its service at station j_ℓ. So what occurs at stations $j_{\ell+1},\ldots,j_m$ does *not* affect the process at stations $j_1,\ldots,j_\ell$. Now consider sequence $j_\ell,j_{\ell-1},\ldots,j_1$ with a departure process which is probabilistically equivalent to the departure process of $j_1,\ldots,j_\ell$. But in sequence $j_\ell,\ldots,j_1$ a customer served at station j_ℓ never can be blocked by preceding customers served at subsequent stations. So the rate at which the customers go through the system is the reciprocal of the expected service time of station j_ℓ. □

We may conclude from the above that the capacity does not depend on the sequence if one or more of the following statements hold:

(i) All stations have deterministic service times but are not necessarily identical,

(ii) All but one station have deterministic service times and are identical,

(iii) The service time of one station is larger than the service times of any other station with probability one.

5. Zero Waiting Room Cases Where the Capacity Does Depend on the Sequence

In this section the following notation will be used:

S_i^h = service time of customer i at station h.

B_i^h = total time customer i occupies station h.

T_i^h = time epoch customer i enters station h.

$X_i^h = T_i^1 - T_{i-h+1}^h$

In Theorem 3 it was shown that with one stochastic station and m - 1 identical deterministic stations the system capacity is independent of the sequence. Consider now two stochastic stations, not necessarily identical, but both with a mean service time of one, and m - 2 identical deterministic stations, also with mean one. For the two stochastic stations distributions will be considered with density functions which are symmetric around the mean; so f(1+x) = f(1-x) where f is the density function. Examples of distributions which satisfy these conditions are (see Figure 1 in [5]):

(i) the Uniform distribution,

(ii) the Normal distribution truncated at 0 and at 2.

Theorem 5. The capacity of a system with m - 2 identical deterministic stations with service times of one time unit and 2 nonidentical stochastic stations, both with mean one and symmetric density functions, is maximized when either one of the stochastic stations is set up at the beginning and the other one at the end of the sequence.

Proof. The proof is organized as follows. We show:

(i) the theorem for m = 3,

(ii) that the capacity increases in m when using the sequence of the theorem,

(iii) the theorem for arbitrary m.

(i) Consider one deterministic and two stochastic stations. The capacity is determined by the expected time a customer remains in the first station. In general, for three stations the following relationship holds:

$$B_i^1 = \max(S_i^1,\ S_{i-1}^2,\ S_{i-2}^3 - (T_i^1 - T_{i-2}^3)) = \max(S_i^1,\ S_{i-1}^2,\ S_{i-2}^3 - X_i^3)$$

This expression has the following explanation: Customer i starts at station 1 at the same time epoch (T_i^1) customer $i-1$ starts at station 2, customer $i-2$, however, may have started earlier at station 3($T_{i-2}^3 \leq T_i^1$). From the Reversibility Property it follows that to prove the theorem for $m=3$ it suffices to compare a sequence where the deterministic station is set up in front with the sequence obtained after performing a pairwise switch between the deterministic station in front and the stochastic station in the middle. The first sequence will be referred to as Sequence I; the sequence after doing the pairwise switch as Sequence II. Interchanging stations 1 and 2 does not affect the first two terms within the parentheses of the expression for B_i^1. The interchange only affects $T_i^1 - T_{i-2}^3 = X_i^3$. The arguments (I) and (II) in the following expressions correspond to Sequences I and II. The service time of customer i at a stochastic station may be written either as $1 + Y_i$ or $1 - Y_i$ where Y_i is a random variable with mean zero, symmetric probability density and bounded by -1 and +1. So

$$B_i^1(I) = \max(1+Y_i(I),\ 1,\ S_{i-2}^3 - X_i^3(I))$$

After a pairwise switch between the first two stations we obtain

$$B_i^1(II) = \max(1,\ 1+Y_{i-1}(II),\ S_{i-2}^3 - X_i^3(II))$$

Instead of comparing $B_i^1(I)$ with the above expression we will compare $B_i^1(I)$ with

$$B_i^1(II) = \max(1,\ 1-Y_{i-1}(II),\ S_{i-2}^3 - X_i^3(II))$$

This change does not alter the process in the probabilistic sense as Y_{i-1} has a symmetric p.d.f. Now we are ready to compare $B_i^1(I)$ with $B_i^1(II)$. Based on common values for S_i^3, and taking $Y_i(I) = Y_{i-1}(II)$, we will show that

$$X_i^3(I) \geq X_i^3(II) \quad i = 3,4,\ldots.$$

by induction on i. It can be checked easily that for $i = 3$

$X_i^3(I) = X_i^3(II)$. For $i > 3$ it is clear that $Y_i(I) \leq 0$ implies $X_{i+1}^3(I) = X_{i+1}^3(II) = 0$. When $Y_i(I) > 0$

$$X_{i+1}^3(I) = (Y_i(I) - (S_{i-2}^3 - X_i^3(I) - 1)^+)^+$$

and

$$X_{i+1}^3(II) = (Y_i(I) - (S_{i-2}^3 - X_i^3(II) - 1 + Y_i(I))^+)^+$$

Clearly

$$X_i^3(I) \geq X_i^3(II) \quad i = 3,4,...$$

It can easily be shown that with positive probability $X_i^3(I) > X_I^3(I)$ (strictly) for $i = 4,5,....$ So

$$E(B_i^1(I)) < E(B_i^1(II))$$

for $i = 4,5,....$ and the theorem is proved for $m = 3$.

(ii) We now show that the capacity of a system of k stations, $k \geq 3$, with a stochastic station set up in front, a stochastic station set up at the end and with $k-2$ deterministic stations in between, increases strictly in k. Consider two systems, one of k stations, say system I, and one of $k+1$ stations, say system II. We show that the capacity of system I is strictly smaller than the capacity of system II. Condition on $S_i^j(II)$, the service time of customer i at station j in system II. Now choose

$$S_i^1(I) = S_{i+1}^1(II)$$

and

$$S_i^k(I) = S_i^{k+1}(II) \quad \text{for } i = 1,2,...$$

From the fact that

$$B_{i+1}^k(II) = \max(1, S_1^{k+1}(II) - (T_{i+1}^k(II) - T_i^k(II)))$$

and

$$B_i^k(I) = S_i^k(I) = S_i^{k+1}(II)$$

follows that when $S_i^k(I)$ $(=B_i^k(I)) \geq 1$

$$B_i^k(I) \geq B_{i+1}^k(II)$$

and when $S_i^k(I) \leq 1$

$$B_{i+1}^k(II) = 1.$$

When $S_i^k(I) \leq 1$ and $B_{i+1}^k(II) = 1$ customer i at station k of system I and customer i + 1 at station k of system II do not cause any blocking of the subsequent customers i+1,i+2,... in preceding stations. The case where $S_i^k(I) > 1$ and $B_{i+1}^k(II) \geq S_i^k(I)$ may cause blocking of subsequent customers at preceding stations. Observe that $B_1^k(II) = 1$, so that customer 1 at station k does not cause any blocking of customer 2 at station k - 1. Also note that $B_2^1(II) = \max(S_2^1(II), 1)$ while $B_1^1(I) = S_1^1(I) = S_2^1(II) \leq B_2^1(II)$. However this difference between $B_1^1(I)$ and $B_2^1(II)$ does not affect further comparisons as the value of $B_1^1(I)$ has no further effect on the process and therefore may be changed to $B_2^1(II)$. So system II can be replaced by a k station system with the same capacity where the service time of customer i + 1 at station k is $B_{i+1}^k(II)$. Comparing customer i + 1 in this new system with customer i in system I it follows immediately that system I has a smaller capacity as $S_i^k(i) \geq B_{i+1}^k(II)$ when $S_i^k(I) \geq 1$. It can be shown easily that the capacity of system I is strictly smaller than the capacity of system II as there is a positive probability that $B_i^1(I) > B_{i+1}^1(II)$. This completes the proof of (ii).

(iii) We now show that any sequence $j_1,\ldots,j_m$ which does not have the stochastic stations in the first and last position can be improved through a series of changes which results in having the stochastic stations first and last. Assume the stochastic stations to be in positions j_k and j_ℓ.

Consider first the case where j_k and j_ℓ are nonadjacent; i.e., there is one or more deterministic stations in between them. We will reason that the capacity of system $j_k,\ldots,j_\ell$ is the same as the capacity of system $j_1,\ldots,j_m$. That stations $j_{\ell+1},\ldots,j_m$ do not cause any additional blocking at stations $j_1,\ldots,j_{\ell-1}$ follows from a similar argument as the one used in the proof of Theorem 3. So system $j_1,\ldots,j_\ell$ has the same capacity as system $j_1,\ldots,j_m$. Because of the Reversibility Property system $j_1,\ldots,j_\ell$ has the same capacity as $j_\ell,\ldots,j_1$. Again, deleting deterministic stations $j_{k-1},\ldots,j_1$ from system $j_\ell,\ldots,j_1$ does not change the capacity of the system. Sequence $j_\ell,\ldots,j_k$, or equivalently $j_k,\ldots,j_\ell$, has the same capacity as $j_1,\ldots,j_m$. Now, because of (ii) the capacity of $j_k,\ldots,j_\ell$ can be increased by inserting additional deterministic stations in between $j_k,\ldots,j_\ell$.

Consider now the case where stations j_k and j_ℓ are adjacent. The capacity of this system can be increased by deleting stations $j_{\ell+1},\ldots,j_m$. Reversing the sequence we obtain $j_\ell,\ldots,j_1$ with the same capacity. The capacity does not change by deleting stations $j_{k+2},\ldots,j_1$. So three stations remain, namely j_ℓ, j_k, j_{k+1}. By (i) we know that sequence j_ℓ, j_{k+1}, j_k has a larger capacity. By (ii) we know that inserting all remaining deterministic stations in between stations j_ℓ and j_k increases the capacity once more. This completes the proof of the theorem. □

In Theorem 4 we found that the capacity of a system, where the service time of one particular station is larger than the service time of any other station with probability one, does not depend on the sequence. Consider now the case of two stations with distributions F and m - 2 stations with distributions G. Service times drawn from distribution F are larger than service times drawn from distribution G with probability one. For this case we have

Theorem 6. The capacity of a system with $m-2$ identical stations with service time distribution G and 2 nonidentical stations with distributions F_1 and F_2, where service times drawn from F_1 or F_2 are larger than service times drawn from G with probability one, is maximized when one of the slow stations is set up first and the other one last.

Proof. Let stations j_k and j_ℓ denote the two slow stations in sequence $j_1,\ldots,j_m$. It is clear that customers at stations $j_{\ell+1},\ldots,j_m$ never can block a customer at station j_ℓ as the service times at stations $j_{\ell+1},\ldots,j_m$ are shorter than the service times at station j_ℓ with probability one. So the capacity of sequence $j_1,\ldots,j_m$ is the same as the capacity of $j_1,\ldots,j_\ell$, which is again equivalent to its reverse. Considering the reverse sequence $j_\ell,\ldots,j_1$, we can again argue that stations $j_{k-1},\ldots,j_1$ do not cause any blocking at stations $j_\ell,\ldots,j_k$. So sequence $j_1,\ldots,j_m$ has the same capacity as sequence $j_k,\ldots,j_\ell$. Now we show that the capacity of a system of k stations, $k \geq 2$, with two slow stations set up in front and at the end and $k-2$ stations set up in between, increases strictly in k. The argument used to show this is similar to the one used in (ii) of Theorem 5. Consider again two systems, one of k stations, say system I, and one of $k+1$ stations, say system II. We show that the capacity of system I is strictly smaller than the capacity of system II. Condition on $S_i^j(II)$, the service time of customer i at station j in system II. Now choose

$$S_i^j(I) = S_i^j(II) \quad \text{for } i = 1,2,\ldots \text{ and } j = 1,2,\ldots,k-2$$

and

$$S_{i+1}^k(I) = S_i^{k+1}(II) \quad \text{for } i = 1,2,\ldots$$

Furthermore, choose

$$S_1^k(I) = S_1^k(II)$$

Assigning the value of $S_1^k(II)$ to $S_1^k(I)$ violates the condition that $S_1^k(I)$ is a random variable drawn from distribution F. But assigning a smaller value to $S_1^k(I)$ can only enlarge the capacity of system I, so if we show that the capacity of system I is smaller than the capacity of system II when $S_1^k(I) = S_1^k(II)$ then it is certainly true when $S_1^k(I) > S_1^k(II)$. It suffices again to show that $B_i^k(I) \geq B_i^k(II)$ for $i = 1,2,\ldots.$ The following relationships hold:

$$B_{i+1}^k(I) = S_{i+1}^k(I) = S_i^{k+1}(II)$$

and

$$B_{i+1}^k(II) = \max(S_{i+1}^k(II),\ S_i^{k+1}(II) - (T_{i+1}^k(II) - T_i^{k+1}(II)))$$

So indeed

$$B_{i+1}^k(I) \geq B_{i+1}^k(II) \quad \text{for } i = 1,2,\ldots$$

which completes the proof. □

6. Discussion

The duality relationship between the infinite waiting room model discussed in Section 3 and the model treated in [5] makes us believe that optimal sequences in the infinite waiting room case have bowl shapes. This means that if the stations can be ordered according to their expected service times, sequences which order the stations initially in increasing order of their expected service times and afterwards in decreasing order of their expected service times should be optimal or near optimal. If the expected service times are equal and the stations can be ordered according to their variances, sequences which order the stations initially in decreasing order of their variances should be optimal or near optimal.

The zero waiting room case appears to be quite different. Theorem 5 shows some similarity with Theorem 1, but a comparison between Theorem 6 and Theorem 2 indicates that the structure of optimal sequences in the zero waiting room case is different from the structure of optimal sequences in the infinite waiting room case. Consider a zero waiting room system with $2m-1$ stations, of which m have service time distribution F and $m-1$ have service time distribution G and F is nonoverlappingly larger than G. There is some evidence that indicates that the optimal sequence *alternates* between stations with large expected service times and small expected service times, while putting at the beginning as well as at the end of the sequence stations with large expected processing times. The special case where we have three stations with exponential service time distribution with mean one and two stations with zero service time distribution can be handled easily through balance equations. It turns out that the "saw-tooth" sequence, which puts the stations with zero processing times in the second and fourth position, is indeed optimal. The optimality of this "saw-tooth" sequence when the stations can be ordered according to their expected service times is rather intuitive. But, suppose now we have $2m-1$ stations all with expected service time of one time unit, m of these stations having an identical symmetric p.d.f. with a high variance and $m-1$ of these stations having an identical symmetric p.d.f. with a low variance. We believe that the optimal sequence alternates between stations with a low variance and stations with a high variance, while putting at the beginning as well as at the end of the sequence stations with a high variance. The optimality of this "saw-tooth" sequence is not very intuitive. Gibbons [3] made a simulation study of these systems. The following system is an example of the kind of systems he considered: Seven stations of which four have the service time distribution

$P(S=10) = 0.2$

$P(S=15) = P(S=5) = 0.4$

and three have the service time distribution

$P(S=10) = 0.8$

$P(S=15) = P(S=5) = 0.1$

Denote the first four stations by a T and the last three by a U. It appeared from the simulation results that the saw-tooth shaped sequence TUTUTUT has the largest capacity.

These saw-tooth shaped sequences appear to be difficult to analyze. Futher understanding of this model might require extensive simulation work.

7. References

[1] A. Avi-Itzhak, "A Sequence of Service Stations with Arbitrary Input and Regular Service Times", Management Sci., 11, 565-573, (1975).

[2] E. Dattatreya, "Tandem Queueing Systems with Blocking", Ph.D. dissertation, IEOR Dept., University of California, Berkeley, (1978).

[3] J. Gibbons, "Simulation of Tandem Queues with Blocking", M.Sc. thesis, School of Industrial and Systems Engineering, Georgia Institute of Technology.

[4] E. Muth, "The Reversibility Property of Production Lines", Management Sci., 25, 152-159, (1979).

[5] M. L. Pinedo, "Minimizing the Makespan in a Stochastic Flowshop", Oper. Res., 30, (1982).

[6] S. V. Tembe and R. W. Wolff, "The Optimal Order of Service in Tandem Queues", Oper. Res., 22, 824-833, (1974).

School of Industrial and Systems Engineering, Georgia Institute of Technology, Atlanta, Georgia 30332

SCHEDULING STOCHASTIC JOBS ON PARALLEL MACHINES
TO MINIMIZE MAKESPAN OR FLOWTIME

Richard R. Weber

Abstract

A number of identical machines operating in parallel are to be used to complete the processing of a collection of jobs so as to minimize the jobs' makespan or flowtime. The amount of processing required to complete the jobs have known probability distributions. It has been established by several researchers that when the required amounts of processing are all distributed as exponential random variables, then the strategy (LEPT) of always processing jobs with the longest expected processing times minimizes the expected value of the makespan, and the strategy (SEPT) of always processing jobs with the shortest expected processing times minimizes the expected value of the flowtime. We prove these results and describe a more general instance in which they are also true: when the jobs have received differing amounts of processing prior to the start, their total processing requirements are identically distributed, and the common distribution of total processing requirements has a monotone hazard rate. Under the stronger assumption that the distribution of the total processing requirements has a density whose logarithm is concave or convex, LEPT and SEPT minimize the makespan and flowtime in distribution.

1. Scheduling to Minimize Makespan or Flowtime

A number of identical machines operating in parallel are to be used to complete the processing of a collection of jobs so as to minimize the jobs' makespan or flowtime. Makespan and flowtime are two criteria commonly used to evaluate scheduling strategies. Suppose that there are n jobs and they they are completed at times $C_1,\ldots,C_n$. The makespan $\max\{C_i\}$ is the time at which the last job is completed. The flowtime ΣC_i is the sum of all the times at which jobs are completed. We suppose that preemptive scheduling is permitted, so that any job may be instantaneously removed from a machine and another job processed instead. A job's processing requirement is the length of time for which it must to be processed by a single machine in order to be completed. We show that when the jobs have processing requirements that are distributed as exponential random variables, then the optimal strategies have simple forms. The expected value of the makespan is minimize by a strategy (LEPT) of always processing jobs with the longest expected processing times. The expected value of the flowtime is minimized by a strategy (SEPT) of always processing jobs with the shortest expected processing times. We prove these results and describe a more general instance in which they are also true: when the jobs are identical but have received differing amounts of processing prior to the start and the distribution of the jobs' total processing requirements has a monotone hazard rate.

It is well known that the LEPT and SEPT strategies are optimal when jobs have known processing requirements. McNaughton (1959) has shown that amongst preemptive scheduling strategies the makespan is minimized by always processing those jobs with the greatest amounts of remaining processing, (so that either all jobs finish together, or the makespan is equal to the processing requirement of the longest job). The flowtime is minimized by always processing those jobs with the least amounts of remaining processing (see Conway, Maxwell and Miller (1967) and Schrage (1968)).

In certain circumstances these results are true when the processing requirements are unknown. Several authors have investigated the case in which processing requirements are distributed as exponential random variables with differing means. Glazebrook (1976 and 1979) has shown that SEPT minimizes the expected value of the flowtime. Bruno (1976) has proved this for just two machines, and Weiss and Pinedo (1979) for any number. Bruno and Downey (1977) have shown that LEPT minimizes the expected value of the makespan for two machines. Bruno, Downey and Frederickson (1981) have shown that LEPT and SEPT are optimal for any number of machines, as has Van der Heyden (1981) for LEPT.

These authors have proved their results by using dynamic programming equations to examine the difference in the expected values of the makespan and flowtime resulting from two strategies which differ only in the jobs scheduled at the start. Complicated notations make the proofs difficult to follow and none can be generalized to processing requirements which are not exponentially distributed. In the following section we present a new proof of the optimalities of LEPT and SEPT in the case of exponentially distributed processing requirements in order to illustrate the method of proof which is used in Weber (1982) to generalize these results to other distributions of processing requirements. Restricting attention to exponentially distributed processing requirements is the best way to put across the flavour of the method. It should become clear that the results may be generalized in several directions. These are discussed in the final section.

2. Proofs When Processing Requirements have Exponential Distributions

Suppose that n jobs are to be completed on m identical machines operating in parallel. All jobs are available for processing from the start. They have processing requirements which are distributed as exponential random variables with means $1/\lambda_1, 1/\lambda_2, \ldots, 1/\lambda_n$.

Theorem 1a. LEPT minimizes the expected value of the makespan.

Proof. Notation will be simplified, but the proof sufficiently illustrated, if we carry it through for the case of just two machines. The proof is essentially the same when there are more than two machines, but the notation is more complicated. The proof is by induction on the number of jobs to be processed. Suppose that the theorem is true whenever there are less than n jobs to process. We will show that it is true when the number of jobs to be processed is n. It is clear that the optimal strategy is non-preemptive, and that if it is optimal to process jobs i and j at time t then it is optimal to continue to process them until one of them is completed. Suppose $\lambda_1 < \lambda_2 < \cdots < \lambda_n$. Let U^I denote the expected value of the remaining time needed to ensure that all n jobs are completed, given that an optimal strategy is employed and that the jobs in the list $I \equiv i_1, \ldots, i_\ell$ have already been completed. Let V^I denote the same quantity when the strategy used to complete the jobs is LEPT. By the inductive hypothesis LEPT is optimal once one job has been completed and so $U^I = V^I$ when I is a list of at least one job. By conditioning on the event that occurs at the first job completion we obtain

$$U = \min_{i \neq j}\{(1 + \lambda_i V^i + \lambda_j V^j)/(\lambda_i + \lambda_j)\}. \tag{1}$$

It is simple to check that (1) is equivalent to

$$0 = \min_{i \neq j}\{1 + \lambda_i(V^i - V) + \lambda_j(V^j - V) + (\lambda_i + \lambda_j)(V - U)\}. \tag{2}$$

In (1) and (2) the minimums are attained by the same pair $\{i,j\}$. Since λ_1 and λ_2 are the two smallest values of λ_1 and $V \geq U$, the fourth term on the right hand side of (2) is minimized by $\{i,j\} = \{1,2\}$. Hence to show that LEPT is optimal it is sufficient to show that $\{i,j\} = \{1,2\}$ also minimizes the sum of the second and third terms. We define

$$V_i = \lambda_i(V^i - V) \quad \text{and} \quad D_{ij} = V_i - V_j.$$

It is interesting to note that $\delta V_i + o(\delta)$ may be interpreted as the amount by which the expected value of the makespan would change from V if when employing LEPT we were to give job i an extra amount δ of processing just before the start. It is the result of theorem 1b that if $\lambda_i < \lambda_j$ then D_{ij} is less than or equal to zero. Hence the sum of the second and third terms on the right hand side of (2), $V_i + V_j$, is minimized by $\{i,j\} = \{1,2\}$ and the induction is complete.

Throughout the rest of this section we shall consider V, V_i and D_{ij} as functions of the variables $\lambda_1, \ldots, \lambda_n$. V_i^I and D_{ij}^I are defined similarly to V_i and D_{ij} when $i,j \notin I$. For example, V_i^I is $\lambda_i(V^{Ii} - V^I)$, where Ii denotes the list I with job i appended.

Theorem 1b. Suppose $\lambda_i < \lambda_j$, $\lambda_1 < \cdots < \lambda_n$. Then

$$D_{ij} \leq 0, \tag{3}$$

and

$$dD_{12}/d\lambda_1 \geq 0. \tag{4}$$

Proof. The proof is by induction on n, the number of jobs to be processed. When $n = 2$, $D_{ij} = (\lambda_i/\lambda_j) - (\lambda_j/\lambda_i)$ and the theorem is true trivially. If i and j are the two smallest indices not in the list I then jobs i and j will be processed first. Conditioning on the event that occurs at the first job completion we have $(\lambda_i+\lambda_j)V^I = 1 + \lambda_i V^{Ii} + \lambda_j V^{Ij}$. By using this fact, and the definition of V_i^I, we can derive the following identities.

$$(\lambda_1+\lambda_2+\lambda_3)V_1 = \lambda_1(\lambda_1+\lambda_2+\lambda_3)V^1 - \lambda_1(\lambda_1+\lambda_2+\lambda_3)V$$

$$= \lambda_1(1+\lambda_1V^1+\lambda_2V^{12}+\lambda_3V^{13}) - \lambda_1(1+\lambda_1V^1+\lambda_2V^2+\lambda_3V)$$

$$= \lambda_1(\lambda_3V^{13}-\lambda_3V^1) + \lambda_2(\lambda_1V^{12}-\lambda_1V^2) + \lambda_3V_1$$

$$(\lambda_1+\lambda_2)V_1 = \lambda_1V_3^1 + \lambda_2V_1^2.$$

We can establish the following similarly.

$$(\lambda_1+\lambda_2)V_2 = \lambda_1V_2^1 + \lambda_2V_3^2.$$

$$(\lambda_1+\lambda_2)V_i = \lambda_1V_i^1 + \lambda_2V_i^2, \quad i = 3,\ldots,n.$$

Combining these we have,

$$D_{12} = \frac{\lambda_1}{\lambda_1+\lambda_2} D_{32}^1 + \frac{\lambda_2}{\lambda_1+\lambda_2} D_{13}^2, \tag{5}$$

and

$$D_{2i} = \frac{\lambda_1}{\lambda_1+\lambda_2} D_{2i}^1 + \frac{\lambda_2}{\lambda_1+\lambda_2} D_{3i}^2, \quad i = 3,\ldots,n. \tag{6}$$

The inductive hypothesis states that (3) and (4) are true when there are less than n jobs to complete, and this hypothesis for (3) implies that both $D_{13}^2 \leq 0$ and $D_{23}^1 \leq 0$ are true when there are n jobs to complete. The hypothesis for (4) similarly implies that $dD_{13}^2/d\lambda_1 \geq 0$. By integrating this with respect to λ_1 we have $D_{13}^2 \leq D_{23}^1 = -D_{32}^1 \leq 0$, and thus remembering that λ_1 is less than λ_2 we can check that (5) is nonpositive. The inductive hypothesis also implies that D_{2i}^1 and D_{3i}^2 are nonpositive, and thus (6) is nonpositive. Since (5) and (6) have been shown to be nonpositive this establishes the inductive step for (3). The inductive step for (4) is established by differentiating the right hand side of (5) with respect to λ_1 and then using the inductive hypothesis to check that every term is nonnegative.

Theorem 2a. SEPT minimizes the expected value of the flowtime.

Proof. The proof is similar to that of theorem 1a. We suppose that $\lambda_i > \cdots > \lambda_n$, and redefine U and V in terms of the expected value of the flowtime. Equation (2) becomes

$$0 = \min_{i \neq j}\{n + \lambda_i(V^i - V) + \lambda_j(V^j - V) + (\lambda_i + \lambda_j)(V - U)\}.$$

The proof is completed using theorem 2b along the same lines as theorem 1a.

Theorem 2b. Suppose $\lambda_j > \lambda_i$, $\lambda_1 > \cdots > \lambda_n$. Then

$$-1 \leq D_{ij} \leq 0, \tag{7}$$

and

$$dD_{12}/d\lambda_1 \leq 0. \tag{8}$$

Proof. The proof is by induction and similar to that of theorem 1b. When $n = 2$, $D_{12} = 0$ and the theorem is true trivially. Instead of (4) and (5) we get

$$D_{12} = \frac{\lambda_1}{\lambda_1 + \lambda_2}(D^1_{32} - 1) + \frac{\lambda_2}{\lambda_1 + \lambda_2}(D^2_{13} + 1), \tag{9}$$

and

$$D_{2i} = \frac{\lambda_1}{\lambda_1 + \lambda_2} D^1_{2i} + \frac{\lambda_2}{\lambda_1 + \lambda_2}(D^2_{3i} - 1), \quad i = 3, \ldots, n. \tag{10}$$

Using (9) and (10) and the inductive hypotheses it is now easy to check the inductive steps for (7) and (8).

3. Generalizations

There are a number of directions in which the results of the previous section may be generalized by using proofs similar to those of theorem 1a and 1b. We shall not give the details of the proofs here, but refer the reader to Weber (1980 and 1982), where they are proved within a framework of optimal control theory and continuous dynamic programming.

(a) By redefining U and V it can be shown that LEPT and SEPT minimize the makespan and flowtime in distribution. This means that, for any γ, the probabilities that the makespan and flowtime are greater than γ are minimized by LEPT and SEPT respectively. It is no longer obvious that the optimal strategy is non-preemptive, but this can be established in the proof.

(b) As Weiss and Pinedo (1979), we may consider non-identical machines which process jobs at rates $s_1 \geq \cdots \geq s_m$. If job i is processed on machine j its instantaneous hazard rate is $\lambda_i s_j$. The expected value of the makespan is now minimized by a version of LEPT which always processes the job of least λ on machine 1, the job of second least λ on machine 2, and so on. The expected valued of the flowtime is minimized by a version of SEPT which always processes the job of greatest λ on machine 1, the job of second greatest λ on machine 2, and so on. The proof is along the lines of the previous section.

(c) For processing requirements that are not distributed as exponential random variables, LEPT and SEPT are still optimal in the following circumstance. Assume the jobs have identical total processing requirements, but have received different amounts of processing prior to the start. A job that has received an amount of processing x has an instantaneous hazard rate of $\rho(x)$. Suppose that $\rho(x)$ is a <u>monotone hazard rate</u> that is increasing or decreasing in x. In these circumstances LEPT and SEPT are still optimal. The proof is along the lines of the previous section. For example, (4) is reformulated in terms of dD_{12}/dx_1, where x_1 is the amount of processing job 1 has so far received. [Note that if several jobs have received equal amounts of processing, LEPT or SEPT may require a 'sharing' of machine effort. For example, if the hazard rate is increasing and exactly three jobs, which have had equal amounts of processing, are to be completed on 2 machines, then LEPT is realized by processing each job at 2/3 the full rate of one machine until one job is completed. In practice, sharing is approximated by frequently changing the set of jobs that are being processed, so that the amounts of processing the jobs have received remain nearly equal.]

This result includes the result of Pinedo and Weiss (1979) who proved that LEPT and SEPT are expected value optimal when the processing requirements are distributed as different mixtures of two exponential distributions. Their model corresponds to a decreasing hazard rate model of the above form. The result also extends the work of Nash (1973) who proved the optimality of SEPT for the case in which all jobs have received identical amounts of processing at the start and the distribution of total processing requirement has an increasing hazard rate.

(d) If in addition to the conditions of (c) the distribution of total processing requirement has a density whose logarithm is concave or convex, then LEPT and SEPT minimize the makespan and flowtime in distribution. A density whose logarithm is concave or convex is said to be <u>sign-consistent of order two</u> (SC_2). Karlin (1968) has made a detailed study of sign-consistent densities. He and other authours have described their importance in areas of statistical theory, reliability, game theory and mathematical economics. The uniform, gamma, hyperexponential, folded-normal and Weibull distributions all have SC_2 densities. The proof is along the same lines as above, with, for example, (4) reformulated in terms of $d\{D_{12}/\rho(x_1)\}/dx_1$.

(e) When the processing requirements are distributed according to any of the models of this paper, the results are still true even if the number of available machines is a non-decreasing function of time. This follows from the fact that in order to carry out the induction in theorem 1b we only needed to be sure that in (5) job 3 could take the role of jobs 1 or 2 in an application of the inductive hypothesis for (4). This will be the case if the number of machines is non-decreasing in time. A stronger result can be proved if the distribution of total processing requirement has a SC_2 density. In this case LEPT minimizes the makespan in distribution even if the number of available machines is an arbitrary function of time and some jobs are not present at the start, but only arrive later according to a stochastic process.

(f) For any of the models of this paper, LEPT also maximizes in distribution the time at which the number of machines first exceeds the number of uncompleted jobs. Thus, if a system requires m components to operate, the length of time for which it can be kept running by using a stock of $n > m$ components is maximized in distribution by LEPT. When $m = 2$ this is the 'lady's nylon stocking problem' of Cox (1959), for which he hypothesised LEPT optimality in the case of monotone hazard rates. Weber and Nash (1979) give further details.

It does not appear possible to find simple strategies minimizing the expected values of the makespan or flowtime if the processing requirements are not distributed according to one of the above models. When the distribution of total processing requirement has a non-monotone hazard rate LEPT and SEPT are generally not optimal.

4. References

[1] Bruno, J. (1976), "Sequencing Tasks with Exponential Service Times on Parallel Machines," Technical report, Department of Computer Science, Pennsylvania State University.

[2] Bruno, J. and Downey, P. (1977), "Sequencing Tasks with Exponential Service Times on Parallel Machines," Technical report, Department of Computer Sciences, University of California at Santa Barbara.

[3] Bruno, J., Downey, P. and Frederickson, G. N. (1981), "Sequencing Tasks with Exponential Service Times to Minimize the Expected Flowtime or Makespan," J. Assoc. Comput. Mach., 28, 100-113.

[4] Conway, R. W., Maxwell, W. L. and Miller, L. W. (1967), The Theory of Scheduling, Addison-Wesley, Reading, Massachusetts.

[5] Cox, D. R. (1959), "A Renewal Problem with Bulk Ordering of Components," J. Roy. Statist. Soc. Ser. B, 21, 180-189.

[6] Glazebrook, K. D. (1976), "Stochastic Scheduling," Ph. D. Thesis, University of Cambridge.

[7] Glazebrook, K. D. (1979), "Scheduling Tasks with Exponential Service Times on Parallel Processors," J. Appl. Probab., 16, 685-689.

[8] Karlin, S. (1968), Total Positivity, Vol. I., Stanford University Press, Stanford.

[9] McNaughton, R. (1959), "Scheduling with Deadline and Loss Functions," Management Sci., 6, 1-12.

[10] Nash, P. (1973), "Optimal Allocation of Resources to Research Projects," Ph. D. Thesis, University of Cambridge.

[11] Pinedo, M. and Weiss, G. (1979), "Scheduling Stochastic Tasks on Two Parallel Processors," Naval Res. Logist. Quart., 27, 528-536.

[12] Schrage, L. E. (1968), "A Proof of the Shortest Remaining Process Time Discipline," Oper. Res., 16, 687-689.

[13] Van der Heyden, J. (1981), "Scheduling Jobs with Exponential Processing and Arrival Times on Identical Processors so as to Minimize the Expected Makespan," Math. Oper. Res., 6, 305-312.

[14] Weber, R. R. and Nash, P. (1979), "An Optimal Strategy in Multi-Server Stochastic Scheduling," J. Roy. Statist. Soc. Ser. B, 40, 323-328.

[15] Weber, R. R. (1980), "Optimal Organization of Multi-Server Systems," Ph. D. Thesis, University of Cambridge.

[16] Weber, R. R. (1982), "Scheduling Jobs with Stochastic Processing Requirements on Parallel Machines to Minimize Makespan and Flowtime," J. Appl. Probab., 19, (to appear).

[17] Weiss, G. and Pinedo, M. (1979), "Scheduling Tasks with Exponential Service Times on Non-Identical Processors to Minimize Various Cost Functions, J. Appl. Probab., 17, 187-202.

Cambridge University, Engineering Department, Control and
Management Systems Division, Mill Lane, Cambridge CB2 1RX, ENGLAND.

Discussant's Report on "Scheduling Stochastic Jobs on Parallel Machines," by Richard Weber

The problem of optimally sequencing a collection of independent jobs with known processing times on parallel machines has a long history in the deterministic scheduling literature [3]. The simplest of these results, regarding the "shortest job first" policy for flowtime [4] and the "longest remnant first" policy for preemptive makespan [10], are so well known that they have become pedagogical "standards", used to convince students that there is something to all this business of scheduling after all.

A few years ago, investigators began to consider whether similar optimal policies exist for jobs whose runtime is described by a random variable. Perhaps forgivably, they first attacked that noble distribution that seems to occur exponentially in the literature [6, 12, 14, 1], allowing for differing mean times for different jobs. The deterministic results carried over (with the adjective "expected" appropriately inserted), but was this merely a trick of our (noble but forgetful) distribution? Pinedo and Weiss [12] were able to push the result to the case of two machines where each job's time is chosen as a different hyperexponential "mix" of the same two fixed exponentials. Nash [11] showed that if jobs shared a common distribution with increasing failure rate function then "shortest expected processing time first" minimized the expected flowtime.

Enter Richard Weber. He has shown that these classic scheduling policies are optimal for fundamental reasons having little to do with the detailed structure of the service time distribution of jobs. Suppose that the service time of each job is chosen independently from the same underlying distribution, but different jobs have "aged" different amounts at the start of scheduling (so that they begin at different points in the failure rate curve for the underlying distribution). All the above problems can be reinterpreted from this new point of view. Weber's discovery is that, assuming scheduling is preemptive, that processor-sharing is allowed, and that the failure rate curve is monotone increasing or decreasing throughout, then "shortest expected processing time first" (SEPT) minimizes expected flowtime and "longest expected processing time first" (LEPT) minimizes expected makespan [13]. From these results, the SEPT policy is reinterpreted as that policy which schedules the job with largest instantaneous failure rate; LEPT schedules the job with smallest instantaneous failure rate; ties must share processors equally.

More is true. LEPT and SEPT are policies with are optimal in distribution (not just in expectation) if the more stringent (but still mild) assumption is made that the processing time density is "sign-regular of order 2" [8]. Such an assumption implies that the failure rate curve is monotone [9].

Thus we have in Weber's work the pleasure of seeing several distinct results suddenly pieced together in a novel way, be shown to be "cases" of a fundamental result, and to obtain a vast generalization of the class of distributions to which the result is applicable. Frankly, results this satisfying are few and far between, and worth the wait when they arrive.

The paper in the current proceedings provides a clear exposition of Weber's proof technique for a particular distribution (the exponential). Shorn of the details necessary to obtain the results of greatest generality, this paper provides a useful introduction to the more elaborate proofs in [13].

Of course, the settling of a question merely causes the bystanders to suggest new questions, and so I will do what is required. There are some possible directions for investigation suggested by both the literature on this problem and by "discrete analogues" from deterministic scheduling theory.

One direction was pursued by Weiss and Pinedo [14] for the exponential distribution. They define a large class of cost functions for schedules (of which makespan and flowtime are but special cases) and show conditions under which LEPT and SEPT minimize expected cost. Does their theorem go through in the more general setting of aged jobs from a monotone failure rate distribution?

So far we have spoken only of scheduling with preemptions. When the failure rate curve is decreasing, then LEPT is nonpreemptive; similarly for SEPT when the failure rate curve is increasing. In these two cases the results yield simple optimal policies for nonpreemptive schedules. From deterministic scheduling results, we know that sequencing a set of jobs with known processing times on two machines to minimize makespan is an NP-complete problem (for three machines it is "strong" NP-complete [5]). It is likely then that no optimal policy exists which is simple enough to be useful.

In the terms of Weber's paper [13], a set of deterministic jobs with processing times $t_1 < t_2 < \cdots < t_n$ can be thought of as "aged" points on an increasing failure rate curve that is zero at the origin with a unit step at t_n. Thus we are trying to minimize makespan when failure rates are increasing while disallowing preemption and processor sharing. This shows that the current results cannot be extended to the nonpreemptive case for all monotone failure rate distributions. But is there a "reasonable" class of distributions for which a "simple" nonpreemptive scheduling rule is optimal?

Another direction of generalization is to introduce precedence constraints among the (still stochastically independent) jobs. On the deterministic side, Hu [7] showed that unit processing time jobs constrained to be sequenced according to an "in-tree" form of partial order (Figure 1) can be finished in the shortest time on m processors if scheduled "highest level first" (HLF) among all those remaining (the levels of jobs are shown in Figure 1). Chandy and Reynolds [2] have shown that if job processing times are random variables with a common exponential or Erlang distribution, and jobs are constrainted according to an "in-tree", then HLF is still optimal to minimize expected makespan on two processors. The result does not extend to three processors: in Figure 1, HLF schedules jobs 1, 2 and 3 first while the optimal policy must pick 1, 2 and 4 first, assuming processing times are exponential [2].

The question in all this is: does the Chandy and Reynolds result carry over for a more general class of distribution? Why does the HLF policy break down for three processors? What, if any, is an optimal policy?

In many scheduling applications (e.g., computer job scheduling), one has mixtures of some short jobs and some long ones with few of middling length. Such a distribution causes the failure rate curve to be "basin" shaped--at first decreasing and then increasing. Based on the insights gained from this paper, what can be said about the relative performance of SEPT against the optimal policy minimizing flowtime? In this context SEPT is a suboptimal "heuristic", but commonly employed with good results.

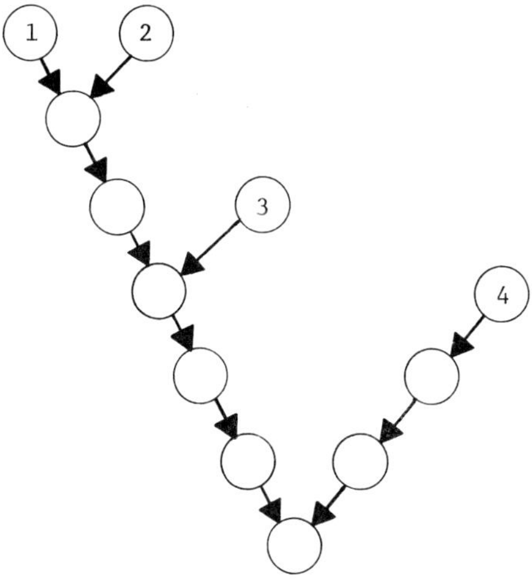

Figure 1. "In-tree" procedence constraint showing levels.

References

[1] Bruno, J., P. Downey and G. N. Frederickson (1981), "Sequencing Tasks with Exponential Service Times to Minimize the Expected Flow Time or Makespan," J. Assoc. Comput. Mach., 28, pp. 100-113.

[2] Chandy, K. M. and P. F. Reynolds (1975), "Scheduling Partially Ordered Tasks with Probabilistic Execution Times," Proceedings of the 5th Symposium on Operating Systems Principles, ACM, 169-177.

[3] Coffman, E. G., Jr. (ed.) (1976), Computer and Job-Shop Scheduling Theory, John Wiley and Sons, New York.

[4] Conway, R. W., W. L. Maxwell and L. W. Miller (1967), Theory of Scheduling, Addison-Wesley Publishing Co., Reading, MA.

[5] Garey, M. R. and D. S. Johnson (1979), Computers and Intractability: a Guide to the Theory of NP-Completeness, W. H. Freeman and Co., San Francisco, CA.

[6] Glazebrook, K. D. (1976), "Stochastic Scheduling," Ph.D. Thesis, Cambridge University.

[7] Hu, T. C. (1961), "Parallel Sequencing and Assembly Line Problems," Oper. Res., 9, pp. 841-848.

[8] Karlin, S. (1968), "Total Positivity, Vol. 1," Stanford University Press, Stanford, CA.

[9] Kaufmann, A., D. Grouchko and R. Crunon (1977), Mathematical Models For the Study of the Reliability of Systems, Academic Press, New York.

[10] McNaughton, R. (1959), "Scheduling with Deadlines and Loss Functions, Management Sci., 16, pp. 1-12.

[11] Nash, P. (1973), "Optimal Allocation of Resources to Research Projects," Ph.D. Thesis, Cambridge University.

[12] Pinedo, M. and G. Weiss (1979), "Scheduling Stochastic Tasks on Two Parallel Processors," Naval Res. Logist. Quart., 26, pp. 527-535.

[13] Weber, R. (1981), "Scheduling Jobs with Stochastic Processing Requirements on Parallel Machines to Minimize Makespan or Flowtime," J. Appl. Probab., 18, to appear.

[14] Weiss, G. and M. Pinedo (1980), "Scheduling Tasks with Exponential Service Times on Non-Identical Processors to Minimize Various Cost Functions, J. Appl. Probab., 17, pp. 187-202.

Discussant: Dr. Peter J. Downey, Department of Computer Science, The University of Arizona, Tucson, Arizona 85721.

AN ADAPTIVE-PRIORITY QUEUE

H. G. Badr[†]
I. Mitrani[††]
and
J. R. Spirn[†††]

Abstract

Priority scheduling is used frequently in computer systems; however, such schedules usually have a "fail-safe" provision to prevent high-priority requests from causing unlimited delay to low-priority ones. In this paper, a queue and server is considered which has two sources of Poisson-arrival customers, types a and b respectively. Non-preemptive priority is granted to type a customers, except that whenever the number of waiting type b customers exceeds a specified threshold, class b receives priority. Service times are assumed exponentially distributed with a common mean, but it is possible to greatly relax this restriction. The intent is to grant priority to class a most of the time, while bounding the mean waiting time of class b customers under heavy class a load. This particular fail-safe mechanism is shown to have the property that for system states in which class b is below the threshold, the occupancy probability is exactly the same as if class a always had high priority.

1. Introduction

This paper presents a non-preemptive, two-customer-class M/M/1 queueing system in which there is dynamic alternation of the priority setting between the two classes. The aim is to model a situation wherein one accords non-preemptive priority to one class, class a, but not at the cost of unlimited penalty to the other class, class b. This penalty is guaged directly in terms of class b's queue length--and so, indirectly, in terms of its (mean) waiting time in queue. Whenever the number of class b customers in queue exceeds a certain predetermined threshold level N_b, $N_b \geq 1$, priority switches to favor class b at the expense of class a until the number of class b customers in queue drops to the threshold level. By suitable choice of the threshold level, the intent is to accord high priority service to class a customers "most of the time" while still bounding the mean waiting time (as a function of system load) of the other class.

We shall initially present the model and outline the intended method of analysis. This analysis is aimed at producing equilibrium distribution transforms for the number of customers in queue and for the waiting times of class a. Note that if the threshold level is set at $N_b = \infty$, we obtain the classical situation [4, 7, 10] in which class a always has priority over b; for $N_b = 0$, class b would have priority all the time. With these situations as "benchmarks", we shall present results to characterize the behavior of the queueing system at various threshold settings. Finally, as an example application, we shall use the system to model some of the features of the memory accessing strategy of the BCC-500 computer system [3]. In the BCC-500, drum accesses to main memory do not receive unqualified priority over CPU requests to the memory. Instead, CPU requests are favored except when there is such a number of drum requests pending that imminent loss of the drum record is threatened.

In this presentation, we shall make very simple service-time assumptions to reduce mathematical complexity--both customer classes will share the same exponential distribution. Under these assumptions we will show the model to be exactly decomposable, meaning that the equilibrium occupancy probabilities of states in which class a has priority are exactly the same as if no priority switch were allowed. In fact, this result can be shown [2] to apply in the much more general M/G/1 case in which each customer class has a distinct and arbitrary service-time distribution, and for preemptive as well as non-preemptive priority.

2. The Model

The system is presented schematically in Figure 1. The arrival process to each of the two classes is an independent Poisson stream. In particular, the distributions of interarrival times for class a and class b are, respectively, given by:

$$A_a(t) = 1 - e^{-\lambda_a t}, \quad A_b(t) = 1 - e^{-\lambda_b t}, \quad t \geq 0,$$

giving Poisson streams with mean rates λ_a and λ_b. The collective arrival process is thus also Poisson, with mean rate $\lambda = \lambda_a + \lambda_b$. Service times for the two classes are taken from the same exponential distribution, given by $B(t) = 1 - e^{-\mu t}$, $t \geq 0$. The mean service time is thus $1/\mu$, irrespective of class. We shall use ρ_a, ρ_b, ρ to denote the load factors for, respectively, class a, class b and the system as a whole, with

$$\rho_a = \lambda_a/\mu; \quad \rho_b = \lambda_b/\mu; \quad \rho = \rho_a + \rho_b.$$

In equilibrium, the system state is characterized by the ordered pair (n_a, n_b), where n_a is the number of class a customers in queue (not including the one, if any, in service); similarly for n_b. We restrict

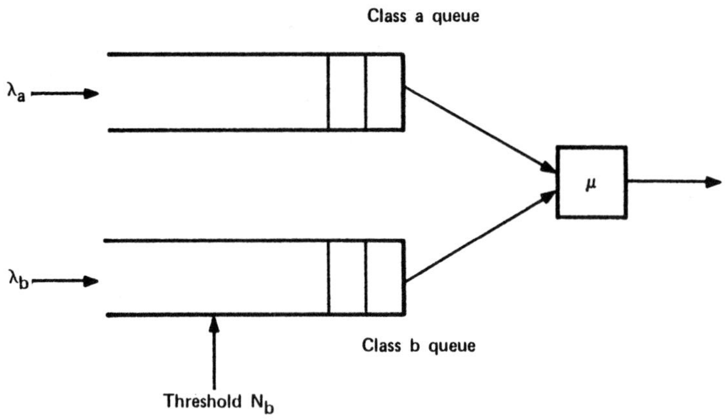

# of class b requests in queue	Priority	
	Class a	Class b
$\leq N_b$	High	Low
$> N_b$	Low	High

Figure 1. The Model

state $(n_a, n_b) = (0, 0)$ to mean that both queues are empty but that the server is busy serving a customer (of unspecified class), and we introduce the state I (idle) to denote that the system as a whole, both server and queues, is empty. Priority switches to class b whenever $n_b > N_b$, where $N_b \geq 1$ is the fixed threshold value; otherwise class a has priority.

The system is thus characterized[1] in equilibrium by the state I, together with the ordered pairs (n_a, n_b). It is clear that we may model this system as a time-homogeneous Markovian process with a discrete, countably infinite state space in two dimensions. The state transition diagram is given in Figure 2. Note that at column N_b and to the left of it, state transitions due to a departing customer flow upward where possible--i.e., in a direction that decreases the first component of the states (n_a, n_b). Since class a has priority in this region, the

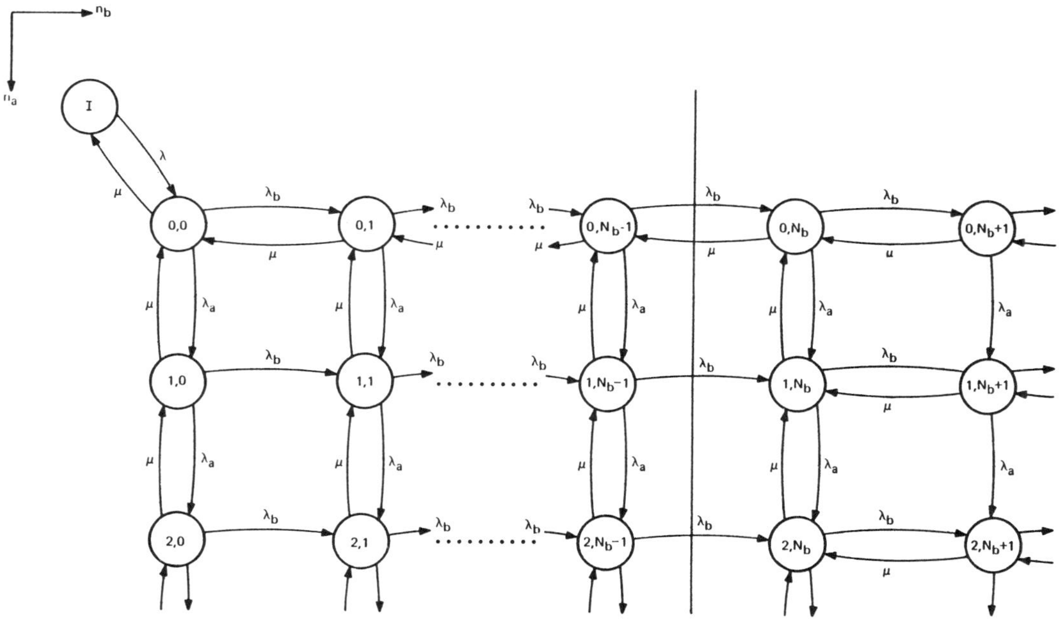

Figure 2. Markov Transition Diagram for Model in Figure 1

customer replacing the one leaving the server will be a class a customer who departs the queue to enter service. To the right of column N_b, transitions due to a departure from service flow leftward--in a direction decreasing the n_b component of state (n_a, n_b)--since class b has priority in this region. For purposes of analysis, we divide the state diagram into two parts by means of a (conceptual) "barrier" between columns $N_b - 1$ and N_b (although the switch in priority takes place after column N_b, the change is reflected in the state balance equations for the states in column N_b itself). The barrier marks, with the exception of state $(0, N_b)$, the points at which the structure of the state balance equations alters due to the switch in priority.

Let P(y, z) be the bivariate generating function of the equilibrium state probabilities. Then P(y, z) is given by

$$P(y, z) = \sum_{i=0}^{\infty} \sum_{j=0}^{\infty} p(i, j) y^i z^j$$

where p(i, j) = Pr[system is in state $(n_a, n_b) = (i, j)$]. Note that we do not include p(I) = Pr[system is in idle state I] in the definition of P(y, z). If we do not distinguish between customer classes, the system is equivalent to a simple M/M/1 queue, so that $p(I) = 1 - \rho$ [5, 6, 8, 9].

We may rewrite P(y, z) as follows:

$$P(y, z) = \sum_{j=0}^{N_b - 1} \left(\sum_{i=0}^{\infty} p(i, j) y^i \right) z^j + \sum_{i=0}^{\infty} \sum_{j=N_b}^{\infty} p(i, j) y^i z^j$$

where the two terms on the RHS contain what lies to the left and right of the barrier, respectively.

Denote the generating function for the equilibrium probabilities of states in the jth column of Figure 2 by $C_j(y) = \sum_{i=0}^{\infty} p(i, j) y^i$, $j \geq 0$, and let S(y, z) be the generating function for the state probabilities to the right of the barrier "shifted" left by N_b columns; i.e.,

$$S(y, z) = \sum_{i=0}^{\infty} \sum_{j=0}^{\infty} p(i, N_b + j) y^i z^j. \tag{1}$$

With this notation, we may now write,

$$P(y, z) = \sum_{j=0}^{N_b - 1} C_j(y) z^j + z^{N_b} S(y, z) \tag{2}$$

We shall analyze the model by considering, first, the portion to the left of the barrier independently of the one to the right of it. The portion to the right of the barrier will then be analyzed to obtain the transform $S(y, z)$. The justification for this approach is embodied in the following theorem which is fundamental to the analysis:

Theorem. The solution of the state balance equations for the states lying to the left of the barrier is not a function of the state balance equations for states lying to the right of the barrier.

Proof. We show that it is possible to solve for the equilibrium probabilities of all states lying to the left of the barrier by starting with the equation for state $p(I)$ and working consistently rightwards, in such a manner that the equilibrium probabilities for states in the jth column (together with the equilibrium probability for state $(0, j + 1)$) are obtained before we move on to states in the $(j + 1)^{st}$ column. The proof is by induction on the vertical sections shown in Figure 3.

Basis (vertical section 0): From the state equation for $p(I)$ we have that $\lambda p(I) = \mu p(0, 0) \Rightarrow p(0, 0) = \rho p(I) = \rho(1 - \rho)$.

Now suppose that the equilibrium probabilities for all states in the vertical sections up to and including section $n - 1$ have been solved where $1 \leq n \leq N_b$.

Step: Consider the states in vertical section n. From the state balance equations for states in column $n - 1$ we form the following equation for $C_{n-1}(y)$ in the usual way:

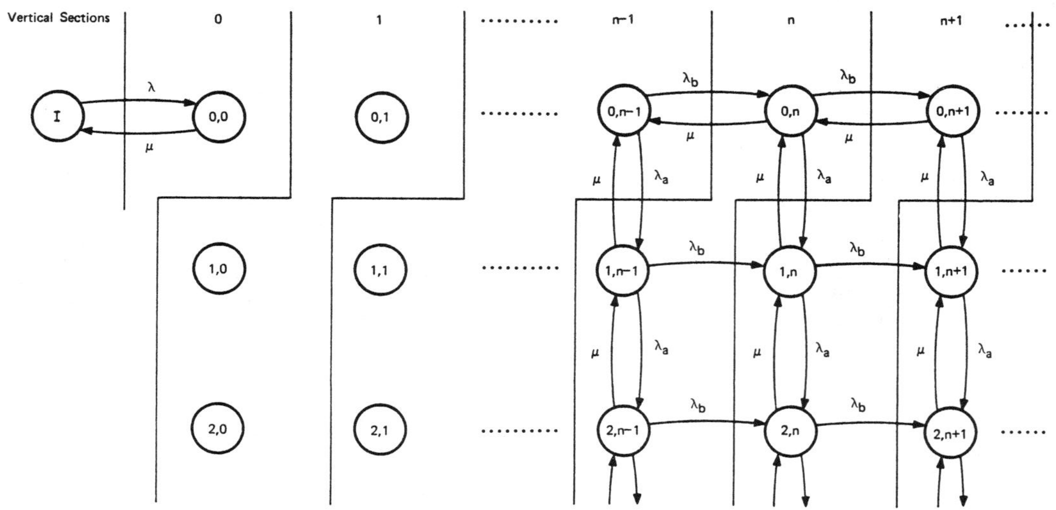

Figure 3. Vertical Sections of Figure 2

$$(\lambda + \mu)C_{n-1}(y) = \lambda_b C_{n-2}(y) + \lambda_a y C_{n-1}(y)$$

$$+ \frac{\mu}{y}(C_{n-1}(y) - p(0, n - 1)) + \mu p(0, n) \qquad (3)$$

$$\therefore\ C_{n-1}(y) = \frac{p(0, n - 1) - y(\rho_b C_{n-2}(y) + p(0, n))}{\rho_a y^2 - (1 + \rho)y + 1}$$

where, for n = 1, we define $C_{-1}(y) = \rho P(I)/\rho_b$. The denominator of expression (3) has a root $y = y_0$ in the open interval (0, 1) for $0 < \rho_a, \rho_b, \rho < 1$, as is easily seen. Since $C_{n-1}(y_0)$ must be finite, the numerator must also be zero:

$$p(0, n - 1) - y_0(\rho_b C_{n-2}(y_0) + p(0, n)) = 0.$$

By the inductive hypothesis, $C_{n-2}(y_0)$ and p(0, n - 1) are known; hence we can find p(0, n). Therefore all terms on the RHS of the expression for $C_{n-1}(y)$ above are now known. We thus have p(0, n) and a transform for the probabilities of the remaining states in section n. Q.E.D.

Note that the proof shows that we can obtain a solution for $p(0, N_b)$ by considering only what is to the left of the barrier. In fact, although state $(0, N_b)$ lies to the right of the barrier, its state equation is not structurally different from those of states $(0, n_b)$, $0 \leq n_b < N_b$. We conclude this section with the following corollary.

Corollary. The solutions to the state balance equations for states (n_a, n_b) with $0 \leq n_a < \infty$, $0 \leq n_b < N_b$ (i.e., the states to the left of the barrier) in our model are the same as those of a 2-class fixed-priority system (with the same arrival and service parameters) in which class a always has priority.

Proof. Immediate from the theorem and from the fact that, for the range of states given, the set of state balance equations in the one model is the same as in the other.

3. Analysis to the Left of the Barrier

We are, in particular, seeking expressions for the column transforms $C_j(y)$ for $j = 0, 1, \ldots, N_b - 1$. In accordance with the corollary, we do this by considering the fixed-priority model of the statement of the corollary which, as already mentioned in the introduction, is equivalent to our model with a threshold set at $N_b = \infty$. The state transition diagram for such a model is given in Figure 4.

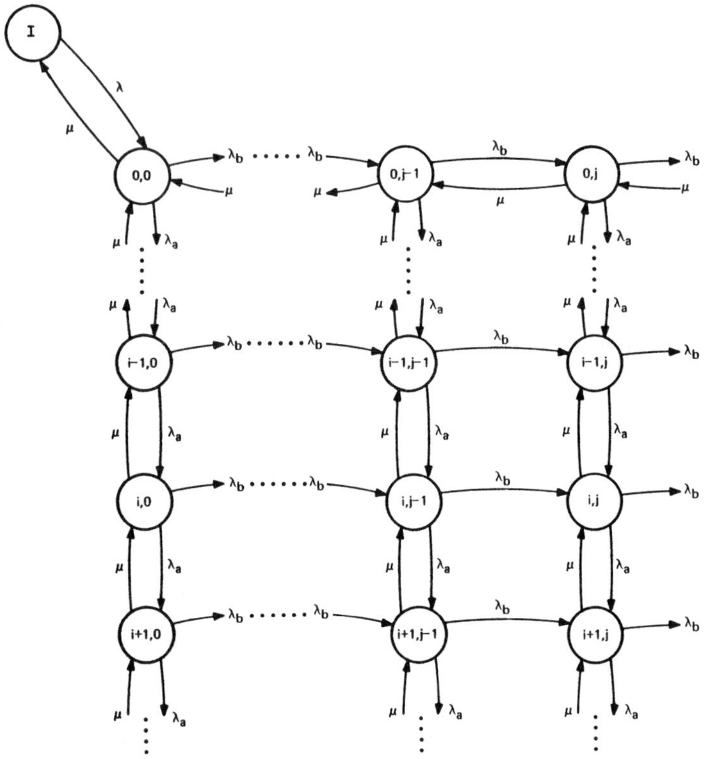

Figure 4. Fixed Priority Model, Left of Barrier

Expression (3) yields a means of evaluating the column transforms for the model of Figure 4. We rewrite (3) in the following form:

$$a(y)\ C_n(y) = b(y)\ C_{n-1}(y) + d_n(y) \qquad n = 0,\ 1,\ 2,\ \ldots\ N_b - 1 \tag{4}$$

where

$$a(y) = \rho_a y^2 - (1 + \rho)y + 1$$

$$b(y) = -\rho_b y$$

$$d_n(y) = p(0,\ n) - yp(0,\ n + 1).$$

By successive substitution, (4) can be solved to yield:

$$C_n(y) = \frac{b(y)^{n+1}\ C_{-1}(y) + \sum_{i=0}^{n} a(y)^i\ b(y)^{n-i}\ d_i(y)}{a(y)^{n+1}}$$

or after manipulation:

$$C_n(y) = \frac{q_0(y) + \sum_{i=1}^{n+1} q_i(y)\ p(0,\ i)}{a(y)^{n+1}} \tag{5}$$

where, recalling that $p(0,\ 0) = \rho P(I) = \rho(1 - \rho)$:

$$q_0(y) = (-1)^n\ \rho(1 - \rho)\rho_b^n y^n(1 - y)$$

$$q_i(y) = (-1)^{n-i}\ \rho_b^{n-i}\ y^{n-i}[a(y)^i + \rho_b y^2\ a(y)^{i-1}] \qquad 1 \leq i \leq n$$

$$q_{n+1}(y) = -ya(y)^n. \tag{6}$$

The denominator of (5) has a zero of order $n + 1$ at $y = y_0$, so that L'Hospital's rule must be applied $n + 1$ times to determine the limit at y_0. Thus, the numerator of (5), together with its first n derivatives, must vanish at $y = y_0$. This yields a triangular (hence easily solved) system of $n + 1$ linear equations in the $n + 1$ unknowns $p(0,\ 1),\ \ldots,\ p(0,\ n + 1)$. In particular, solving this system with $n = N_b - 1$ yields $p(0,\ 1),\ \ldots,\ p(0,\ N_b)$, which can be used in (5) to obtain the column transforms $C_0(y), \ldots, C_{N_b-1}(y)$.

The only apparent difficulty with this approach is in evaluating the first $N_b - 1$ derivatives of the numerator of (5), with $n = N_b - 1$. But it is straightforward to obtain any number of derivatives of a polynomial, and the functions $q_i(y)$ in (6) are easily manipulated to polynomial form upon application of the multinomial formula:

$$a(y)^n = \sum_{i=0}^{n} \sum_{j=0}^{n-i} \frac{n!}{i!j!(n-i-j)!} (-1)^j \rho_a^i (1+\rho)^j y^{2i+j}$$

An alternative approach, given in [1], is to use complex integration to directly invert the bivariate generating function corresponding to the model of Figure 4. This yields closed-form expressions for the column transforms $C_i(y)$. The closed form expressions include definite integrals, however these can be evaluated exactly in time $O(N_b)$.

4. Analysis to the Right of the Barrier

We seek an expression for the shifted probability generating function $S(y, z)$ defined in (1). With respect to Figure 2, let $Q_i(z)$ be the row transform, shifted left by N_b columns, for the equilibrium probabilities of states lying in row $i \geq 0$ to the right of the barrier; i.e., $Q_i(z) = \sum_{j=0}^{\infty} p(i, N_b + j)z^j$. Consider the balance equations for states lying in row 0:

$$(1+\rho)\, p(0, N_b) = \rho_b p(0, N_b - 1) + p(0, N_b + 1) + p(1, N_b), \quad j = 0;$$

$$(1+\rho)\, p(0, N_b + j) = \rho_b p(0, N_b + j - 1) + p(0, N_b + j + 1), \quad j \geq 1.$$

Multiplying the jth equation by z^j and summing, we have

$$(1+\rho)Q_0(z) = \rho_b p(0, N_b - 1) + \rho_b z Q_0(z) + 1/z(Q_0(z) - p(0, N_b)) + p(1, N_b),$$

$$\Rightarrow Q_0(z) = \frac{p(0, N_b) - zp(1, N_b) - \rho_b z p(0, N_b - 1)}{\rho_b z^2 - (1+\rho)z + 1}.$$

Similarly, for row $i \geq 1$ we obtain

$$Q_i(z) = \frac{p(i, N_b) - zp(i+1, N_b) - \rho_b zp(i, N_b - 1) - \rho_a zQ_{i-1}(z)}{\rho_b z^2 - (1+\rho)z + 1}, \quad i \geq 1.$$

Noting that $S(y, z) = \sum_{i=0}^{\infty} Q_i(z)y^i$ we get, after algebraic manipulation,

$$S(y, z) = \frac{(y - z)C_{N_b}(y) - \rho_b yzC_{N_b-1}(y) + zp(0, N_b)}{\rho_b yz^2 + (\rho_a y - \rho - 1)yz + y} .$$

Taking the vertical cutset between columns $N_b - 1$ and N_b, we have that $p(0, N_b) = \rho_b \sum_{i=0}^{\infty} p(i, N_b - 1) = \rho_b C_{N_b-1}(1)$. So,

$$S(y, z) = \frac{(y - z)C_{N_b}(y) - \rho_b yzC_{N_b-1}(y) + \rho_b zC_{N_b-1}(1)}{\rho_b yz^2 + (\rho_a y - \rho - 1)yz + y} .$$

The denominator of $S(y, z)$, considered as a quadratic in z, has a root given by,

$$z_0(y) = [(1 + \rho - \rho_a y) - \sqrt{(\rho_a y - \rho - 1)^2 - 4\rho_b}]/2\rho_b, \tag{7}$$

where $|z_0(y)| \leq 1$ for $|y| \leq 1$, i.e., the root lies within the region of convergence of $S(y, z)$. Since the denominator vanishes at the $z = z_0(y)$, the numerator of $S(y, z)$ must approach 0 as $z \to z_0(y)$, yielding, after manipulation,

$$C_{N_b}(y) = \frac{\rho_b z_0(y)(yC_{N_b-1}(y) - C_{N_b-1}(1))}{(y - z_0(y))} .$$

Substituting for $C_{N_b}(y)$ in the expression for $S(y, z)$, and after algebraic manipulation,

$$S(y, z) = \frac{z_0(y) - z}{\rho_a yz + \rho_b z^2 - (1 + \rho)z + 1} \cdot A(y), \tag{8}$$

where

$$A(y) = \frac{\rho_b(yC_{N_b-1}(y) - C_{N_b-1}(1))}{y - z_0(y)} \quad . \tag{9}$$

The function $z_0(y)$ is given in (7), and $C_{N_b-1}(y)$, $C_{N_b-1}(1)$ may be obtained from (5). Recall that $P(y, z)$ is given in (2) by

$$P(y, z) = \sum_{j=0}^{N_b-1} C_j(y)z^j + z^{N_b}S(y, z). \tag{10}$$

All terms on the RHS of this expression are now known with $C_j(y)$ defined in (5) for $0 \leq j \leq N_b-1$, and $S(y, z)$ defined in (8).

5. Queue Lengths and Waiting Times

Let L_a and L_b be the random variables giving the number of class a and class b customers, respectively, in queue. Noting that the idle probability $p(I)$ does not contribute to the queue lengths[2], we have that

$$E[L_b] = \left.\frac{\partial P(y, z)}{\partial z}\right|_{y=z=1} = \sum_{j=0}^{N_b-1} jC_j(1) + N_b\, S(y, z)\Big|_{y=z=1} + \left.\frac{\partial S(y, z)}{\partial z}\right|_{y=z=1} \tag{11}$$

But evaluation of $S(y, z)$ and $\frac{\partial S(y, z)}{\partial z}$ at $y = z = 1$ in (11) will require application of l'Hospital's rule, differentiating with respect to y. We choose, instead, the following approach.

First, from the definition of $P(y, z)$, we have that $P(y, z)|_{y=z=1} = 1 - p(I) = \rho$. Using this fact in (10), which we evaluate at $y = z = 1$

$$S(y, z)|_{y=z=1} = \rho - \sum_{j=0}^{N_b-1} C_j(1). \tag{12}$$

Next, we evaluate the first factor on the RHS of (8) at $y = z = 1$ by first letting $y \to 1$, evaluating the limit, and then letting $z \to 1$.

From (7), $z_0(1) = 1$. Therefore,

$$\left.\frac{z_0(y) - z}{\rho_a yz + \rho_b z^2 - (1+\rho)z + 1}\right|_{y=z=1} = \left.\frac{1-z}{(z-1)(\rho_b z - 1)}\right|_{z=1} = \frac{1}{1-\rho_b} \tag{13}$$

Evaluating (8) and using (12) and (13),

$$\frac{1}{(1-\rho_b)} \cdot \lim_{y\to 1} A(y) = \rho - \sum_{j=0}^{N_b-1} C_j(1)$$

$$\Rightarrow \lim_{y\to 1} A(y) = (1-\rho_b)(\rho - \sum_{j=0}^{N_b-1} C_j(1)). \tag{14}$$

Now

$$\frac{\partial S(y, z)}{\partial z} = \frac{(\rho_b z^2 - 1) - z_0(y)(\rho_a y + 2\rho_b z - (1+\rho))}{(\rho_a yz + \rho_b z^2 - (1+\rho)z + 1)^2} \cdot A(y).$$

Evaluating $\frac{\partial S(y, z)}{\partial z}$, first as $y \to 1$ using (14), then as $z \to 1$,

$$\left.\frac{\partial S(y, z)}{\partial z}\right|_{y=1,\ z=1} = \{\left.\frac{(\rho_b z^2 - 1) - (2\rho_b z - (1+\rho_b))}{(\rho_b z^2 - (1+\rho_b)z + 1)^2}\right|_{z=1}\} \cdot$$

$$(1-\rho_b)(\rho - \sum_{j=0}^{N_b-1} C_j(1))$$

$$= \{\left.\frac{\rho_b (z-1)^2}{(\rho_b z - 1)^2 (z-1)^2}\right|_{z=1}\} \cdot (1-\rho_b)(\rho - \sum_{j=0}^{N_b-1} C_j(1))$$

$$\Rightarrow \left.\frac{\partial S(y, z)}{\partial z}\right|_{y=z=1} = \frac{\rho_b}{1-\rho_b}(\rho - \sum_{j=0}^{N_b-1} C_j(1)). \tag{15}$$

Using (12) and (15) in (11), and after manipulation,

$$E[L_b] = \sum_{j=0}^{N_b-1} jC_j(1) + \frac{N_b - (N_b - 1)\rho_b}{1-\rho_b}(\rho - \sum_{j=0}^{N_b-1} C_j(1)). \tag{16}$$

Recalling that, if we do not distinguish between customer classes, our model is just an $M|M|1$ system whose mean queue length is given by $\frac{\rho^2}{1-\rho}$ [5, 6, 8, 9] we have that $E[L_a] + E[L_b] = \frac{\rho^2}{1-\rho}$; hence,

$$E[L_a] = \frac{\rho^2}{1-\rho} - E[L_b], \tag{17}$$

where $E[L_b]$ is defined in (16).

Let T_a and T_b be the random variables giving the waiting times in queue for, respectively, class a and class b customers. Each class' queue forms a distinct subsystem to which Little's result may be independently applied [6, 8, 11]. Thus,

$$E[T_a] = \frac{E[L_a]}{\lambda_a} \quad \text{and} \quad E[T_b] = \frac{E[L_b]}{\lambda_b} ; \tag{18}$$

where $E[L_a]$ and $E[L_b]$ are given, respectively, in (16) and (17). Expression (16) or (18) may be used to choose a vlaue for N_b, to implement the following policy: maximize class a performance provided that mean class b queue length (or waiting time) does not exceed a specified value. Observing that $C_j(1)$ is not a function of N_b, by the theorem of Section 2, we can compute $C_j(1)$ for all j less than some N, and then quickly use (16) or (18) to calculate class b performance for all $N_b \leq N$.

We can find higher moments (in fact, the Laplace transform) of waiting time in queue for class a customers, only, assuming FCFS service within queue a. This is similar to a well-known method for finding the waiting time Laplace transform of an FCFS M/G/1 queue [5, 6, 8, 9], and depends on the fact that class a arrivals subsequent to that of a tagged class a customer have no effect on the tagged customer's waiting time.

The class a queue forms a distinct subsystem which, in particular, changes state by unit jumps only. Further, the arrival process to this subsystem is Poisson (with rate λ_a). Let

$q_a(n_a)$ = Pr (class a arrival to the queue finds $n_a \geq 0$ class a customers already in queue),

$q_d(n_a)$ = Pr (class a departure from queue leaves $n_a \geq 0$ class a customers behind it in queue),

$q_t(n_a)$ = Pr ($n_a \geq 0$ class a customers are in queue at an arbitrary time instant t)

$= \sum_{j=0}^{\infty} p(n_a, j)$.

It is well known [8, 9] that $q_d(n_a) = q_a(n_a) = q_t(n_a) = \sum_{j=0}^{\infty} p(n_a, j)$. Let $W_a(t)$ be the class a queueing time distribution (i.e., waiting time in queue), with Laplace transform $W_a^*(s) = \int_0^\infty e^{-st} dW_a(t)$. Note that any class a customers left behind in queue by a tagged class a customer departing the queue (to enter service) must have all arrived during the latter's residence time in queue. Since such arrivals have no effect on the waiting time of the tagged customer, we may write,

$$q_d(i) = \sum_{j=0}^{\infty} p(i, j) = \int_0^\infty \frac{(\lambda_a t)^i}{i!} e^{-\lambda_a t} dW_a(t), \tag{19}$$

which, after multiplication by y^i and summation over j, yields:

$$P(y, 1) = \int_0^\infty e^{-(1-y)\lambda_a t} dW_a(t) = W_a^*(\lambda_a - \lambda_a y).$$

Thus,

$$W_a^*(\lambda_a - \lambda_a y) = P(y, 1) = \sum_{j=0}^{N_b - 1} C_j(y) + \frac{(z_0(y) - 1)}{\rho_a(y - 1)} \cdot A(y) \tag{20}$$

where we have used (8) and (10) to evaluate P(y, 1) and where $C_j(y)$ and $z_0(y)$ are given in (5) and (7) respectively, and A(y) in (9). Expression (20) can be differentiated with respect to $s = \lambda_a - \lambda_a y$ to obtain moments of T_a, but the algebra is tedious. Recall that we have already determined $E[T_a]$ in (17) and (18).

An argument similar to the one used above to obtain (20) cannot be applied to the class b queue to obtain its queueing time distribution. This is because class b arrivals subsequent to a tagged class b arrival to the queue have a direct effect on the tagged customer's waiting time, in that they partially determine the priority of queue b. Hence, the analogous relation to (19) for queue b does not hold. Of course, we have determined the mean waiting time for class b in (18).

Finally, it is of interest to determine the frequency with which each queue has high priority. We have that $S(y, z) = \sum_{j=N_b}^{\infty} C_j(y) z^j$. Taking $z = 0$ in (8):

$$C_{N_b}(y) = S(y, 0) = A(y), \tag{21}$$

where $A(y)$ is defined in (9). Further, taking vertical cutsets between columns $N_b + i$ and $N_b + i + 1$, $i \geq 0$, in Figure 2, it is straightforward to show that

$$C_j(y) = \rho_b^{\,j-N_b} C_{N_b}(y) = \rho_b^{\,j-N_b} A(y), \quad j \geq N_b \tag{22}$$

Expressions (5), for $0 \leq j < N_b$, and (22) together give $C_j(y)$ for all j. In particular:

$$\Pr\,(\text{class a has high priority} \mid \text{server busy}) = \frac{1}{\rho} \sum_{j=0}^{N_b} C_j(1); \tag{23}$$

and, either from (20), using (13) and (19), or from the fact that $\Pr\,(\text{server busy}) = \rho$,

$$\Pr\,(\text{class b has high priority} \mid \text{server busy}) = \frac{1}{\rho} \sum_{j=N_b+1}^{\infty} C_j(1)$$

$$= 1 - \frac{1}{\rho} \sum_{j=0}^{N_b} C_j(1). \tag{24}$$

The two results (23) and (24) give us a measure of the mutual interference between the two classes.

6. Discussion and Application

We may use Kleinrock's Conservation Law for M/G/1 nonpreemptive, work-conserving (in the sense that no work is either created or destroyed in the system) priority systems [10], to obtain, for our system:

$$\lambda_a E[T_a] + \lambda_b E[T_b] = \frac{\rho^2}{1-\rho}, \quad \lambda_a, \lambda_b, \lambda < \mu \tag{25}$$

For a given total load ρ, the total work in the system is a constant and so any reduction of one class's mean waiting time in queue may be achieved only at the cost of increasing the other class's mean. In particular, a unit reduction in $E[T_a]$ will cause an increase in $E[T_b]$ of λ_a/λ_b. Similarly, a unit reduction in $E[T_b]$ will cause $E[T_a]$ to increase by $\frac{\lambda_b}{\lambda_a}$ units. The relative values of λ_a and λ_b are thus central to the determination, in any given case, of the offset cost paid in terms of an increase in one class's mean queueing time due to a reduction of the other class's mean queueing time.

In order to characterize the system's behavior (in terms of mean queueing times) and to place the above discussion on the conservation law in perspective, we plot the results for two cases in Figures 5 and 6. The system in Figure 5 has $\lambda_a = .15$, $\lambda_b = .6$, $\mu = 1$, $\frac{\lambda_a}{\lambda_b} = 0.25$. In Figure 5a we plot the mean queueing times for each class at various settings of the threshold level N_b, starting with the "benchmark" case $N_b = 0$ (class b has priority always) and ending with the other "benchmark" case $N_b = \infty$ (class a has priority always). The results for the benchmark cases are obtained from Cobham's model [4, 7, 10]. In Figure 5b we plot the quantities:

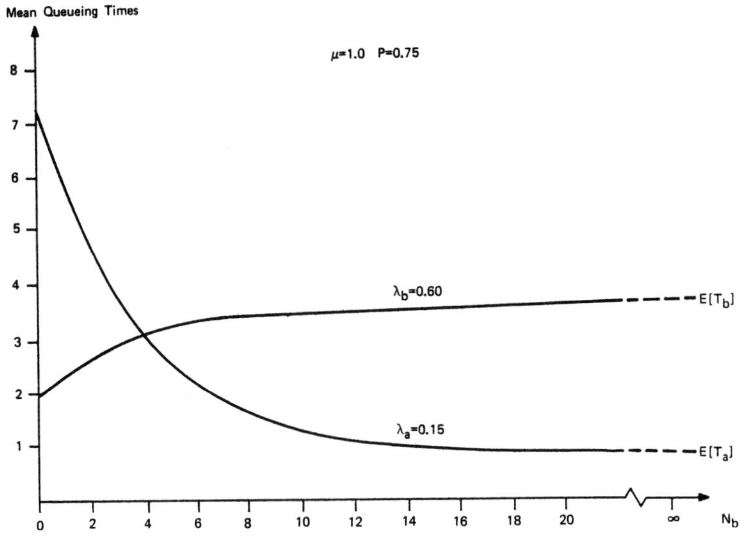

Figure 5a

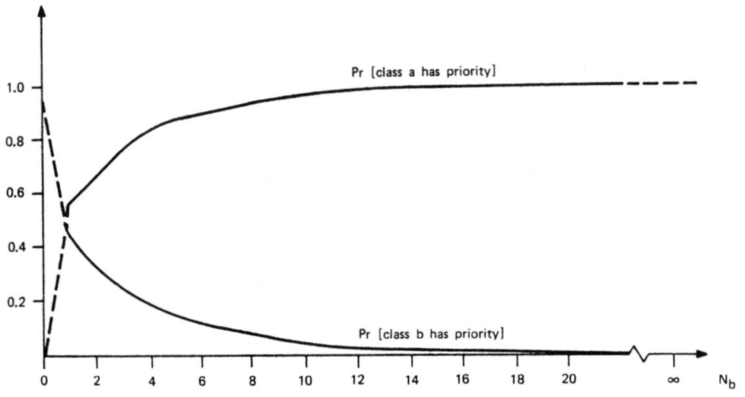

Figure 5b

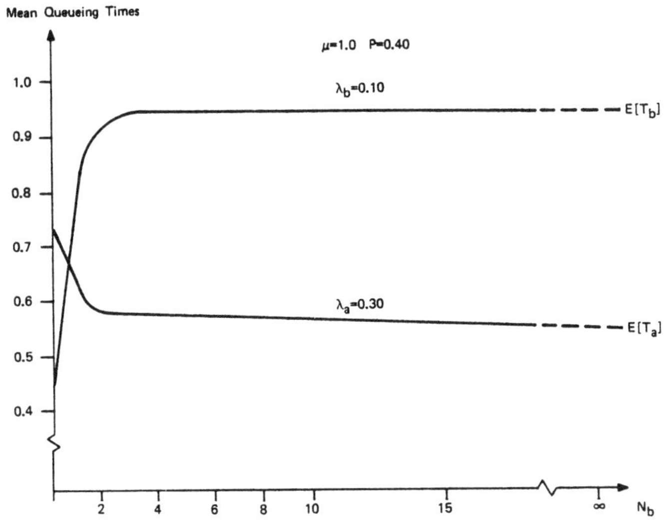

Figure 6a

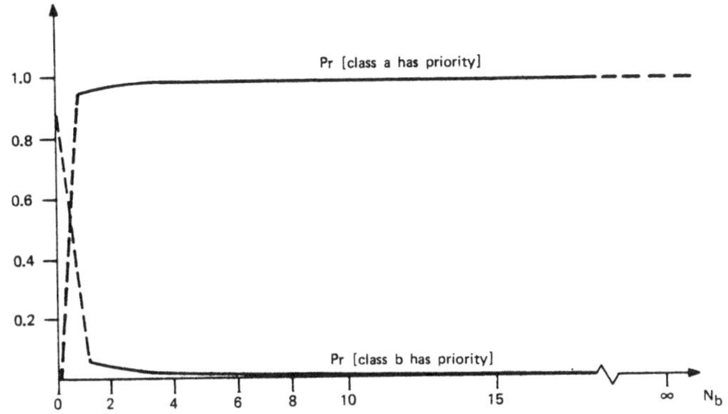

Figure 6b

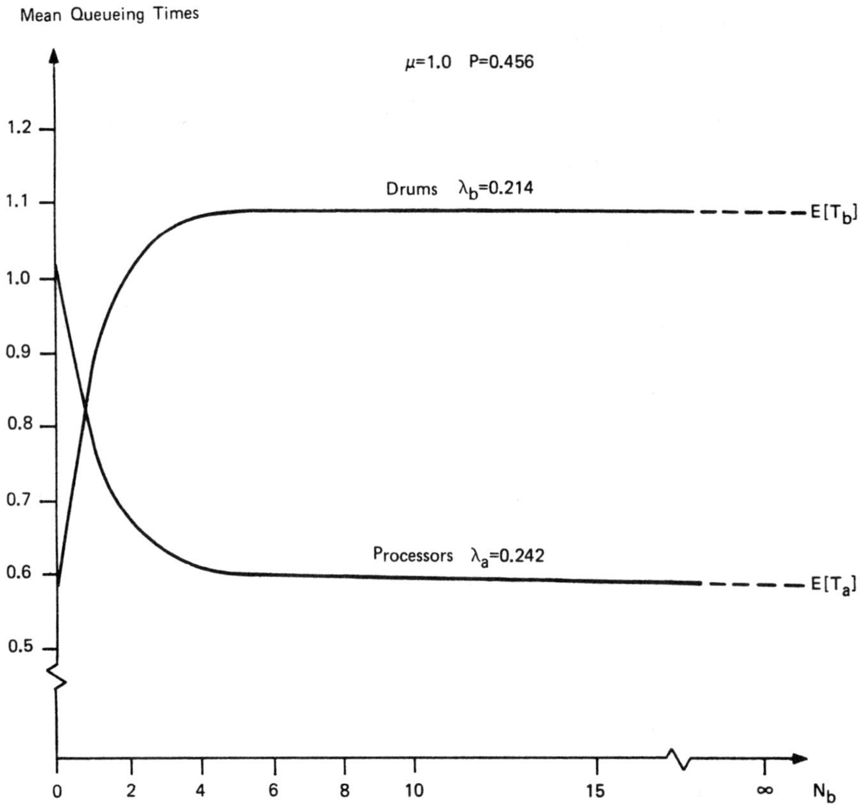
Mean Queueing Times
μ=1.0 P=0.456
1.2
1.1
1.0
0.9
0.8
0.7
0.6
0.5
Drums λb=0.214
E[Tb]
Processors λa=0.242
E[Ta]
0
2
4
6
8
10
15
∞
Nb

Figure 7a

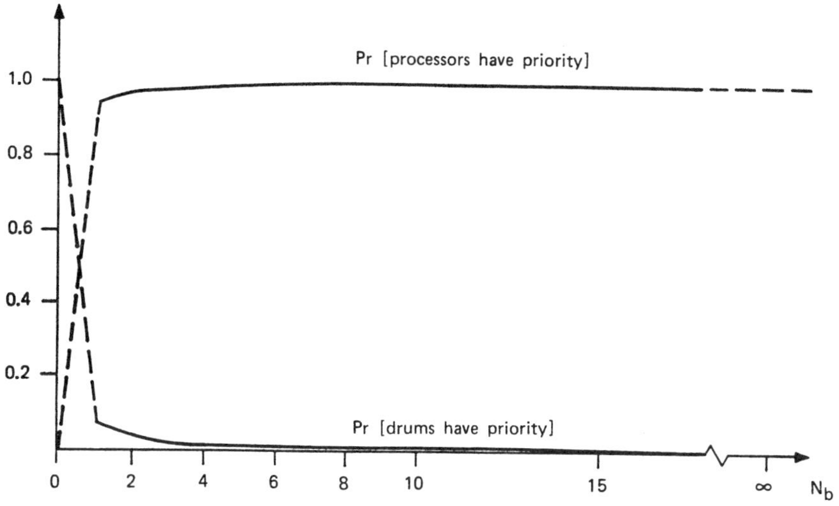

Figure 7b

Threshold setting N_b	Processor requests' mean waiting time in queue	Percentage improvement over standard case (N_b = 0)
0	1.0665	------
1	0.79492	25.46
2	0.68115	36.13
3	0.63446	40.51
4	0.61526	42.31
5	0.60732	43.06
10	0.60166	43.591
15	0.60159	43.592
20	0.60158	43.593
∞	0.60158	43.593

Figure 7c

Proportion of time class b has priority in a busy period =

$$\Pr\ (\text{class b has priority} \mid \text{busy period})$$

and

Proportion of time class a has priority in a busy period =

Pr (class a has priority|busy period)

as a measure of the mutual interference between the classes. In Figures 6a and 6b we plot analogous results for the system with $\lambda_a = .3$, $\lambda_b = .1$, $\mu = 1$, $\frac{\lambda_a}{\lambda_b} = 3$.

We now consider an example application of our model. We have assumed service times for both customer classes to be identically distributed; thus the model is not well suited to situations in which service times are class-dependent. However, there are instances in computer systems in which all customers can be assumed identical with respect to service time distribution, but in which there is a partition of customers into classes on which scheduling is to be performed.

We shall consider a memory system architecture similar in many respects to that of the BCC-500 computer system [3]. This architecture attempts to reduce the interference of drum requests to main memory upon central processor requests to the memory.

The BCC-500 system has 2 central processors and 1μs cycle-time main memory, interleaved by a factor of 8. We will model a single memory module. Drum requests to main memory are issued at a peak-load rate of 0.214 requests|memory module|memory cycle. Drum requests are predicted (as consecutive locations within a page) and pre-issued by the drum controller so that relatively long waiting times are tolerable for such requests. Hence, processor requests (customer class a) are normally accorded higher priority than drum requests (customer class b). But if the queue of drum requests backs up to the point that loss of the drum record is imminent, then such requests are immediately accorded high priority. (Although it is still possible to lose the record, this is a recoverable condition as the record need only be re-transferred.)

Let the unit of time be one memory access time, so that $\mu = 1$, and

$\lambda_b = 0.214$. Suppose $\lambda_a = 0.242$, which yields an average of 2 pending requests/processor (over all modules) when $N_b = 0$ (processor has unconditional priority). In Figure 7a we plot the mean queueing times $E[T_a]$ and $E[T_b]$ for increasing values of N_b. Note that the benchmark case $N_b = 0$ represents a "conventional" strategy in which drum requests always have priority over processor requests. Observe that $E[T_a]$ and $E[T_b]$ are mean waiting times for a request to begin service on the module, after that request is initially made, and do not include the service time itself. Figure 7b gives a measure of the potential interference between processor and drum requests, from (23) and (24). Figure 7c tabulates the relative improvement in the mean waiting times of processor requests as compared with the case $N_b = 0$ (drum always has priority). This improvement is achieved at the cost of increasing the mean size of the queue of drum requests.

7. Conclusions and Acknowledgements

Priority scheduling has certain advantages, but it must be applied carefully in heavy-load situations. Low priority requests may be arbitrarily delayed in such a case. In practice, many priority schemes do incorporate a mechanism (priority increase with time, for example) to attempt to guarantee at least a stated minimum level of service to every request. Such schemes do not always work as intended, however. For example, if priorities are simply increased with time, there is the well-known possibility of "priority inflation" under heavy load, in which no request is granted except at higher-than-initial priority, and the schedule differs little from one of fixed priority. The method we have proposed appears immune to such problems, although under heavy lower-class load, the lower class may actually receive priority service most of the time.

A further advantage of this scheme, as compared with other proposed modified-priority schemes, is that there is only a single, integer-valued parameter (N_b) to determine. The (exact) decomposition approach of the Corollary in Section 2 provides one means for determining N_b: one can apply known results for fixed-priority queues to calculate performance below the N_b barrier, choosing N_b to achieve a desired value. As previously noted, this decomposition technique applies to the model even under M/G/1 assumptions and with different service-time distributions for the two classes. Although this technique does not appear to have especially wide application to other kinds of models, there may be others for which it can be used.

We are grateful to Donald Rung for his helpful suggestions about contour integration.

8. References

[1] Badr, H. G. and J. R. Spirn, "An Adaptive-Priority Queue," The Pennsylvania State University Computer Science Technical Report CS-79-38.

[2] Badr, H. G. and J. R. Spirn, "An M/G/1 Adaptive-Priority Queue," to appear.

[3] Chattergy, R. and W. W. Lichtenberger, "BCC 500: A Distributed Function Multiprocessor System," University of Hawaii Electrical Engineering Technical Report, 1978.

[4] Cobham, A., "Priority Assignment in Waiting Line Problems," Oper. Res., Vol. 2, pp. 70-76, 1954.

[5] Cohen, J. W., The Single Serve Queue, North Holland (Amsterdam), 1969.

[6] Coffman, E. G., and P. J. Denning, Operating Systems Theory, Prentice Hall, Englewood Cliffs, NJ, 1973.

[7] Conway, R. L., W. L. Maxwell, and L. W. Miller, Theory of Scheduling, Addison-Wesley, Reading, Mass., 1967.

[8] Gross, D., and C. M. Harris, Fundamentals of Queueing Theory, Wiley-Applied Statistics, New York, 1974.

[9] Kleinrock, L., Queueing Systems, Vol. I: Theory, Wiley-Interscience, New York, 1975.

[10] Kleinrock, L., Queueing Systems, Vol. II: Computer Applications, Wiley-Interscience, New York, 1975.

[11] Little, J. D. C., "A Proof of the Queueing Formula L = λW," Oper. Res., Vol. 9, pp. 383-387, 1961.

9. Endnotes

[1] Although the characterization is not complete in as much as the class of customer currently in service is unknown, it is sufficient to permit an analysis to be undertaken.

[2] Recall that P(I) is not included in transform P(y, z).

† Department of Computer Science, SUNY at Stony Brook, Stony Brook, New York 11796

†† Computing Laboratory, The University of Newcastle, Newcastle upon Tyne, UNITED KINGDOM

††† Digital Equipment Corporation, Maynard, Massachusetts 01754

MARKOV CHAIN MODELS IN PERFORMANCE ANALYSIS
Matt Sobel, Chairman

P. Tzelnic

THE LENGTH OF PATH FOR FINITE MARKOV CHAINS AND ITS APPLICATION TO MODELLING PROGRAM BEHAVIOUR AND INTERLEAVED MEMORY SYSTEMS

Percy Tzelnic

Abstract

The distribution of the <u>number of distinct states</u> visited along a Markov chain path is obtained. This random variable is hereby called <u>length of path</u>, to be distinguished from the well known first passage time (the number of steps taken firstly to reach a given state). The length of path is related to a notion of capacity in potential theory for Markov chains.

The use of these results is warranted in a variety of possible applications, wherever probabilistic walks on graphs may be benefically described not only in terms of the number of steps taken, but also by the number of distinct nodes traversed.

In this paper, two different areas of application are considered. One concerns modelling program behaviour in virtual memory systems by a Markov chain model. In this context, the paging rate of the Least Recently Used paging algorithm is obtained. The other concerns the memory bandwidth in interleaved memory systems with saturated demand, where the stream of memory requests is Markovian. An approach to computing the mean memory bandwidth is proposed.

1. Introduction

1.1 The original motivation of this paper is in modelling of computer programs' behavior (PB) as streams of successive memory (page) references, in (paged) virtual memory systems. A model of PB is a conveniently defined stochastic process which is amenable to the calculation of performance measures of these computer systems. The simplest such process, a sequence of i.i.d. variables -- or, a zero order Markov chain -- was termed the Independent Reference Model (IRM) and used by Denning and Schwartz ([5]) to obtain the first moment of the Working Set size (see section 3.4 below). In obtaining this result a sample path variable of the PB process played a central role. This variable is the first return time, in this context usually termed the inter-reference interval. Although used in many subsequent works, it was never made clear whether the inter-reference interval is intrinsically important, or whether another path variable may be better suited to a different model.

When attempting to generalize the IRM to a first order Markov chain model (Tzelnic [16]), I realized that a sample path characterization by a different variable is needed. This variable is the _number of distinct states_ visited along the path, especially for first passage paths. Furthermore, the investigation of such paths that avoid a certain subset of states (_taboo set_) appeared necessary.

1.2 This variable I termed _length of path_, to distinguish it from the well-known (Kemeny and Snell [12]) _first passage time_, or, to use another ad-hoc term -- _path duration_. See Figure 1 for an illustration of these concepts, in a special case of a Markov chain whose 6 states are, for convenience, denoted as i, j, j_1, j_2, j_3, j_4, and a path whose end states are i and j. The arcs joining the states visited are numbered in the chronological sequence of visits.

The length of path (related to a notion of capacity in the potential theory for Markov chains -- Syski [13], [14]) appears not to have received any formal consideration. It seems to warrant a certain conceptual interest, extending beyond the scope of the modelling effort that led to it -- various applications where probabilistic walks on graphs are to be described by their length, and not only by their duration, come to mind (probabilistic automata, computational complexity, etc.).

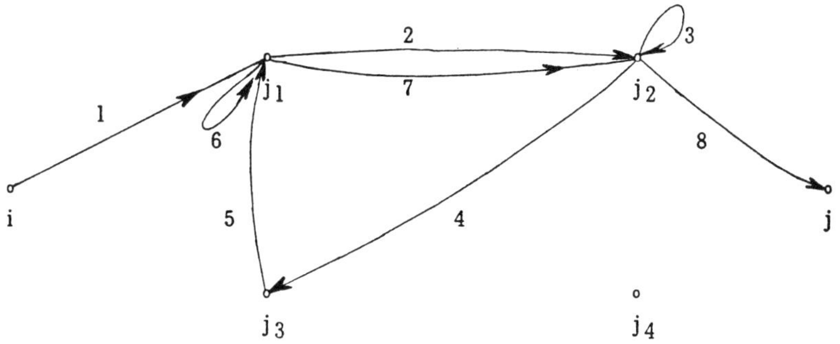

Figure 1. An illustration of the length of path concept
T(path duration) = 8
D(length of path) = 3

In section 2, the probability mass function (pmf) of the length of path is derived under various conditions. The tools used to establish the main result (Proposition 2.1) are routine probabilistic techniques (effective conditioning, taboo probabilities, the combinatorial inclusion-exclusion principle). What renders them methodologically important is the demonstration of how an approach based on path counting leads to closed form expressions.

1.3 Several applications of these results are presented in section 3. First, within the modelling context that has motivated this excursion into Markov chain theory -- the Markov Chain Model (MCM) of program behaviour. In section 3.1 the performance of the Least Recently Used paging algorithm (LRU) is evaluated for the MCM; the numerical complexity of this result is analyzed in section 3.2, and a generalization of the LRU is proposed and investigated in section 3.3. Second, an application to an unrelated problem -- the modelling of interleaved memory systems -- is presented in section 3.4. Another application to the MCM, namely the computation of the Working Set size distribution and of its first moments, is treated in Hofri and Tzelnic ([11]).

2. Length of Path

2.1 Let $X = \{X(t),\ t=1,2,\ldots\}$ be a Markov chain defined over a canonic probability space, with values in the finite set of states $N = \{1,2,3\ldots n\}$. Its one-step transition probability matrix, P, has elements:

$$p_{ij} = P[X(t+1)=j \mid X(t)=i], \quad i,j \in N. \tag{2.1}$$

Let $H \subset N$ be a fixed set of states, and $H^c = N - H$. The s-step transition probability with taboo set H (briefly, the taboo probability) is defined for all $i,j \in N$ (Syski [13]):

$$_Hp_{ij}^{(s)} = P[X(2) \in H^c,\ldots X(s) \in H^c, X(s+1)=j \mid X(1)=i], \quad s > 1. \tag{2.2}$$

For $s=1$, $_Hp_{ij}^{(1)} = p_{ij}$ and for $s=0$ we define $_Hp_{ij}^{(0)} = \delta_{ij}$. Sample paths for which H is a taboo set are called H-avoiding paths.

Taboo probabilities satisfy the Chapman-Kolmogorov equations:

$$_Hp_{ij}^{(s+r)} = \sum_{k \in H^c} {_Hp_{ik}^{(s)}}\ {_Hp_{kj}^{(r)}}. \tag{2.3}$$

The first entrance ("hit") time to H is T_H (if no possibility of confusion arises, T will be used for brevity), defined as:[1]

$$T_H = \inf\{t \mid t>1,\ X(t) \in H\}. \tag{2.4}$$

The probability of first entrance to H in s steps, starting from state i, is:

$$_H h_i(s) = P[X(T)\varepsilon H,\ T{=}s{+}1 \mid X(1){=}i], \quad i \in N,\ s \geq 1 \tag{2.5}$$

where the compound random variable X(T) represents the position of the first entrance to H ("hitting place"). The joint distribution of X(T) and T is:

$$_H h_{ij}(s) = P[X(T){=}j,\ T{=}s{+}1 \mid X(1){=}i]. \tag{2.6}$$

Note that, by equations (2.2), (2.5), and (2.6):

$$_H h_{ij}(s) = {_H p_{ij}^{(s)}}, \quad i \in n,\ j \in H,\ s \geq 1, \tag{2.7}$$

and

$$_H h_i(s) = \sum_{j\varepsilon H} {_H h_{ij}(s)}, \quad i \in N,\ s \geq 1. \tag{2.8}$$

In the special case when H consists of a single state, say j, it is customary to substitute the notation $f_{ij}(s)$ for $_{\{j\}}h_{ij}(s)$. By (2.7) this is the probability of "first passage" (Kemeny and Snell [12]) from i to j at the s-th step:

$$f_{ij}(s) = P[X(2)\neq j, \ldots X(s)\neq j,\ X(s{+}1){=}j \mid X(1){=}i], \quad i,j \in N,\ s \geq 1. \tag{2.9}$$

The probability of first passage from i to j is then obtained as:

$$f_{ij} = \sum_{s=1} f_{ij}(s), \quad i,j \in N. \tag{2.10}$$

For $j = i$, f_{ii} is the first return probability to state i; if $f_{ii} < 1$, i is a transient state, and i is recurrent if $f_{ii} = 1$. A Markov chain whose states are all recurrent (as is the case for all the chains considered in this paper) is termed <u>regular</u>.

2.2 Matrix notation (Syski [13]). Write P in the partitioned form:

$$P = \begin{array}{c} H^c \\ \hline H \end{array} \overset{\begin{array}{cc} H^c & H \end{array}}{\left(\begin{array}{c|c} P_{11} & P_{12} \\ \hline P_{21} & P_{22} \end{array}\right)} \tag{2.11}$$

Define the matrix projector operators I_1 and I_2:

$$I_1 = \begin{array}{c|c|c} & H^c & H \\ \hline H^c & I & 0 \\ H & 0 & 0 \end{array}, \qquad I_2 = \begin{array}{c|c|c} & H^c & H \\ \hline H^c & 0 & 0 \\ H & 0 & I \end{array}, \tag{2.12}$$

where I and 0 stand for the unit and the null matrices with the corresponding dimensions. I_1 and I_2 in (2.12) represent a partition of unity: $I = I_1 \oplus I_2$.

The taboo matrix ${}_HP^{(s)}$, defined by elements in equation (2.2), can be now expressed as:

$${}_HP^{(s)} = P(I_1P)^{s-1} = (PI_1)^{s-1}P = \begin{pmatrix} P_{11}^s & P_{11}^{s-1}P_{12} \\ P_{21}P_{11}^{s-1} & P_{21}P_{11}^{s-2}P_{12} \end{pmatrix}$$

for $s > 1$, and

$${}_HP^{(1)} = P. \tag{2.13}$$

Equation (2.13) follows from the Chapman-Kolmogorov equations (2.3).

Ultimately our interest is focused on taboo probabilities of only those H-avoiding paths that reach states from H; therefore, instead of ${}_HP^{(s)}$ from (2.13), the form ${}_HP^{(s)}I_2$ will be preferred (for computational convenience).

From the transition matrix P we produce n matrices P_j. Such a matrix is the $n \times n$ zero matrix, with the exception of column j, which is column j of P. The algebraic sum of r such matrices: $P_{j_1},\ldots P_{j_r}$ $(r = 1,\ldots n)$ will be further designated by P_{J_r}, where $J_r = \{j_1,\ldots j_r\}$. A notation like P_{iJ_r}, signifying $P_i + P_{J_r}$, will sometimes be used. We also find use for the n matrices $\underline{P}_j$, defined as $\underline{P}_j = P - P_j$, i.e., matrices identical to P with the exception of a zero column j.

As a matter of notational consistency, we will use $\hbar$ to denote joint probability mass functions (pmf) of length of path and path duration (analogously to the above use of h for pmf related to path duration); somewhat less consistently, f will be used to denote marginal pmf of length of path.

2.3 After this brief excursion through general Markov chain concepts, with an emphasis on first entrance probabilities, we come to the point of this section. Consider an H-avoiding path of X, which starts from state i and first enters the taboo set H by the s-th step, at the hitting place j, $j \in H$. Along this path some states of H^c are visited at least once, while others not at all. The number of distinct states of the first category, excluding both ends of the path, is hereafter termed length of path.

Clearly, the length of a path whose duration is s successive steps is at most $s - 1$. There is a distinction to be made between "H-open" paths ($j \in H$, while $i \in H^c$; obviously $j \neq i$) and "H-closed" paths (both $i, j \in H$; either $j \neq i$, or $j = i$). If $j \neq i$, in the first case the sample path can return several times to i before hitting H, whereas in the second case this is forbidden[2].

We make this discussion more precise:

Definition 2.1. Let $L = \{X(2), X(3), \ldots X(T-1)\}$, $X(T) \in H$, and $X(t) \in H^c$ for all $1 < t < T$, be an H-avoiding sample path of the regular Markov chain X. L is a random set, whose elements are the values taken by the above $T - 1$ chain variables. Let us denote by $L - X(1)$ the above set, from which the element whose value is X(1) is excluded (if such an element exists). D, the length of the path L, is a random variable defined as the size of this set:

$$D = |L - X(1)|. \qquad (2.14)$$

In order to formalize the above mentioned distinction between H-open and H-closed paths, we may introduce another random variable, D':

Definition 2.2. The length of path for an H-closed path, D', is the random variable:

$$D' = \begin{cases} D, & X(1) \in H, \\ \text{undefined}, & X(1) \in H^c. \end{cases} \tag{2.15}$$

We now present the main results. Firstly, we give the joint taboo pmf of D', T, and X(T), conditional upon the starting state. From this, the joint conditional pmf of D, T and X(T) is obtained. Next, the marginal pmf of D and X(T) is derived. The proof of the first result is given in Appendix A (the others follow a similar pattern).

Proposition 2.1. The joint conditional pmf of D', T, and X(T), for given X(1), i.e., the probability that an H-closed sample path of X starting from state $i \in H$ firstly returns to the taboo set H at the hitting place $j \in H$, in s steps, after having visited k distinct states in H^c, is:

$${}_H\hbar'_{ij}(s,k) = \sum_{r=1}^{k} (-1)^{k-r} \binom{n-m-r}{k-r} \sum_{J_r \subset H^c} [P_{J_r}^{s-1} P_j]_{ij}, \tag{2.16}$$

for $1 \leq k \leq n-m$, $s \geq k+1$, $m = |H|$, $r = |J_r|$.

Note that equation (2.16) can be extended for $k = 0$ (a path that returns to H by the first step):

$${}_H\hbar'_{ij}(1,0) = p_{ij}, \quad i,j \in H. \tag{2.17}$$

Proposition 2.2. The joint conditional pmf of D, T, and X(T), for given X(1), i.e., the probability that an H-open sample path of X starting from state $i \in H^c$ first enters the taboo set H at the hitting place $j \in H$, in s steps, after having visited k distinct states in H^c, is:

$$ {}_{H}\hbar_{ij}(s,k) = \sum_{r=0}^{k} (-1)^{k-r}\binom{n-m-r-1}{k-r} \sum_{J_r \subset H^c - i} [P_{iJ_r}^{s-1} P_j]_{ij}, \tag{2.18} $$

for $1 \leq k \leq n-m-1$, $s \geq k+1$, $m = |H|$, $r = |J_r|$.

Note that equation (2.18) too can be extended for $k = 0$ (the sample path loops $s-1$ times around state i before hitting j at the s-th step):

$$ {}_{H}\hbar_{ij}(s,0) = p_{ii}^{s-1} p_{ij}. \tag{2.19} $$

An important special case is defined by $H = \{j\}$. In this case, we obtain by direct substitution in (2.16)-(2.19) (dropping the pre-subscript "H"):

$$ \hbar'_{jj}(s,k) = \sum_{r=1}^{k} (-1)^{k-r}\binom{n-r-1}{k-r} \sum_{J_r \subset N-j} [P_{J_r}^{s-1} P_j]_{jj}, \tag{2.20} $$

$$ j \in N,\ 1 \leq k \leq n-1,\ s \geq k+1, $$

$$ \hbar'_{jj}(1,0) = p_{jj}, $$

and

$$ \hbar_{ij}(s,k) = \sum_{r=0}^{k} (-1)^{k-r}\binom{n-r-2}{k-r} \sum_{J_r \subset N-i-j} [P_{iJ_r}^{s-1} P_j]_{ij} \tag{2.21} $$

$$ i,j \in N,\ 0 \leq k \leq n-2,\ s \geq k+1. $$

2.4 Example. For $N = \{1,2,3\}$, $j = 1$, equation (2.20) yields:

$$ \hbar'_{11}(1,0) = p_{11}, \qquad s = 1, $$

$$ \hbar'_{11}(s,1) = [P_2^{s-1} P_1]_{11} + [P_3^{s-1} P_1]_{11} $$

$$ = p_{12} p_{22}^{s-2} p_{21} + p_{13} p_{33}^{s-2} p_{31}, \qquad s \leq 2, $$

$$ \hbar'_{11}(s,2) = -[P_2^{s-1} P_1]_{11} - [P_2^{s-1} P_1]_{11} $$

$$ + [(P_2 + P_3)^{s-1} P_1]_{11}, \qquad s \geq 3. \tag{2.22} $$

For instance, when $s = 3$ we have:

$$ \hbar'_{11}(3,2) = p_{12} p_{23} p_{31} + p_{13} p_{32} p_{21}. $$

For $i=1$, $j=2$, equation (2.21) yields:

$$\hbar_{12}(s,0) = p_{11}^{s-1}p_{12}, \qquad s \geq 1,$$

$$\hbar_{12}(s,1) = -[P_1^{s-1}P_2]_{12} + [(P_1+P_3)^{s-1}P_2]_{12}, \qquad s \geq 2. \tag{2.23}$$

When $s=3$:

$$\hbar_{12}(3,1) = p_{11}p_{13}p_{32} + p_{13}p_{31}p_{12} + p_{13}p_{33}p_{32}.$$

2.5 Far more interesting than the above (in view of the subsequent applications) is the joint pmf of the length of path and hitting place. We obtain this pmf as a marginal distribution from the corresponding one derived in section 2.3. To do this, we have to sum over all the possible path durations, s, in the equations (2.16)-(2.19). A lemma from (Kemeny and Snell [12]) is of use here: if A is a $n \times n$ sub-stochastic matrix, then $I-A$ is invertible, and the following equation holds:

$$(I-A)^{-1} = \sum_{s=0}^{\infty} A^s.$$

The lemma assures the existence of the matrices $(I-P_{J_r})^{-1}$, for all $r<n$. With the notation ${}_Hf_{ij}(.)$ for the marginal pmf derived from ${}_H\hbar_{ij}(s,.)$, the following result is then obtained:

<u>Proposition 2.3</u>: The joint conditional pmf of D' and X(T), and of D and X(T), for given starting state X(1) are respectively:

$${}_Hf'_{ij}(k) = \sum_{r=1}^{k} (-1)^{k-r}\binom{n-m-r}{k-r} \sum_{J_r \subset H^c} [P_{J_r}^k (I-P_{J_r})^{-1}P_j]_{ij},$$

$$i,j \in H, \ 1 \leq k \leq n-m; \tag{2.24}$$

$${}_Hf_{ij}(k) = \sum_{r=0}^{k} (-1)^{k-r}\binom{n-m-r-1}{k-r} \sum_{\substack{J_r \subset H^c, \\ i \notin J_r}} [P_{iJ_r}^k (I-P_{iJ_r})^{-1}P_j]_{ij},$$

$$i \in H^c, \ j \in H, \ 1 \leq k \leq n-m-1,$$

where $m = |H|$, $r = |J_r|$.

3. Applications

3.1 The Markov chain model (MCM) of program behavior (PB) is defined by specifying a finite Markov chain, as in section 2.1, whose sample paths are realizations of all the program's reference strings (streams of successive page references). Such models have been considered in (Aho et al. [1], Freiberger et al. [8], Franklin and Gupta [7], Glowacki [10], Courtois [3], Courtois and Vantilborgh [4]). More recently (Hofri and Tzelnic [11] and Tzelnic [16]) have shown how the pmf of the Working Set size can be derived for the MCM, using some of the techniques developed in section 2. Here it is presented a derivation of the asymptotic miss rate of the paging algorithm Least Recently Used (LRU). This result is computationally different, and numerically superior, to that obtained in (Glowacki [10]).

Let us first define the conceptual modelling framework for the MCM of PB.

A memory state is a collection of a program's pages resident in main memory. For a fixed memory allocation to a program of, say, size c, a memory state can be modelled at time t by a subset $S(t)$ of N (the set of all the program pages). If the page reference at t, represented by $X(t)$, is such that $X(t) \notin S(t)$, a miss (page fault) occurs. The page $X(t)$ has to be brought into memory, taking the place of one of the c pages in $S(t)$. (The identity of the page which is to be removed is established by a control mechanism that models the paging algorithm). Thus, a new memory state is created, say, $S(t+1)$. If $X(t) \in S(t)$, then $S(t+1) = S(t)$.

The derived process $S = \{S(t),\ t = 1,2,\dots\}$, is the model of the memory evolution, as determined by the reference string and the paging algorithm. It is important to note that the above description corresponds to page fetching _on demand_ (as opposed to prefetching, i.e., bringing pages into memory prior to their reference) and this is how most paging algorithms work.

The cost of a paging algorithm is usually evaluated as the number of misses accumulated over the life of a program, or, locally, by the miss rate function (the ratio of page faults to the number of page references, within some time interval).

One of the most widely used paging algorithms is the LRU. For its description, it is convenient to model the memory evolution by $\underline{S}$, the _memory stack process_, rather than by S. The memory stack at t, $\underline{S}(t)$, is defined as a vector whose elements are those of S(t), sorted in the order of the most recent reference. Thus, X(t-1) is always the first element of $\underline{S}(t)$. When a page fault occurs at t, the memory stack is updated by the LRU algorithm in the following simple way:

(i) X(t) becomes the first element of $\underline{S}(t+1)$;

(ii) all the elements of $\underline{S}(t)$ are pushed down one position in $\underline{S}(t+1)$;

(iii) The c-th element of $\underline{S}(t)$ is the page replaced from main memory.

It can then be proven that the bivariate process $(X,\underline{S})$ is a regular Markov chain. Its states are $(i,\underline{s})$, where $i \in N$, and $\underline{s} = (s_1,s_2,\ldots s_c)$ are realizations of the memory stack process--briefly: memory stacks. If the equilibrium probability of this bivariate process is denoted by $\pi_{(i,\underline{s})}$, for all the possible pairs $(i,\underline{s})$, then it is easily seen that the asymptotic probability of a miss is given by:

$$F(\text{LRU},c) = \sum_{(i,\underline{s})} \pi_{(i,\underline{s})} [1 - \sum_{j=1}^{c} p_{is_j}], \tag{3.1}$$

where p_{is_j} is the transition probability from state i to state s_j in X.

F(LRU,c) in equation (3.1) is, by ergodicity, the asymptotic miss rate of the LRU paging algorithm, for an MCM of a program running in a fixed memory of size c. Following this approach, Glowacki ([10]) has obtained a complicated expression for F(LRU,c). The numerical complexity of that formula is essentially related to the tremendous size of the state space of $(X,\underline{S})$: $c!\binom{n}{c}$.

We follow here a different approach, based on the concept of length of path. Instead of dealing with the bivariate process $(X,\underline{S})$, we concentrate upon the chain X, and reason as follows. In the steady state of X, the probability of a miss at time t is the probability that the page reference X(t) is not in the memory stack $\underline{S}(t)$:

$$F(LRU,c) = \lim_{t\to\infty} P[X(t)\neq s_1(t),\ldots X(t)\neq s_c(t)]$$

$$= \sum_{i=1}^{n} (\lim_{t\to\infty} P[X(t)\neq s_1(t),\ldots X(t)\neq s_c(t)\,|\,X(t)=i])\pi_i, \qquad (3.2)$$

where π_i is the equilibrium probability that the chain X is in state i, and obvious convergence and limit commutativity properties were used.

Let us note now that the paranthesized expression in the RHS of equation (3.2) is, in point of fact, the probability of first return to state i along a path whose length is at least c; this follows directly from the structure of the algorithm LRU, as well as from renewal properties of the Markov chain X. Thus:

$$F(LRU,c) = \sum_{i=1}^{n} \pi_i \sum_{k=c}^{n-1} f'_{ii}(k). \qquad (3.3)$$

Using equation (2.24) in equation (3.3), we obtain the following:

<u>Proposition 3.1</u>: The asymptotic miss rate of the paging algorithm LRU, for the MCM of program behavior X, for a fixed memory allocation of size c, is:

$$F(LRU,c) = \sum_{i=1}^{n} \pi_i \sum_{k=c}^{n-1} \sum_{r=1}^{k} (-1)^{k-r} \binom{n-r-1}{k-r} \circ$$

$$\sum_{J_r \subseteq N-i} [P_{J_r}^{k} (I-P_{J_r})^{-1} P_i]_{ii}. \qquad (3.4)$$

From equation (3.3) it follows that $F(LRU,c) \geq F(LRU,c+1)$; this suggests an iterative computational procedure for obtaining F(LRU,c) for all the possible values of c $(c = 1,2,\ldots n)$:

Proposition 3.2:

$$F(LRU,n) = 0,$$

$$F(LRU,c) = F(LRU,c+1) + \sum_{i=1}^{n} \pi_i f'_{ii}(c), \qquad c = n-1, n-2, \ldots 1. \qquad (3.5)$$

3.2 Numeric complexity. When performing the iterative procedure of equation (3.5), the steady state distribution $\underline{\pi}$ of X is required -- which can be obtained in $O(n^3)$ multiplications -- as well as the pmf of the length of path for the first return to state i, for all $i = 1,\ldots n$, $c = 1,\ldots n-1$. In addition, $O(n(n-1))$ term by term multiplications of these two distributions are required.

To compute all the $f'_{ii}(c)$ the procedure given in Appendix B may be followed. The number of multiplications needed to carry it out are given by the expression

$$\left(\frac{7n^4}{3} - 4n^3 + 7n^2 - 6n\right)2^{n-4}. \qquad (3.6)$$

The additional $\frac{n^3}{8} + n(n-1)$ multiplications mentioned before are, obviously, negligible when compared to the above figure. Table 1 summarizes the numbers of multiplications needed for the computation of F(LRU,c), for several accessible values of n, for the computational procedure of Franklin and Gupta ([7]), for Glowacki's formula ([10]) and for ours -- equation (3.5).

If we take 10^{-6} sec to be a representative time for a

Table 1. The number of multiplications required for the computation of F(LRU,c)

	# multiplications \ n	5	10	15	20	25
FRANKLIN & GUPTA	$\sum_{c=1}^{n-1} \frac{1}{3} [\frac{n!}{(n-c)!}]^3$	$6x10^5$	$2x10^{19}$	$8x10^{35}$	$5x10^{54}$	$1x10^{75}$
GLOWACKI	$\sum_{c=1}^{n-1} \frac{n!}{(n-c)!} [\frac{c^3}{3c!} + c + 3]$	$2x10^3$	$7x10^7$	$4x10^{14}$	$9x10^{19}$	$7x10^{26}$
EQ. (3.5)	exp. (3.6)	$3x10^3$	$2x10^6$	$3x10^8$	$3x10^{10}$	$2x10^{12}$

multiplication in a modern computer, the LRU miss rate for a program whose virtual space includes 22 pages is computable with equation (3.5), in contrast to a maximum size of 12 pages for Glowacki's formula, or at most 6 following the approach of Franklin and Gupta.

3.3 The algorithm LRU with fixed kernel (LRUFK). The paging algorithm LRU can be significantly generalized specifying a hierarchy of algorithms that permanently keep in memory a subset of "non-pageable" pages H ($|H| = m < c$), and select the replacement page from among the $c - m$ "pageable" pages. At one end of this hierarchy is LRU ($m = 0$), at the other is a class of algorithms that include A_0 (Aho et al. [1]) at its steady state ($m = c - 1$).

This may be compared with the hierarchy of the RPPL paging algorithms of (Gelenbe [9]), which employ instead random selection with probability $1/(c - m)$ for choosing the replacement page from among the pageable ones.

Actually the LRUFK paging algorithms have a direct practical significance. A virtual system using LRUFK answers the need for freezing in memory some pages that are either in the process of information transfer (I/O in IBM 370 OS/VS, or WANG VS), or are very popular/important (the permanently resident segments in MULTICS).

The approach of section 3.1 can be followed in order to compute the asymptotic miss rate for the algorithms LRUFK, although some extensions are needed. The argument proceeds as follows: the memory stack is redefined over the $c - m$ pageable pages (excluding the pages in H). By analogy with equations (3.2), (3.3), it is established that:

$$F(\mathrm{LRUFK}, c, H) = \sum_{i \in H^c} \pi_i \sum_{k=c-m}^{n-m-1} {}_H g_{ii}(k), \qquad (3.7)$$

where

$$\sum_{k=c-m}^{n-m-1} {}_Hg_{ii}(k) = P[X(t)\neq s_1(t),\ldots X(t)\neq s_{c-m}(t)\,|\,X(t)=i], \tag{3.8}$$

and ${}_Hg_{ii}(k)$ in equations (3.7), (3.8) is the probability of the first return to state i along a path that visits exactly k distinct states of H^c (a Markov subchain, ${}_HX$, may be thus defined by restricting X to the subset H^c), and any number of states in H (references to pages in H do not cause misses, neither do they change the memory stack as redefined above). Note that in fact this is the pmf of the length of path for the Markov subchain ${}_HX$. The following result can be obtained, along the same lines as for the derivation of Proposition 2.3:

$$ {}_Hg_{ii}(k) = \sum_{r=0}^{k} (-1)^{k-r}\binom{n-m-r-1}{k-r} \circ $$

$$ \sum_{J_r\subset N-H-i} [(P_H+P_{J_r})^k(I-P_H-P_{J_r})^{-1}P_i]_{ii}. \tag{3.9}$$

To establish equation (3.9), equation (2.16) is used, substituting $H+i$ for H (i is "taboo" too) and, correspondingly, $m+1$ for m. Also, P_H in the summand accounts for the possibility of an indeterminate number of visits to H (the probability of first return to i in s steps through states of J_r and H only is -- similarly to equation (4.2) -- $[(P_H+P_{J_r})^{s-1}P_i]_{ii}$). Finally, the summation over r starts at $r=0$, since the corresponding term is not null anymore and reflects the possibility of a return path through states of H only.

Then, in the same way Proposition 3.1 was established, we obtain:

<u>Proposition 3.2</u>: The asymptotic miss rate of the paging algorithm LRUFK, for the MCM of program behavior X, for fixed kernel H and a fixed memory allocation of size c is:

$$F(LRUFK,c,H) = \sum_{i\in N-H} \pi_i \sum_{k=c-m}^{n-m-1} \sum_{r=0}^{k} (-1)^{k-r} \binom{n-m-r-1}{k-r} \cdot$$

$$\sum_{J_r \subset N-H-i} [(P_H+P_{J_r})^k (I-P_H-P_{J_r})^{-1} P_i]_{ii}, \tag{3.10}$$

for $m = |H|$.

Example. Let $N = \{1,2,3,4,5\}$, $H = \{4,5\}$ $(m=2)$. Then, for $i=1$ equation (3.8) yields:

$$\begin{aligned}
{}_H g_{11}(0) &= [(I-P_4-P_5)^{-1}P_1]_{11},\\
{}_H g_{11}(1) &= -2[(P_4+P_5)(I-P_4-P_5)^{-1}P_1]_{11}\\
&\quad + [(P_4+P_5+P_2)(I-P_4-P_5-P_2)^{-1}P_1]_{11}\\
&\quad + [(P_4+P_5+P_3)(I-P_4-P_5-P_3)^{-1}P_1]_{11},\\
{}_H g_{11}(2) &= [(P_4+P_5)^2(I-P_4-P_5)^{-1}P_1]_{11}\\
&\quad - [(P_4+P_5+P_2)^2(I-P_4-P_5-P_2)^{-1}P_1]_{11}\\
&\quad - [(P_4+P_5+P_3)^2(I-P_4-P_5-P_3)^{-1}P_1]_{11}\\
&\quad + [(P_4+P_5+P_2+P_3)^2(I-P_4-P_5-P_2-P_3)^{-1}P_1]_{11},
\end{aligned} \tag{3.11}$$

and similarly for $i = 2,3$.

Then, using equation (3.10):

$$F(LRUFK,3,\ 4,5) = \sum_{i=1}^{3} \pi_i [{}_H g_{ii}(1) + {}_H g_{ii}(2)]$$

$$\sum_{i=1}^{3} \pi_i [1 - {}_H g_{ii}(0)] = 1 - \sum_{i=1}^{3} \pi_i\, {}_H g_{ii}(0),$$

$$F(LRUFK,4,\ 4,5) = \sum_{i=1}^{3} \pi_i\, {}_H g_{ii}(2),$$

$$F(LRUFK,5,\ 4,5) = 0.$$

It is also seen that:

$$F(LRUFK,2,\ 4,5) = \sum_{i=1}^{3} \pi_i \sum_{k=0}^{2} {}_H g_{ii}(k) = \sum_{i=1}^{3} \pi_i = 1 - \pi_4 - \pi_5,$$

which obviously corresponds to the fact that when all c resident pages are fixed, all references to other pages are misses. Note though that this situation does not correspond to a realizable operation (at least one resident page should always be pageable, for the virtual memroy system to function!)

3.4 Interleaved memory systems. This is a storage system consisting of n memory modules such that each module can serve one memory request during a memory cycle. Denoting the set of memory modules by $N = \{1,2,...n\}$, a model of the system is built that is driven by the model of the stream of consecutive memory requests. This may be assumed to be a Markov chain, as in (Burnett and Coffman [2], and Török [15]) -- let us denote it by $X = \{X(t),\ t=1,2,...\}$, with values in the state set N; t is more conveniently viewed as an index of successive memory requests, rather than the time of request generation. Refer to Figure 2 for a schematic representation of an interleaved memory system.

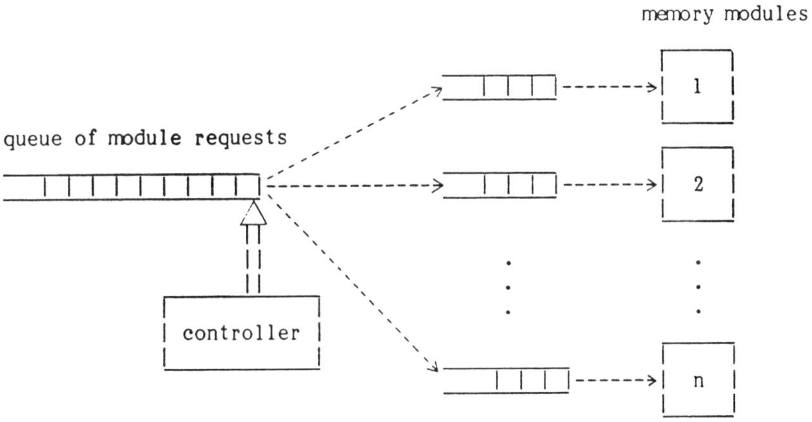

Figure 2. Interleaved memory system

The requests to be served during each memory cycle are taken from an auxiliary buffer, in the order of arrival (increasing t). When the first repetition occurs, or n distinct requests have been taken, the process of selection stops. (The queues in front of each memory modules, in Figure 2, are thus always kept empty.) It is further assumed that the system is saturated: the buffer always contains enough pending requests so that one of the above conditions holds.

The random process which is the number of requests served during each memory cycle is the bandwidth (BW) of the interleaved memory system. The pmf and first moment of BW are the objects of analytical interest. An approach toward their computation is suggested in (Török [15]). A pair renewal process, associated with X, is firstly defined as $(Y,T) = \{(Y(s),T(s)),\ s=1,2,\ldots\}$, as follows:

$$Y(s) = X(T(s)), \tag{3.12}$$

$$T(s) = \min\ \{t>T(s-1)\,|\,X(t)=X(t'),\ t'\varepsilon[T(s-1),\ t)\}.$$

In equation (3.16), T(s) is the index of that memory request X(T(s)) which, being a repetition of one of the requests already selected for the current memory cycle, stops the selection process.

The Markov chain $(Y,D) = \{(Y(s),D(s)),\ s=1,2,\ldots\}$ is then derived from (Y,T), by putting:

$$D(s) = T(s+1) - T(s). \tag{3.13}$$

D is thus the process of "time" intervals between repetitions -- or, more to the point, the process of successive bandwidths. It is shown that the events $\{D(s) = k\}$, when conditioned upon the events $\{Y(s-1) = i\}$, depend only on i, so that state dependent renewal times can be defined for each state $i\,\varepsilon\,N$. The following distributions are further introduced:

$$\begin{aligned} q_{ij}(k) &= P[Y(s)=j,\ D(s)=k\,|\,Y(s-1)=i], \qquad k=1,\ldots n,\\ q_{ij} &= \sum_{k=1}^{n} q_{ij}(k), \qquad i,j\ \varepsilon\ N, \end{aligned} \tag{3.14}$$

so that q_{ij} is the transition probability of the imbedded Markov chain Y, and:

$$h_i(k) = \sum_{j=1}^{n} q_{ij}(k) = P[D(s)=k|Y(s-1)=i], \quad i,k \in N, \tag{3.15}$$

so that $h_i(k)$ is the pmf of the renewal time dependent on the state at time s. The stationary probability of Y is denoted by $\underline{f}$, the fixed point vector of the stochastic matrix $Q = (q_{ij})$. The asymptotic pmf of the bandwidth is then easily obtained as:

$$P[BW=k] = \lim_{s\to\infty} [D(s)=k] = \sum_{i=1}^{n} f_i h_i(k). \tag{3.16}$$

In order to apply equation (3.16) to the computation of the bandwidth's pmf, the matrices $Q(k) = (q_{ij}(k))$ are needed. These can be obtained based on the distribution of the length of path, and are given here without proof (which is fairly direct):

$$q_{ij}(k) = \sum_{m=1}^{k-1} \sum_{J_{m-1}\subset N-i-j} \sum_{\text{Permutations}(J_{m-1})} (P_{j_1}\cdots P_{j_{m-1}} P_j)_{ij} \circ$$

$$\sum_{r=1}^{k-m-1} (-1)^{k-m-r-1} \binom{n-m-r-1}{n-k} \circ$$

$$\sum_{J_r\subset N-J_{m-1}-i-j} (P_{J_r}^{k-m-1} P_j)_{jj}, \quad i\neq j,\ i,j \in N,\ k=1,\ldots n,$$

(3.17)

$$q_{jj}(k) = \sum_{r=0}^{k-1} (-1)^{k-r-1} \binom{n-r-1}{n-k} \sum_{J_r\subset N-j} (P_J^{k-1} P_j)_{jj},$$

$$j \in N,\ k=1,\ldots n.$$

Refer to Figure 3 for sample paths that start in i and pass in k + 1 steps through k distinct states, such that j is the first repetition, and (a) and (b) correspond to the two cases in equation (3.21): j≠i and j=i, respectively.

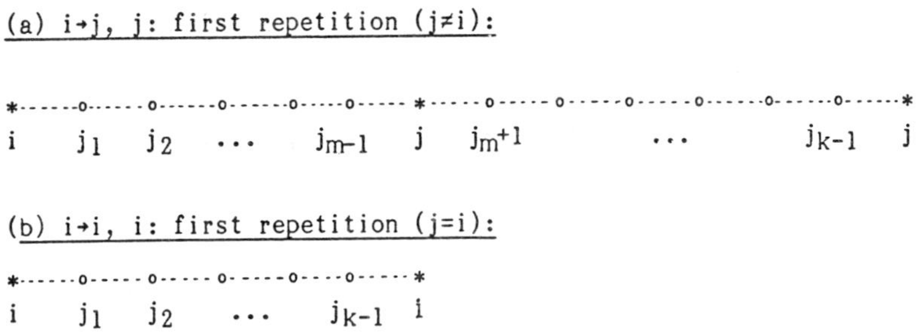

Figure 3. The bandwidth of an interleaved memory system (sample selections of requests to be served in a memory cycle)

4. Appendix A

Proof of Proposition 2.1: Refer to Figure 4; i and j are used to designate the exit and end state of a sample path, respectively.

1 Consider the sets $J_r \subset H^c$, $J_r = \{j_1,\ldots j_r\}$, for $r = 0,\ldots n-m$, and the sets $H_r = J_r^c$, $H_r = \{i_1,\ldots i_{n-r}\}$, $H \subset H_r$. Even though through this proof various indexings of the states in N are used, the states of H are invariantly denoted: $H = \{i_1,\ldots i_m\}$, and also: $i_1 = j$.

The method of proof is to compute the probabilities of all the paths that leave H at i and first return to it at j, after having visited only states of J_r. We build the sets J_r so that the family of their complementary sets H_r have the common intersection H. This approach warrants the use of the inclusion-exclusion principle.

The projection operators I_1 and I_2 (defined as in section 2.2) will be related to the sets $H_r^c = J_r$ and H_r, respectively. The matrix of first return to H_r in s steps is, as in equation (2.13):

$$(PI_1)^{s-1}PI_2 = (P_{j_1}+\ldots+P_{j_r})^{s-1}(P_{i_1}+\ldots+P_{i_{n-r}}). \tag{4.1}$$

Since we are interested only in the j column of the matrix in equation (4.1), corresponding to the first return to H_r at the hitting place $j \in H$, we disregard other columns. Thus, we define the probability of

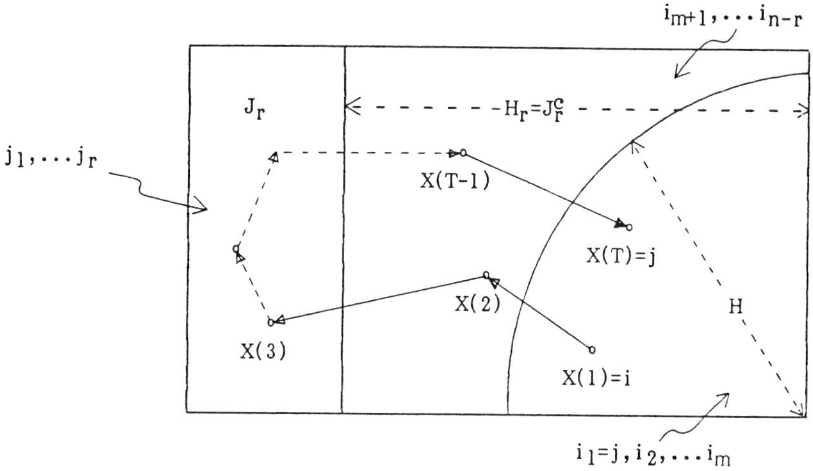

Figure 4. Illustration for the proof of Proposition 2.1

first passage to j, under taboo H_r, having started in i:

$$_{H_r}h_{ij}(s) = (P_J^{s-1}P_j)_{ij}, \quad i,j \in H. \tag{4.2}$$

2 Let T_{i_z} be the time of first passage to each $i_z \in N$ $(z=1,\ldots n)$, and let us define the events:

$$A_{i_z} = \{T_{i_z} > s+1\} \cap \{T_H = s+1\} \cap \{X(T_H)=j\} \quad \text{(of course, } A_{i_1}=\emptyset\text{)}, \tag{4.3}$$

i.e., A_{i_z} is the event: the state i_z is not visited along a path of first return to H (at the hitting place j), whose duration is s steps. The event of first passage of j under the taboo set H_r in s steps can be expressed through the events A_{i_z}, as follows:

$$\{T_{H_r}=s+1\} \cap \{X(T_{H_r})=j\} = \bigcap_{i_z \in H_r - j} \{T_{i_z}>s+1\} \cap \{T_{H_r}=s+1\} \cap \{X(T_{H_r})=j\}$$

$$= [\bigcap_{z=m+1}^{n-r} \{T_{i_z}>s+1\}] \cap [\bigcap_{z=2}^{m} \{T_{i_z}>s+1\} \cap \{T_{H_r}=s+1\} \cap \{X(T_{H_r})=j\}].$$

On the other hand, since $H = \{j, i_2, \ldots i_m\} \subset H_r$, we obtain:

$$\bigcap_{z=2}^{m} \{T_{i_z} > s+1\} \cap \{T_{H_r} = s+1\} \cap \{X(T_{H_r}) = j\} = \{T_H = s+1\} \cap \{X(T_H) = j\},$$

and therefore the following equation holds:

$$\{T_{H_r} = s+1\} \cap \{X(T_{H_r}) = j\} = \bigcap_{z=m+1}^{n-r} \{T_{i_z} > s+1\} \cap \{T_H = s+1\} \cap \{X(T_H) = j\}$$

$$= \bigcap_{z=m+1}^{n-r} A_{i_z} .$$

By conditioning the last equation upon the event $\{X(1)=i\}$, we obtain the conditional probability:

$$_{H_r}h_{ij}(s) = P[\bigcap_{z=m+1}^{n-r} A_{i_z} | X(1)=i]. \tag{4.4}$$

3 For each r, $1 \leq r \leq n-m$, consider the following expressions:

$$s_{ij}(r,s) = \sum_{J_r \subset H^c} {}_{H_r}h_{ij}(s), \tag{4.5}$$

$$s_{ij}(n-m,s) = 1.$$

Using equation (4.4), the interpretation of $s_{ij}(r,s)$ for all $r \leq n-m$ is: the sum of the probabilities of joint realization of the events $\{A_{h_1},\ldots A_{h_{n-m-r}}\} \subset \{A_{i_{m+1}},\ldots A_{i_n}\}$, i.e.:

$$s_{ij}(r,s) = \sum_{\substack{\{h_1,\ldots h_{n-m-r}\} \\ \subset \{i_{m+1},\ldots i_n\}}} P[\bigcap_{q=1}^{n-m-r} A_h | X(1)=i]. \tag{4.6}$$

Note: $s_{ij}(r,s)$ is not a probability (it may be greater than 1) for $r > 0$, but $s_{ij}(0,s)$ is, and in fact it is $P[T_{H_r} > s,\ T_H = s,\ X(s) = j | X(1) = i] = 0.$

4 Let us now consider the n-m events $A_{i_{m+1}},\ldots A_{i_n}$, and let Y be the random variable defined as the number of such events that are true:

$$Y = \sum_{z=m+1}^{n} I_{A_z}, \tag{4.7}$$

where I_A is the indicator function associated with the event A. Using the inclusion-exclusion principle (Feller, [6], p. 106), and denoting by P[Y=u] the probability that precisely u events are true (u = 0,1,..n-m), we obtain:

$$P[Y=u] = \sum_{v=0}^{n-m-u} (-1)^v \binom{u+v}{u} S_{u+v}, \tag{4.8}$$

where:

$$S_w = \sum_{\substack{\{h_1,\ldots h_w\} \\ \subset\{i_{m+1},\ldots i_n\}}} P[\bigcap_{q=1}^{w} A_{h_q}], \qquad w > 0, \text{ and:} \tag{4.9}$$

$$S_0 = 1.$$

5 Comparing the equations (4.6) and (4.9), it becomes evident that $s_{ij}(r,s)$ is S_u, if we condition the probabilities of which S_u is the sum, upon the event {X(1)=i}, and identify w with n-m-r. Moreover, effecting the substitution of indices u=n-m-k, v=k-r in equation (4.8), we obtain:

$$P[Y=n-m-k|X(1)=i] = \sum_{r=0}^{k} (-1)^{k-r} \binom{n-m-r}{k-r} s_{ij}(r,s) \tag{4.10}$$

6 {Y=n-m-k} is the event: precisely n-m-k distinct states of H^c are avoided by the path that exiting state i, enters state $j \in H$ after s steps; or, in other words: precisely k distinct states of H^c are visited along the path of first return in s steps to H.

Using now the equations (4.2) together with (4.10), and according to the Note following equation (4.6) (i.e., $s_{ij}(0,s)=0$), the desired equation (2.16) is immediately obtained.

5. Appendix B

The following procedure may be used for the computation of $f'_{ii}(c)$, as defined in equation (2.24) (H = {i}), for all the values of

$i=1,\dots n$, $c=1,\dots n-1$. It is written in pidgin ALGOL, and to the right of every statement representing a computational step, is given the number of multiplications or iterations required.

```
for i=1 till n do                                    [n times]
    begin
    for c=1 till n-1 do                              [n-1 times]
        begin
        for all the tuples J_c do                    [(n-1 choose c) times]
            begin
            compute the c significant rows,
            j_1,...j_c of (I-P_{J_c})^{-1};          [c^3/3 multipl.]
            postmultiply by the c significant
            columns, j_1,...j_c of P_i;              [c^2 multipl.]
            compute P^c_{J_c};                       [c-1 multipl.]
            for r=c till n-1 do                      [n-c times]
                begin
                compute P^r_{J_c} (given P^{r-1}_{J_c});  [c^2 multipl.]
                compute i-th row of
                [P^r_{J_c} (I-P_{J_c})^{-1} P_i];    [c multipl.]
                end;
            end;
        end;
    end
```

Thus, the number of multiplications required is:

$$n \sum_{c=1}^{n-1} \binom{n-1}{c} \left[(c^3/3)+c^2+(c-1)c^2+(n-c)(c^2+c)\right]$$

$$= \left(\frac{7n^4}{3} - 4n^3 + 7n^2 - 6n\right)2^{n-4}.$$

6. Acknowledgements

A substantial part of this paper parallels parts of the author's thesis (Tzelnic [16]), and the advisor's -- M. Hofri -- valuable contribution in shared insight is gratefully acknowledged. In a personal communication, R. Syski ([14]) has suggested a formal approach, followed here, which is substantially more elegant than that of a first draft.

7. References

[1] Aho, A. V., P. J. Denning, J. D. Ullman, "Principles of Optimal Page Replacement," J. Assoc. Comput. Mach., 18, (1971).

[2] Burnett, G. J., E. G. Coffman, "A Combinatorial Problem Related to Interleaved Memory Systems," J. Assoc. Comput. Mach., 20, (1973), 34-95.

[3] Courtois, P. J., Decomposability, Academic Press, 1977.

[4] Courtois, P. J., H. Vantilborgh, "A Decomposable Model of Program Paging Behaviour," Acta Inform., 6, (1976), 251-275.

[5] Denning, P. J., S. C. Schwartz, "Properties of the Working Set Model," Comm. ACM, 15, (1972), 191-198.

[6] Feller, W., An Introduction to Probability Theory and its Application, I, 3rd ed., John Wiley, 1968.

[7] Franklin, M. A., R. K. Gupta, "Computation of Page Fault Probabilities from Program Transition Diagrams," Comm. ACM, 17, (1974), 186-191.

[8] Freiberger, W. F., U. Grenander, P. D. Sampson, "Patterns in Program Behaviour," IBM J. Res. Develop., 19, (1975), 230-243.

[9] Gelenbe, E., "A Unified Approach to the Evaluation of a Class of Replacement Algorithms," IEEE Trans. Comput., C-226, (1973), 611-618.

[10] Glowacki, C., "A Closed form Expression of the Page Fault Rate for the LRU Algorithm in a Markovian Reference Model of Program Behaviour," in Proc. Int. Comp. Symp., E. Morlet, D. Ribbens (Eds.), North Holland, 1977, 315-318.

[11] Hofri, M., P. Tzelnic, "On the Working Set size for the Markov Chain Model of Program Behaviour," in Performance of Comp. Syst., M. Arato et al. (Eds.), North Holland, 1979, 393-405.

[12] Kemeny, J. G., J. L. Snell, Finite Markov Chains, Van Nostrand, 1960.

[13] Syski, R., "Potential Theory for Markov Chains," in Probab. Methods in Appl. Math., III, A. T. Barucha-Reid (Ed.), Academic Press, 1973.

[14] Syski, R., personal communication.

[15] Török, T. L., "Interleaved Memory Systems with Markovian Requests," in Performance of Comp. Syst., M. Arato et al. (Eds.), North Holland, 1979.

[16] Tzelnic, P., "Stochastic Models of Program Behaviour in Paged Virtual Memory Systems," D.Sc. Thesis (Hebr.), Department of Computer Science, Israel Institute of Technology, 1979.

8. Endnotes

[1]Strictly speaking, the first hit time must be defined in equation (2.4) for $t \geq 1$, and so that $X(1) \varepsilon H^c$. Equation (2.4), with $X(1) \varepsilon H$, is in fact the definition of the first return time to H; but the distinction between these two variables is of no import in the sequel.

[2]It is for this reason that H-open and H-closed paths are not the same as, respectively, first hit and first return paths (as mentioned in the note on page 3.)

WANG Laboratories Inc., M/S 1379, One Industrial Avenue, Lowell, MA 01851.

Discussant's Report on
"On Length of Path for Finite Markov Chains, and its Applications to Program Behaviour and Interleaved Memory Systems Modelling,"
by Percy Tzelnic

I'd like to ask a few questions concerning the ultimate practical applicability of the material presented. The given method for computing page fault rates is offered as an alternative to an "obvious" method with inherent computational problems since it is based on a Markov chain whose state space increases exponentially. The alternative cleverly uses a chain with linearly increasing state space but involves summing over an exponentially increasing number of terms. This is vastly superior, and represents a definite theoretical advance, but computations still get unmanageable very quickly.

In addition, it appears difficult to estimate the transition probabilities for the Markov chain giving succesive page requests, regardless of computation method used.

What are the practical limits to the size of the parameters for which numerical results can be obtained? Can such results give insight into the behavior of the page fault rate for a real system, and if not, how can such problems be attacked?

Author's Response to Discussant's Report

There are two different ways of applying the techniques presented in this paper to, say, program behavior modelling (I am picking this application area only because the discussant did so). One relates to gaining insight in the behavior of a given, specific real program, and that implies that the analyst has to measure it, fit the Markov chain model to the data, perform the prescribed computations, contemplate the results. The other relates to using the program behavior model as input to drive the model of a computer system. In this case the analyst is not as concerned with estimating the parameters of the program behavior from a real life program, but rather wants a generic model, indeed one which is postulated according to the analyst's belief in what is representative.

In the first case, one would have to estimate the n^2 parameters of a Markov chain model of program behaviour (for a real life program with n pages n can be of O(100)). This is quite a difficult undertaking. It involves data collection and analysis - statistical hypothesis testing to ascertain the level of confidence that such data fits a Markov chain model. The difficulty resides in the large number of degrees of freedom (n^2) rather than in the methodology used, which is well known to statisticians and has been succesfully employed in the past. A review of such techniques, applied to program behavior modelling, can be found in Jeffrey Spirn's book Program Behavior: Models and Measurements, Elsevier, 1977.

However, once the n^2 parameters are estimated (or postulated, if this is the analyst's way) the computation of the page fault rate is itself bounded to relatively small values of n - as the discussant pointed out. As shown in the paper (Table 1), n = 22 seems to be an upper bound.

Nevertheless, even models with much larger values for n might be amenable to calculation, as I suggested in the paper co-authored with M. Hofri: "The Working Set Size Distribution for the Markov Chain Model of Program Behavior" (to appear in SIAM Journal of Computing in Spring 1982). What makes this possible is, once again, a program's locality of reference. This allows stipulating a "phase" model using Courtois' near decomposability technique so that in many instances all the localities would have under 22 pages. This is an approximative approach, of course, and it is not yet clear whether the degree of accuracy achievable would justify the effort. The point though is that such an approach is feasible, and more research ought to be done on its usefulness.

Note too that in the case of interleaved memory systems the size of the state space is typically below the 22 mark, so that the approach outlined in this paper is clearly feasible. Finally, I believe that there are numerous, though yet unexplored problems waiting for the method of solution proposed in this paper. For such problems (relating to probabilistic walks on graphs that are best characterized by the number of distinct nodes traversed) the dimensionality of the state space may not be an issue at all.

Discussant: Dr. Donald R. Smith, Bell Laboratories, WB-1H316, Holmdel, New Jersey 07733

NETWORKS OF QUEUES, II
Jean Walrand, Chairman

M. I. Reiman
A. Hordijk & N. van Dijk
J. McKenna, D. Mitra,
& K. G. Ramakrishnan

THE HEAVY TRAFFIC DIFFUSION APPROXIMATION FOR SOJOURN TIMES IN JACKSON NETWORKS

Martin I. Reiman

Abstract

Using a heavy traffic limit theorem for open queueing networks, we find the correct diffusion approximation (D.A.) for sojourn times in Jackson networks with single server stations. The D.A. for sojourn times is a function of the D.A. for the queue length process, which is reflected Brownian motion on the nonnegative orthant.

We display a partial differential equation with boundary conditions which is satisfied by the stationary density of the D.A. for the queue length process. The solution of this equation is a product of exponentials when the diffusion is obtained as a limit of Jackson networks. Finally, we derive an expression for the stationary distribution of the D.A. for sojourn times based on the relationship between the D.A.'s for sojourn times and queue lengths. For many special cases, the stationary distribution of the D.A. for sojourn times precisely matches that of the Jackson network.

1. Introduction and Summary

Queueing networks have been receiving a great deal of attention over the last several years for their role as models of computer systems. The study of such networks has been based to a large degree on the pioneering work of Jackson (1957). His model has been greatly extended, incorporating many changes which can be applied to the complexities of computer systems. Jackson's original work, as well as the bulk of the extensions of his model have dealt with the queue length process.

From the point of view of computer system modelling, response or sojourn times are of more interest than queue lengths. Very few results are available for sojourn times in these networks. Reich (1957) studied the sojourn time in a tandem, or series network. His results were extended to acyclic networks with at most one path between any two nodes by Lemoine (1979) and Walrand and Varaiya (1980).

In this paper we give the heavy traffic diffusion approximation for sojourn times in Jackson networks with single server stations. There are no restrictions placed on the structure of the network (such as tandem, or acyclic). The diffusion approximation is justified by a heavy traffic limit theorem for a more general class of queueing networks, proven in Reiman (1982). It is shown in Reiman (1982) that properly normalized sequences of queue length and sojourn time processes converge weakly (in distribution) to a certain diffusion as the network traffic intensity converges to unity. More precisely, the sequence of queue length processes converges weakly to reflected Brownian motion on the nonnegative orthant, and the sequence of sojourn times converges to a function of this process. Reflected Brownian motion on the nonnegative orthant (RBM) was defined and characterized in Harrison and Reiman (1981a). Results related to its distribution are contained in Harrison and Reiman (1981b). Roughly speaking, reflected Brownian motion on the nonnegative orthant behaves like Brownian motion on the interior of its state space (the nonnegative orthant) and reflects instantaneously at the boundary. The direction of reflection is a constant for each boundary hyperplane.

It was shown in Harrison and Reiman (1981b) that the density of the stationary distribution of RBM solves a particular partial differential equation with boundary conditions. They also showed that when the coefficients of the RBM satisfy a particular relationship, the solution of the equation is a product of exponentials. We show in this paper that the condition yielding a product form solution is satisfied for the diffusion arising as the limit of a sequence of Jackson networks. Using the relationship between the diffusion approximation for sojourn times and queue lengths, we derive an expression for the stationary distribution of the diffusion approximation for sojourn times.

In the next section we introduce the queue length and sojourn time processes of interest to us. In Section 3 we give the main result of this paper, the distribution of stationary sojourn times for the diffusion approximation. The remainder of the paper outlines the method by which this result was obtained. Section 4 contains a definition of the limit process. In Section 5 we state the limit theorems which justify the diffusion approximation. Choosing the correct diffusion to approximate a given Jackson network is discussed in Section 6. Section 7 contains the partial differential equation for the stationary density of RBM, as well as its solution for the special case related to Jackson networks.

2. The Queue Length and Sojourn Time Processes

We assume that the reader is familiar with Jackson networks, and merely state the notation we shall be using in this paper. We assume that the queueing network we study has K stations, each of which has a single server. The arrival rates are given by a K-vector λ, where λ_i is the arrival rate from outside the network to station

$i (\lambda_i \geq 0,\ 1 \leq i \leq K)$. Similarly, the service rates are given by a K-vector μ. We let P be the routing matrix, and define $\gamma = \lambda R$, where

$$R = (I - P)^{-1} = \sum_{n=0}^{\infty} P^n. \qquad (1)$$

We assume that the sum in (1) converges, which occurs when the network is open. We define the vector traffic intensity, ρ, for the network by

$$\rho_i = \gamma_i / \mu_i, \quad 1 \leq i \leq K. \qquad (2)$$

We also define

$$\nu_i = \lambda_i + \sum_{j=1}^{K} \mu_j p_{ji}, \quad 1 \leq i \leq K.$$

We let $Q = \{Q(t),\ t \geq 0\}$ denote the vector queue length process. We take $Q(0) \in \mathbb{Z}_+^K$, the set of nonnegative integer K-tuples. The structure of the Jackson network then defines Q as a particular Markov process.

To describe the sojourn times studied here, let $h \in \mathbb{Z}_+^K$. In addition, assume that h is such that it is possible for a customer to enter the network at station k $(1 \leq k \leq K)$, to pass through the network, and visit station i h_i times, for all $1 \leq i \leq K$. We call such an h k-feasible. For a k-feasible h, let $W_{k,h}(t)$ be the total time spent in the network (sojourn time) by the next customer entering station k after time t who has a visit vector h. Note that we are not distinguishing between paths with the same visit vector but with different orders for the visits.

3. The Diffusion Approximation for Stationary Sojourn Times

We can now describe the distribution of stationary sojourn times in a Jackson network. We assume that $\rho_i < 1$ for $1 \leq i \leq K$. As a result, the queue length process possesses a stationary distribution. Start the queue length process in its stationary distribution. For a k-feasible h, let $W_{k,h}$ denote the sojourn time of the next customer to enter the network at station k who has visit vector h. Let $\{\tau_i, 1 \le i \le K\}$ be independent random variables, where τ_j is exponentially distributed with mean $\mu_j - \gamma_j$. The main result of this paper is that when ρ_i is close to one for all $1 \le i \le k$,

$$W_{k,h} \overset{d}{\approx} \sum_{i=1}^{K} h_i \tau_i , \tag{3}$$

where $\overset{d}{\approx}$ means approximately equal in distribution. Several facts can be gleaned from equation (3). First, the right hand side of (3) is actually independent of k in the sense that if h is k'-feasible for $k' \neq k$, both $W_{k,h}$ and $W_{k',h}$ have the same diffusion approximation. The diffusion approximation is also independent of the particular path through the network, and only depends on the visit vector. The form of (3) also implies that a customer's waiting time at station i is the same every time that customer visits station i. It is as if the customer takes a snapshot of the network when he enters and all queues remain at that same value during the customers sojourn throughout the network. This remarkable fact was first pointed out by G. J. Foschini (1980) and is a result of his observation that on the diffusion time scale, customers spend zero time in the network. Since the diffusion is continuous, it does not change in zero time and so the queue lengths seen by the customer are constant during his sojourn through the network. This intuitive idea was made rigorous in Reiman (1982). In the remainder of the paper we sketch out the argument leading to (3).

4. The Approximating Diffusion

The process we propose as an approximation for the queue length and sojourn time processes arises as the limit process in a heavy traffic limit theorem that we will state in Section 5. This process, which we call reflected Brownian motion on the nonnegative orthant, was rigorously defined by Harrison and Reiman (1981a). RBM is defined by way of a mapping; when applied to Brownian motion it yields RBM. A second definition of the mapping, which relates more closely to the queueing application, is given by Reiman (1982). We will merely restate the definition as given in Harrison and Reiman (1981a).

Let C be the space of continuous functions $x: [0,\infty) \to \mathbb{R}^K$ and let C_+ be the subset of C with $x(0) \in \mathbb{R}_+^K = \{w \in \mathbb{R}^K: w_i \geq 0, 1 \leq i \leq K\}$. The following result appeared as Theorem 1 in Harrison and Reiman (1981a).

<u>Proposition 1</u>. For each $x \in C_+$ there exists a unique pair of functions $y \in C$ and $z \in C$ satisfying

$$z_j(t) = x_j(t) + y_j(t) - \sum_{i=1}^{K} p_{ij} y_i(t), \quad t \geq 0 \tag{4}$$

$$z_j(t) \geq 0, \quad t \geq 0 \tag{5}$$

$y_j(\cdot)$ is non-decreasing with $y_j(0) = 0$, and (6)

$y_j(\cdot)$ increases only at those times t where $z_j(t) = 0$ for all $j = 1,\ldots,K$. (7)

Moreover, setting $y = \psi(x)$ and $z = \phi(x)$, we have the following

The restrictions of y and z to $[0,T]$ depend only on the restriction of x to $[0,T]$. (8)

Both ψ and ϕ are continuous mappings $C_+ \to C$. (9)

Definition. Let $X = \{X(t), t \geq 0\}$ be a K dimensional Brownian motion with drift vector c, positive definite covariance matrix A, and $X(0) = 0$. Let $Z = \{Z(t), t \geq 0\}$ be defined by $Z = \phi(X)$, where ϕ is given by Proposition 1. Then Z is a reflected Brownian motion on $\mathbb{R}_+^K$ with drift c, covariance A, and reflection matrix $(I - P)$.

5. The Limit Theorems

The limit theorems justifying the diffusion approximations advanced in this paper apply to normalized versions of the queue length and sojourn time processes. Our aim in this paper is to obtain approximations for the stationary distributions of queue lengths and sojourn times, so we can focus our interest on sequences of stable systems, where $\rho \to 1$ from below. As a result, we can use a slightly different (but equivalent) normalization than was used in Reiman (1982), which makes the choice of the correct approximating diffusion more straightforward.

We assume that we have a sequence of Jackson networks, indexed by $n \geq 1$, with associated sequences of queue length processes and sojourn time processes. Arrival rates, service rates, and the routing matrix may depend on n. K is assumed fixed. Let

$$\alpha(n) = \min_{1 \leq i \leq K} (1 - \rho_i(n)), \quad n \geq 1. \tag{10}$$

We assume

$$\lambda(n) \to \lambda, \tag{11}$$

$$\mu(n) \to \mu \tag{12}$$

$$P(n) \to P, \tag{13}$$

$$\alpha(n) \to 0, \text{ and} \tag{14}$$

$$[\alpha(n)]^{-1}(\nu(n) - \mu(n)) \to c \tag{15}$$

as $n \to \infty$, with all limits assumed finite.

Let

$$\begin{aligned} A_{ii} &= 2\gamma_i(1 - p_{ii}), \quad 1 \leq i \leq K, \text{ and} \\ A_{ij} &= -[\gamma_i p_{ij} + \gamma_j p_{ji}], \quad 1 \leq i < j \leq K. \end{aligned} \tag{16}$$

We define

$$Z^n(t) = \alpha(n) Q^n(t[\alpha(n)]^{-2}), \quad n \geq 1, \quad 0 \leq t \leq 1,$$

and

$$T^n_{k,h}(t) = \alpha(n)W^n_{k,h}(t[\alpha(n)]^{-2}), \quad n \geq 1, \quad 0 \leq t \leq 1,$$

$1 \leq k \leq K$, and h k-feasible.

We denote by D^N the set of functions from $[0,\infty)$ into $\mathbb{R}^N$ which are right continuous and have left hand limits everywhere. The sample paths of both the queue length and sojourn time processes are elements of D^N. Weak convergence, denoted by the symbol $\Rightarrow$ is the generalization of convergence in distribution to stochastic processes.

The following results were proven in Reiman (1982).

Theorem 1. Suppose (10) - (14) hold. Then $Z^n \Rightarrow Z$ in D^K as $n \to \infty$.

Theorem 2. Suppose (10) - (14) hold and h is k-feasible. Then $T^n_{k,h} \Rightarrow \sum_{i=1}^{K} \frac{h_i}{\gamma_i} Z_i$ in D as $n \to \infty$.

6. Approximating Stable Jackson Networks

Given a stable Jackson network, we can use Theorems 1 and 2 to obtain diffusion approximations for queue lengths and sojourn times respectively. The limit theorems can be used to justify the claim that for a queueing network with α "small", the distribution of $Q(t)$ is approximately the same as that of $\alpha^{-1}Z(\alpha^2 t)$, and the distribution of $W_{k,h}(t)$ is approximately the same as that of $\alpha^{-1} \sum_{i=1}^{K} \frac{h_i}{\gamma_i} Z_i(\alpha^2 t)$. The limit theorem does not yield information on how small α must be to obtain a certain degree of accuracy in the approximation. This information would come from results on the rate of convergence. Although the limit theorem does not say how good the approximation is for a given system, it states that the approximation is asymptotically exact in the sense of weak convergence.

To determine which diffusion to use as an approximation for a given network, we simply match up coefficients. From equation (15) we find that

$$c = \alpha^{-1}[\nu - \mu]. \tag{17}$$

The covariance, A, is given by (16), and the reflection matrix is $(I - P)$.

7. The Stationary Distribution of Z

The great appeal of diffusion approximations is that there is an entire theory available which makes it a straightforward task to write equations for the probability densities or distributions for most quantities of interest. In Harrison and Reiman (1981b), equations are presented for the transient density as well as the stationary density. Our interest here is in stationary distributions so we restrict our attention to this case.

Before discussing equations for the stationary density of Z we must deal with the question of determining when Z is positive recurrent. In Harrison and Reiman (1981b) it was shown that if Z is positive recurrent then $(cR)_i < 0$ for all $1 \leq i \leq K$. This condition will present itself in a natural way in the results of this section.

Assume that Z is positive recurrent and has stationary density π. The following partial differential equation with boundary conditions was given for π in Harrison and Reiman (1981b):

$$\frac{1}{2}\sum_{i=1}^{K}\sum_{j=1}^{K}\frac{\partial^2\pi}{\partial x_i \partial x_j}(x) - \sum_{i=1}^{K} c_i \frac{\partial \pi}{\partial x_i}(x) = 0, \quad x \in \mathbb{R}_+^K, \tag{18}$$

$$\frac{1}{2}\sum_{j=1}^{K} B_{ij}\frac{\partial \pi}{\partial x_j}(x) - c_i \pi(x) = 0, \quad x_i = 0, \tag{19}$$

where

$$B_{ij} = 2A_{ij} + A_{ii}(\hat{p}_{ij} - \delta_{ij}),$$

and

$$\hat{p}_{ij} = \begin{cases} 0 & i = j \\ \dfrac{p_{ij}}{1-p_{ii}} & i \neq j \end{cases} .$$

The solution for (18) and (19) is not known in general, but when Z arises as the limit of a sequence of Jackson networks we can solve these equations. For the Jackson network of Section 2, assume $\rho_i < 1$ for $1 \leq i \leq K$, and let q be a vector random variable with the stationary distribution of Q. Define

$$\hat{\pi}(n_1,n_2,\ldots,n_K) = P\{q_1=n_1,q_2=n_2,\ldots,q_K=n_K\} \tag{20}$$

for $n_i \geq 0$, $1 \leq i \leq K$. The main result of Jackson (1957) states that

$$\hat{\pi}(n_1,n_2,\ldots,n_K) = \prod_{i=1}^{K} (1-\rho_i)\rho_i^{n_i} . \tag{21}$$

Let

$$\hat{\Pi}(n_1,\ldots,n_K) = \sum_{j_1 \leq n_1} \cdots \sum_{j_K \leq n_K} \hat{\pi}(j_1,\ldots,j_K). \tag{22}$$

Then

$$\hat{\Pi}(n_1,\ldots,n_K) = \prod_{i=1}^{K} (1-\rho_i^{n_i+1}). \tag{23}$$

A simple calculation shows that (15) is equivalent to

$$[\alpha(n)]^{-1}(1-\rho_i(n)) \to \gamma_i^{-1}(cR)_i \tag{24}$$

as $n \to \infty$, $1 \leq k \leq K$. Let

$$\Pi(x_1,x_2,\ldots,x_K) = \int_0^{x_1}\int_0^{x_2}\cdots\int_0^{x_K} \pi(y_1,\ldots,y_K)\,dy_K\ldots dy_1, \tag{25}$$

for $x_1 \geq 0$, $1 \leq i \leq K$. Taking the limit as $n \to \infty$ in (23) we obtain

$$\Pi(x_1,\ldots,x_K) = \prod_{i=1}^{K} (1-e^{-d_i x_i}), \quad x \in \mathbb{R}_+^K, \tag{26}$$

where

$$d_i = -\gamma_i^{-1}(cR)_i, \quad 1 \leq i \leq K. \tag{27}$$

Differentiating (26),

$$\pi(x) = \prod_{i=1}^{K} d_i e^{-d_i x_i}, \qquad x \in \mathbb{R}_+^K. \tag{28}$$

Substituting (28) into (18) and (19) verifies that it is a solution.

It was shown in Harrison and Reiman (1981b) that (18) and (19) have a solution of the product form if and only if

$$A_{ij} = -\frac{1}{2}[A_{ii}\hat{p}_{ij} + A_{jj}\hat{p}_{ji}], \qquad 1 \le i < j \le K, \tag{29}$$

in which case the solution is (28). It can be seen from (16) that (29) is always satisfied for diffusions arising as limits of sequences of Jackson networks. A solution to (18) and (19) which is not of product form, associated with a two station tandem queue having deterministic service at the first station, is shown in Harrison and Reiman (1981b).

We now calculate the mean queue length for the diffusion approximation, and compare it to the mean queue length for the exact Jackson network model. From Section 6 we know that we are interested in $\alpha^{-1}E[Z]$, where

$$\alpha = \min_{1 \le i \le K} (1 - \rho_i).$$

We assume that the queue length process has a stationary distribution, so that $\alpha > 0$. From (28) we have

$$\alpha^{-1}E[Z_i] = (\alpha d_i)^{-1}. \tag{30}$$

Using (17) and (27),

$$\begin{aligned} \alpha d_i &= \gamma_i^{-1}[(\mu - \nu)R]_i \\ &= \rho_i^{-1}(1 - \rho_i). \end{aligned} \tag{31}$$

Substituting the last line of (31) into (30) yields

$$\alpha^{-1}E[Z_i] = \frac{\rho_i}{1-\rho_i} \tag{32}$$

which is precisely the result obtained for $E(q_i)$ using either (21) or (23). So the diffusion approximation gives the exact value of the mean queue length for all values of the traffic intensity.

Combining equation (28) with Theorem 2 we obtain equation (3). If we restrict our attention to a tandem system, and set $h_i = 1$, $1 \leq i \leq K$, we can obtain the diffusion approximation for the total sojourn time of an arbitrary customer (in steady-state). The distribution function of this sojourn time is precisely the same as that of the original Jackson network. This can be shown for mean queue lengths. The result can also be obtained using Little's theorem. Little's theorem can be used to show that the mean waiting time at each station is the same in both the exact model and the diffusion approximation, since the mean queue lengths are equal. The waiting time distribution is exponential in both cases, so the entire waiting time distribution for each station individually is the same for both the exact model and the diffusion approximation. The result for the entire sojourn time follows from the independence of the waiting times at individual stations. This argument obviously extends to the class of networks considered by Lemoine (1979) and Walrand and Varaiya (1980).

8. References

[1] G. J. Foschini (1980). Personal correspondence.

[2] J. M. Harrison and M. I. Reiman (1981a). Reflected Brownian motion on an orthant. Ann. Probab., 9, 302-308.

[3] J. M. Harrison and M. I. Reiman (1981b). On the distribution of multidimensional reflected Brownian motion. SIAM J. Appl. Math., 41, 345-361.

[4] J. R. Jackson (1957). Networks of waiting lines. Oper. Res. 5, 518-521.

[5] A. J. Lemoine (1979). On total sojourn time in networks of queues. Tech. Report, Systems Control, Inc.

[6] E. Reich (1957). Waiting times when queues are in tandem. Ann. Math. Stat. 28, 768-773.

[7] M. I. Reiman (1982). Open queueing networks in heavy traffic. *Math. Oper. Res.*, to appear.

[8] J. Walrand and P. Varaiya (1980). Sojourn times and the overtaking condition in Jackson networks. *Adv. in Appl. Probab.*, **12**, 1000-1018.

Bell Laboratories, Murray Hill, New Jersey 07974

STATIONARY PROBABILITIES FOR NETWORKS OF QUEUES

A. Hordijk

and

N. van Dijk

Abstract

Using the phase-method, stationary distributions will be obtained for general service-requirements in case of symmetric-service-disciplines. A way of generating results for open networks by results for closed networks is given. In the second section an "invariance-property" is introduced. This property, which is equivalent to "job-local-balance", generalizes Kelly's models with symmetric-service-disciplines. Again stationary distributions are obtained. Some examples of this generalization are given.

1. Introduction

Considered are networks of queues. Customers may travel from one queue to another according to a routing scheme. At each queue a customer demands a service with certain distribution. Many models of such systems have been studied, such as: Communication networks (Kleinrock (1964)); Population models (Kingman (1969)); Networks of service facilities (Jackson (1954)) etc.

Results for these models have been obtained by assuming that the service requirements are negative-exponential distributed and that the behaviour of the system at a moment only depends on the number of customers present at the different nodes. These assumptions have been largely extended by considering general service-requirements with far more dependence upon the state and history of the system.

Our research has been stimulated by the papers of Chandy, et al., Cohen, Kelly and Schassberger (for references see Kelly [7]). By means of classification of customer-routes and service-requirements Kelly derived stationary distributions for a large class of queueing networks. In this class of networks a symmetric-service-discipline is assumed. Under this condition his results are based on time reversibility of the system. In Section 1 analogous results will be derived by using the phase method. Classification of customers is allowed and the routing may be random.

In Section 2 the condition of a symmetric-service-discipline is extended. The behaviour of a customer at a node may depend on several numbers present at that node. Under an "invariance-property", which includes the "symmetric-service-discipline", again stationary distributions are derived. The extension especially concerns a more general specification of a job by its fixed job/class-number and its service-number which is chosen each time a job enters a node. The amount of service capacity given to a job may strongly depend on its own numbers as well as even on those of other jobs present. For instance, in this way it is possible to consider queues with some kind of priority, and/or queues with a general queue-dependent (possibly randomized) capacity function.

Section 3 contains some examples such as: Queues in which the capacity function factorizes to the different types of jobs present; queues in which job/class-numbers and/or service numbers represent priority numbers; queues with Lifo-pre-emptive service-discipline and general queue-dependent capacity function.

In all sections closed as well as open models are considered. Only those references are given which were of direct stimulation for this paper. For further references we refer to Kelly's book.

Section 1

We consider nodes 0, 1, ..., N. At any node there is a service facility with a service-discipline of a rather general type. The facility has characteristic (f,φ,d) if:

1. The service capacity is $f(x)$ when x jobs are present at the facility. Let $f(x)$ be positive whenever x is positive.
2. The fraction of the total capacity when x jobs are present given to the job at the i-th place is $\varphi(x|i)$; $i = 1,\ldots,x$.
3. The probability that when x jobs are present an incoming job gets the i-th place is $d(x|i)$; $i = 1,\ldots,x+1$.

The place numbers of the jobs are always different. When x jobs are present and the job at the i-th place leaves the node, the jobs at places $i+1,\ldots,x$ will get places $i,\ldots,x-1$; when x jobs are present and an incoming job is assigned place i, the jobs at places $i,\ldots,x$ will get places $i+1,\ldots,x+1$.

Many types of service facilities can be modelled with this capacity function f, sharing function φ and placing function d.

In this section we assume the service-discipline to be symmetric, i.e., we assume that $\varphi(x|i) = d(x-1|i)$ (cf. "Reversibility and Stochastic Networks" by F. P. Kelly [5]). There are many service facilities which satisfy this assumption: Processor-sharing, loss models, infinite numbers of servers, last-in-first-out-pre-emptive, etc.

Let us first consider closed networks. Say there are jobs 1, 2, ..., M, which are travelling around via the nodes with routing-matrix $(p_{ij}(k))$ for job k, i.e., $p_{ij}(k)$ is the probability that job k goes to node j when its service is finished at node i.

With respect to the service requirements we consider three cases:

1. Exponential model; the service requirement of job k at node ℓ has a negative-exponential distribution with parameter $\mu_{\ell k}$.
2. Phase-model; the service-requirement of job k at node ℓ is given by a mixture of Erlang distributions. The distribution of service-requirement of job k at node ℓ is:

$$F_{\ell k} = \sum_{i=1}^{m_{\ell k}} h_i^{\ell k} E_{\mu_{\ell k}}^{i}$$

with $E_{\mu_{\ell k}}^{i}$ an Erlang distribution with parameter $\mu_{\ell k}$ and i phases and $h_i^{\ell k}$ is the probability that $F_{\ell k}$ has i phases.
3. General model; the service-requirement of job k at node ℓ has a distribution $F_{\ell k}$ with a finite first moment $\tau_{\ell k}$.

Case 1. The state of the system is given by specifying the numbers of the jobs at the occupied places of the nodes. Let $r_{\ell i}$ be the number of the job at the i-th place of node ℓ. The generic notation of a state is $\{r_{\ell i};\ \ell = 0,1,\dots,N,\ i = 1,\dots,x_\ell\}$ where x_ℓ is the number of jobs at node ℓ.

To give the expression for stationary probabilities we need the following notation. For job k let $\lambda_\ell(k)$, $\ell = 0,1,\dots,N$ be a solution of the equations:

$$\lambda_j(k) = \sum_{i=0}^{N} \lambda_i(k)\, p_{ij}(k), \quad j = 0,1,\dots,N. \tag{1.1}$$

Without restriction of generality it is assumed that for each k, $k = 1,\dots,M$, the routing-matrix $(p_{ij}(k))$ is irreducible, which implies that there exists, after normalization, a unique solution for (1.1). This irreducibility of $(p_{ij}(k))$ together with the positiveness of f(x), also implies the existence of a unique stationary distribution.

Write $\rho_{\ell k}$ for $\lambda_{\ell k}\,\mu_{\ell k}^{-1}$ and assume that node ℓ has characteristic $(f_\ell, \varphi_\ell, d_\ell)$, $\ell = 0,1,\ldots,N$. We will show that the stationary probabilities are then given by:

$$P\{r_{\ell i};\ \ell = 0,1,\ldots,N,\ i = 1,\ldots,x_\ell\} = c \prod_{\ell=0}^{N} \prod_{i=1}^{x_\ell} \frac{\rho_{\ell r_{\ell i}}}{f_\ell(i)} \tag{1.2}$$

where c is a normalizing constant. The proof that these probabilities are stationary is given by showing that there is local-balance for each job k, $k = 1,\ldots,M$. I.e., the stream out of a state due to job k equals the stream into that state due to job k.

Suppose that job k occupies the i-th place at node ℓ i.e., $r_{\ell i} = k$. Let $[r_{jt}]$ denote that state in which job k is at the t-th place of node j and the other jobs are at the same places as in the state $\{r_{\ell i};\ \ell = 0,1,\ldots,N,\ i = 1,\ldots,x_\ell\}$. Let $P[r_{jt}]$ denote the probability on state $[r_{jt}]$ as given by (1.2).

The stream out of state $[r_{\ell i}]$ due to job k is equal to:

$$P[r_{\ell i}]\ f_\ell(x_\ell)\ \varphi_\ell(x_\ell|i)\ \mu_{\ell k}. \tag{1.3}$$

The stream out of state $[r_{jt}]$ into state $[r_{\ell i}]$ is:

$$\begin{cases} P[r_{jt}]\ f_j(x_j+1)\ \varphi_j(x_j+1|t)\ \mu_{jk}\ p_{j\ell}(k)\ d_\ell(x_\ell-1|i) \\ \text{for } j \neq \ell,\quad t = 1,\ldots,x_j+1 & \text{and} \\ P[r_{\ell t}]\ f_\ell(x_\ell)\ \varphi_\ell(x_\ell|t)\ \mu_{\ell k}\ p_{\ell\ell}(k)\ d_\ell(x_\ell-1|i) \\ \text{for } j = \ell,\quad t = 1,\ldots,x_\ell. \end{cases}$$

By summation over j, $j = 0,1,\ldots,N$ and t, $t = 1,\ldots,x_j+1$ for $j \neq \ell$ and $t = 1,\ldots,x_\ell$ for $j = \ell$, we find that the total stream into state $[r_{\ell i}]$ due to job k is given by:

$$\sum_{j\neq\ell,\, t=1,\dots,x_j+1} P[r_{jt}]\, f_j(x_j+1)\, \varphi_j(x_j+1|t)\, \mu_{jk}\, d_\ell(x_\ell-1|i)\, p_{j\ell}(k)$$

$$+ \sum_{t=1,\dots,x_\ell} P[r_{\ell t}]\, f_\ell(x_\ell)\, \varphi_\ell(x_\ell|t)\, \mu_{\ell k}\, d_\ell(x_\ell-1|i)\, p_{\ell\ell}(k). \tag{1.4}$$

Directly from (1.2) we also have, for $j \neq \ell$, $t = 1,\dots,x_\ell+1$:

$$\frac{P[r_{jt}]}{P[r_{\ell i}]} = \frac{\lambda_j(k)\ \mu_{\ell k}\ f_\ell(x_\ell)}{\lambda_\ell(k)\ \mu_{jk}\ f_j(x_j+1)}$$

and for $t = 1,\dots,x_\ell$:

$$\frac{P[r_{\ell t}]}{P[r_{\ell i}]} = 1.$$

By substituting these equations in (1.4) we find that (1.4) is equal to:

$$P[r_{\ell i}]\ f_\ell(x_\ell)\ d_\ell(x_\ell-1|i) \left\{ \sum_{\substack{j=0\\ j\neq\ell}}^{N} p_{j\ell}(k)\ \frac{\lambda_j(k)}{\lambda_\ell(k)} \sum_{t=1}^{x_j+1} \varphi_j(x_j+1|t) \right.$$

$$\left. + \ p_{\ell\ell}(k)\ \frac{\lambda\ell(k)}{\lambda_\ell(k)} \sum_{t=1}^{x_\ell} \varphi_j(x_\ell|t) \right\} \mu_{\ell k}$$

By using the assumption of symmetric-disciplines, i.e., $d_\ell(x_\ell-1|i)$ equals $\varphi_\ell(x_\ell|i)$ and the fact that all capacity is given to the place numbers present i.e.:

$$\sum_{t=1}^{x_j+1} \varphi_j(x_j+1|t) = 1,$$

we find that (1.4) equals:

$$P[r_{\ell i}]\ f_\ell(x_\ell)\,\varphi_\ell(x_\ell|i) \sum_{j=0}^{N} p_{j\ell}(k)\ \frac{\lambda_j(k)}{\lambda_\ell(k)}\ \mu_{\ell k}\ .$$

From relations (1.1), (1.3) and (1.4) follows that the total "stream into" a state due to job k equals the "stream out" of that state due to job k. Since this "job-balance" holds for each k, $k = 1,\dots,M$, also the "global-balance" equations are satisfied. Consequently, the probabilities as given in (1.2) are stationary probabilities.

Case 2. In this case the state of the system is given by specifying the place numbers of the jobs, as well as their residual service-requirements, measured in phases. Let $a_{\ell i}$ denote the number of phases of the residual service-time of the job at the i-th place of node ℓ. Let $\bar{\rho}_{\ell k} := \lambda_\ell(k)\mu_{\ell k}^{-1}$. The stationary probabilities are now given by

$$P\{(r_{\ell i},a_{\ell i});\ \ell=0,1,\ldots,N;\ i=1,\ldots,x_\ell\}$$

$$= c \prod_{\ell=0}^{N} \prod_{i=1}^{x_\ell} \frac{\bar{\rho}_{\ell r_{\ell i}}}{f_\ell(i)} \left[h_{a_{\ell i}}^{\ell r_{\ell i}} + \cdots + h_{m_{\ell r_{\ell i}}}^{\ell r_{\ell i}} \right] \tag{1.5}$$

The proof that these probabilities are stationary is again given by showing "local-balance" for each job k; k = 1,...,M. Suppose that job k occupies the i-th place at node ℓ, i.e., $r_{\ell i} = k$, and requires $a_{\ell i}$ phases of service. Let $[r_{jt},a_{jt}]$ denote the state in which job k is at the i-th place at node j with a_{jt} phases of residual service-requirement and in which the other jobs are at the same places and have the same residual service-requirements measured in phases as given in state $\{(r_{\ell i},a_{\ell i});\ \ell=0,1,\ldots,N;\ i=1,\ldots,x_\ell\}$. Let $P[r_{jt},a_{jt}]$ denote the probability on state $[r_{jt},a_{jt}]$ as given by (1.5).

The stream out of state $[r_{\ell i},a_{\ell i}]$ due to job k is equal to:

$$P[r_{\ell i},a_{\ell i}]\ f_\ell(x_\ell)\ \varphi_\ell(x_\ell|i)\ \mu_{\ell k}\ . \tag{1.6}$$

The stream out of state $[r_{jt},a_{jt}]$ into state $[r_{\ell i},a_{\ell i}]$ is given by:

$$P[r_{jt},1]\ f_j(x_j+1)\ \varphi_j(x_j+1|t)\ \mu_{jk}\ p_{j\ell}(k)\ d_\ell(x_\ell-1|i)\ h_{a_{\ell i}}^{\ell k}$$

for $j \neq \ell$, $t = 1,\ldots,x_j+1$;

$$P[r_{\ell t},1]\ f_\ell(x_\ell)\ \varphi_\ell(x_\ell|t)\ \mu_{\ell k}\ p_{\ell\ell}(k)\ d_\ell(x_\ell-1|i)\ h_{a_{\ell i}}^{\ell k}$$

for $j = \ell$, $k = 1,\ldots,x_\ell$;

$$P[r_{\ell i}, a_{\ell i}+1]\, f_\ell(x_\ell)\, \varphi_\ell(x_\ell|i)\, \mu_{\ell k}$$

for $j=\ell$, $t=i$ and $a_{jt}=a_{\ell i}+1$, and 0 elsewhere.

So by summation over j, $j=0,1,\ldots,N$ and $t=1,\ldots,x_j+1$ for $j\neq\ell$ and $t=1,\ldots,x_\ell$ for $j=\ell$, we find that the total stream into state $[r_{\ell i}, a_{\ell i}]$ due to job k is given by:

$$\sum_{j\neq\ell,t=1,\ldots,x_j+1} P[r_{jt},1]\, f_j(x_j+1)\, \varphi_j(x_j+1,t)\, \mu_{jk}\, p_{j\ell}(k)\, d_\ell(x_\ell-1|i)\, h^{\ell k}_{a_{\ell i}}$$

$$+ \sum_{t=1,\ldots,x_\ell} P[r_{\ell t},1]\, f_\ell(x_\ell)\, \varphi_\ell(x_\ell|t)\, \mu_{\ell k}\, p_{\ell\ell}(k)\, d_\ell(x_\ell-1|i)\, h^{\ell k}_{a_{\ell i}}$$

$$+ \qquad P[r_{\ell i}, a_{\ell i}+1]\, f_\ell(x_\ell)\, \varphi_\ell(x_\ell|i)\, \mu_{\ell k} \;. \tag{1.7}$$

Directly from (1.5) we also have, for $j\neq\ell$, $t=1,\ldots,x_j+1$:

$$\frac{P[r_{jt},1]}{P[r_{\ell i}, a_{\ell i}]} = \frac{\left[h^{jk}_1+\cdots+h^{jk}_{m_{jk}}\right]}{\left[h^{\ell k}_{a_{\ell i}}+\cdots+h^{\ell k}_{m_{\ell k}}\right]}\; \frac{\lambda_j(k)\;\mu_{\ell k}\;f_\ell(x_\ell)}{\lambda_\ell(k)\;\mu_{jk}\;f_j(x_j+1)} \;.$$

For $t=1,\ldots,x_\ell$:

$$\frac{P[r_{\ell t},1]}{P[r_{\ell i}, a_{\ell i}]} = \frac{\left[h^{\ell k}_1+\cdots+h^{\ell k}_{m_{\ell k}}\right]}{\left[h^{\ell k}_{a_{\ell i}}+\cdots+h^{\ell k}_{m_{\ell k}}\right]} \;,$$

and

$$\frac{P[r_{\ell i}, a_{\ell i}+1]}{P[r_{\ell i}, a_{\ell i}]} = \frac{\left[h^{\ell k}_{a_{\ell i}+1}+\cdots+h^{\ell k}_{m_{\ell k}}\right]}{\left[h^{\ell k}_{a_{\ell i}}+\cdots+h^{\ell k}_{m_{\ell k}}\right]} \;.$$

By substituting these equations in (1.7), using $h^{jk}_1+\cdots h^{jk}_{m_{jk}}=1$, we find that (1.7) is equal to:

$$P[r_{\ell i},a_{\ell i}]\, f_\ell(x_\ell)\, d_\ell(x_\ell-1|i)\, \mu_{\ell k}\; h^{\ell k}_{a_{\ell i}} \left[h^{\ell k}_{a_{\ell i}}+\cdots+h^{\ell k}_{m_{\ell k}}\right]^{-1}$$

$$\left\{\sum_{\substack{j=0\\ j\neq 0}} p_{j\ell}(k)\, \frac{\lambda_j(k)}{\lambda_\ell(k)} \sum_{t=1}^{x_j+1} \varphi_j(x_j+1|t) + p_{\ell\ell}(k)\, \frac{\lambda_\ell(k)}{\lambda_\ell(k)} \sum_{t=1}^{x_\ell} \varphi_\ell(x_\ell|t)\right\} +$$

$$P[r_{\ell i},a_{\ell i}]\, f_\ell(x_\ell)\, d_\ell(x_\ell-1|i)\, \mu_{\ell k} \left[h^{\ell k}_{a_{\ell i}+1}+\cdots+h^{\ell k}_{m_{\ell k}}\right]\left[h^{\ell k}_{a_{\ell i}}+\cdots+h^{\ell k}_{m_{\ell k}}\right]^{-1}.$$

Again, by using properties of a symmetric-discipline and the sharing function this expression for (1.7) equals:

$$P[r_{\ell i},a_{\ell i}]\, f_\ell(x_\ell)\, \varphi_\ell(x_\ell|i)\, \mu_{\ell k} \left[h^{\ell k}_{a_{\ell i}}+\cdots+h^{\ell k}_{m_{\ell k}}\right]^{-1}$$

$$\left\{h^{\ell k}_{a_{\ell i}} \sum_{j=0}^{N} p_{j\ell}(k)\, \frac{\lambda_j(k)}{\lambda_\ell(k)} + \left[h^{\ell k}_{a_{\ell i}+1}+\cdots+h^{\ell k}_{m_{\ell k}}\right]\right\}.$$

Hence, from relations (1.1), (1.6) and (1.7) it follows similarly as in case 1: The probabilities given by (1.5) satisfy the "local-balance" equations and therefore are stationary probabilities.

Now let $\tau_{\ell k}$ denote the first moment of the service-requirement of job k at node ℓ, then:

$$\tau_{\ell k} = \sum_{i=1}^{m_{\ell k}} i\, \mu^{-1}_{\ell k}\, h^{\ell k}_i = \sum_{i=1}^{m_{\ell k}} \sum_{j=1}^{i} \mu^{-1}_{\ell k}\, h^{\ell k}_i = \mu^{-1}_{\ell k} \sum_{j=1}^{m_{\ell k}} \left[h^{\ell k}_j+\cdots+h^{\ell k}_{m_{\ell k}}\right]$$

Using this together with (1.5) we conclude that

$$\mu^{-1}_{\ell k}\, \tau^{-1}_{\ell k} \left[h^{\ell k}_{a_{\ell i}}+\cdots+h^{\ell k}_{m_{\ell k}}\right]$$

is the probability that job k at node ℓ has $a_{\ell i}$ phases of residual service-requirement. Since each phase has an exponential distribution with mean $\mu^{-1}_{\ell k}$, the density for a residual service-time of $t_{\ell i}$ is given by:

$$\sum_{i=1}^{m_{\ell k}} \left[h_i^{\ell k}+\cdots+h_{m_{\ell k}}^{\ell k}\right] \tau_{\ell k}^{-1} \mu_{\ell k}^{-1} \frac{d}{dt} E_{\mu_{\ell k}}^{i} (t_{\ell i})$$

$$= \frac{1}{\tau_{\ell k}} \sum_{i=1}^{m_{\ell k}} \sum_{j=i}^{m_{\ell k}} h_j^{\ell k} \tau_{\ell k}^{-1} \mu_{\ell k}^{-1} \frac{(\mu_{\ell k} t_{\ell i})^{(i-1)}}{(i-1)!} e^{-(\mu_{\ell k} t_{\ell i})}$$

$$= \frac{1}{\tau_{\ell k}} \sum_{j=1}^{m_{\ell k}} h_j^{\ell k} \sum_{i=1}^{j} \frac{(\mu_{\ell k} t_{\ell i})^{(i-1)}}{(i-1)!} e^{-(\mu_{\ell k} t_{\ell i})} = \frac{1}{\tau_{\ell k}} \left[1-F_{\ell k}(t_{\ell i})\right] .$$

By using this derivation for each job k, $k = 1,\ldots,M$, we find for the stationary densities:

$$P\{(r_{\ell i}, t_{\ell i});\ \ell = 0,1,\ldots,N;\ i = 1,\ldots,x_\ell\}$$

$$= c \prod_{\ell=0}^{N} \prod_{i=1}^{x_\ell} \frac{\rho_{\ell r_{\ell i}}}{f_\ell(i)} \cdot \frac{\left[1 - F_{\ell r_{\ell i}}(t_{\ell i})\right]}{\tau_{\ell r_{\ell i}}} \tag{1.8}$$

with $\rho_{\ell k} := \lambda_\ell(k) \cdot \tau_{\ell k}$ and $t_{\ell i}$ the residual service-requirement of the job on the i-th place at node ℓ.

Case 3. The expression of the stationary probabilities given in (1.8) does not depend on the fact that the distributions $F_{\ell k}$ are assumed to be mixtures of Erlang distributions. It is well known that any distribution concentrated on the nonnegative real line can be approximated arbitrarily closely by a mixture of Erlang distributions.

A conjecture would then be that the expression (1.8) is true in more general models. A rigorous proof that this conjecture is true under mild assumptions, can be given with results from the theory of weak convergence for probability measures on metric spaces. The assumption is that the technologies of the service facilities are almost sure conditions. There are symmetric facilities which are not continuous. However, since stationary distributions are continuous with respect to the residual service-requirements they become almost sure continuous.

As a consequence we have that the stationary probabilities are valid for general service distributions. (cf. Barbour [1], A. Hordijk and R. Schassberger [6], W. Whitt [9]).

To analyse open networks we consider two cases: a) Generalized quasi-random input b) Poisson input. For case a we start with nodes $1,\dots,N$ and then add an extra node 0 with characteristics (f_0, φ_0, d_0) such that $f_0(x) = x$, $\varphi_0(x|i) = d_0(x-1|i) = \frac{1}{x}$. Then this node produces a generalized quasi-random input with M sources such that source k has idle-time distribution F_{0k}, $k = 1,\dots,M$. For case b with one class of jobs having routing-matrix (p_{ij}) assume the Poisson-input at node ℓ has rate γ_ℓ, $\ell = 1,\dots,N$. Again we add the extra node 0 with characteristic given as in case a.

We assume M jobs which all have a negative-exponential distributed service-requirement with parameter $\mu_0(M)$. For λ_j, $j = 1,\dots,N$ a solution of the equations:

$$\lambda_j = \gamma_j + \sum_{i=1}^{N} \lambda_i \, p_{ij}, \quad j = 1,\dots,N,$$

we define:

$$p_{i0} := (1 - \sum_{j=1}^{N} p_{ij}) \text{ and } \lambda_0 := \sum_{i=1}^{N} \lambda_i \, p_{i0};$$

with

$$\gamma_0 := \sum_{i=1}^{N} \gamma_i, \quad p_{0i} := \gamma_i \gamma_0^{-1} \text{ and } p_{00} = 0.$$

Then we have that $\lambda_0, \lambda_1, \dots, \lambda_N$ is a solution of:

$$\lambda_j = \sum_{i=0}^{N} \lambda_i \, p_{ij}, \quad j = 0,1,\dots,N.$$

We consider now the closed network with M jobs and $\mu_0(M) := \gamma_0 M^{-1}$. Using another theorem from the probability theory on metric spaces we conclude that the output-process from node 0 to node j tends to a Poisson process with rate γ_j, $j = 1,\dots,N$, as M tends to infinity. As a consequence we can obtain stationary probabilities for the open network from (1.8) by taking the appropriate limits. More precisely, the probability that there are x_ℓ jobs at node ℓ having residual service-requirements less than or equal to $t_{\ell i}$, $i = 1,\dots,x$ for $\ell = 1,\dots,N$ becomes:

$$c \prod_{\ell=1}^{N} \rho_\ell^{x_\ell} \prod_{i=1}^{x_\ell} \frac{1}{f_\ell(i)} \cdot \frac{1}{\tau_\ell} \int_0^{t_{\ell i}} (1-F_\ell(u))\, du, \tag{1.9}$$

where c is a normalizing constant, F_ℓ is the distribution of the service-requirement at node ℓ with first moment τ_ℓ and $\rho_\ell = \lambda_\ell \tau_\ell$.

Now (1.9) will be extended to more classes of jobs. Suppose there are Q classes of jobs and each node has a Poisson-input of jobs of class k with rate $\gamma_\ell(k)$; $\ell = 1,\dots,N$, $k = 1,\dots,Q$. Again we add the extra node 0 with characteristic as given in case b. Suppose there are M_k jobs of class k; $k = 1,\dots,Q$. Jobs of class k have routing-matrix $(p_{ij}(k))$. Moreover, their service-requirements are negative-exponential distributed with parameters $\mu_{\ell k}$ at node $\ell \neq 0$ and $\mu_{0k}(M_k)$ at node 0.

Let $\lambda_j(k)$; $j = 1,\dots,N$ be a solution of the equations:

$$\lambda_j(k) = \gamma_j(k) + \sum_{i=1}^{N} \lambda_i(k)\, p_{ij}(k) \qquad k = 1,\dots,Q.$$

Define:

$$p_{i0}(k) := 1 - \sum_{j=1}^{N} p_{ij}(k) \text{ and } \lambda_0(k) := \sum_{i=1}^{N} \lambda_i(k)\, p_{i0}(k).$$

Then with:

$$\gamma_0(k) := \sum_{i=1}^{N} \gamma_i(k) \quad \text{and} \quad p_{0i}(k) := \gamma_i(k)\, \gamma_0(k)^{-1}$$

we have that $\lambda_0(k),\dots,\lambda_N(k)$ for $k = 1,\dots,Q$ is a solution of

$$\lambda_j(k) = \sum_{i=0}^{N} \lambda_i(k)\, p_{ij}(k).$$

Now we consider the closed network with M_k jobs of class k;

$k=1,\dots,Q$ and $\mu_{0k}(M_k) := \gamma_0(k)\, M_k^{-1}$. Using the same arguments as given for one class of jobs we conclude that the output process for jobs of class k from node 0 to node j tends to a Poisson process with rate $\gamma_j(k)$; $j=1,\dots,N$; $k=1,\dots,Q$, as M_k tends to infinity. In particular, by taking $M_k = M$, $k=1,\dots,Q$ we can obtain stationary probabilities for the open network from (1.6) by taking appropriate limits as M tends to infinity. More precisely, the probability that there are $x_\ell(k)$ jobs of class k at node ℓ, having residual service-requirements less than or equal to $t_{\ell i}(k)$, $i=1,\dots,x\ (k)$, $k=1,\dots,Q$, $\ell=1,\dots,N$ is given by:

$$c \prod_{\ell=0}^{N} \Big[\prod_{j=0}^{x_\ell(1)+\cdots+x_\ell(Q)} \frac{1}{f_\ell(j)} \Big]$$

$$\prod_{\ell=1}^{Q} \prod_{i=1}^{x_\ell(k)} \rho_{\ell k} \frac{1}{\tau_{\ell k}} \int_0^{t_{\ell i}(k)} (1 - F_{\ell k}(u))\, du. \qquad (1.10)$$

with $F_{\ell k}$ the distribution of service-requirement of a job of class k at node ℓ with mean $\tau_{\ell k}$, $\rho_{\ell k} = \lambda_\ell(k)\, \tau_{\ell k}$ and c a normalizing constant.

Section 2

In this section each job at a node carries three numbers. For the closed model they are: Its job-number, its service-number and its serial-number. For the open model they are: Its class-number, its service-number and its serial-number. The job-numbers in the closed model as well as the class-numbers in the open model will never change. To each job which arrives at a certain node is assigned a service-number. There are several different protocols possible for changing the service-numbers. We distinguish between the following protocols: With protocol 1 the service-numbers at a node are different. If service-numbers $1,2,\dots,x$ are present at a certain node and to an incoming job is assigned service-number k, $k=1,\dots,x$ then the service-numbers

$k,\dots,x$ are changed to $k+1,\dots,x+1$. If a job with service-number ℓ leaves the node then the service-numbers $\ell+1,\dots,x$ are changed to $\ell,\dots,x-1$. In protocol 2 jobs at a node may have the same service-numbers. The service-number of a job will not be changed during his service at a node.

By representing the place-numbers of section 1 by service-numbers under protocol 1 we get the model of section 1. With service-numbers of protocol 2 it will be possible to introduce a kind of priority.

The serial-numbers are only introduced for notational purpose. We use these numbers to index the job/class-, and service-numbers. Further assume that a job with job/class-number k has a negative-exponential distributed service-requirement with parameter $\mu_{\ell k}$ at node ℓ and travels through the system according routing-matrix $(p_{ij}(k))$.

We introduce the following notation: The job with i-th serial number at node ℓ has job-number $r_{\ell i}$ in the closed model, and class-number $r_{\ell i}$ in the open model. In both models $s_{\ell i}$ will denote the service-number of this job. The state of the system will be an unordered set of number-pairs denoted as $\{(r_{\ell i},s_{\ell i});\ \ell = 0,1,\dots,N;\ i = 1,\dots,x_\ell\}$ when x_ℓ jobs are present at node ℓ.

In this notation the index i is the serial-number. In the sequel we will often, when convenient, change the serial-numbers. Note that although used in the notation of the state the precise information of the serial-numbers is not contained in the information given by the state. In other words: From the state we find for each node ℓ the number of jobs present, their job/class-numbers and service-numbers but not their serial-numbers.

The service-discipline at a node is characterized by (f,φ,δ) where f is the capacity, φ the sharing-function and δ the service-number-assigning function. Whenever jobs/class-numbers r_i with service-numbers s_i, $i=1,\dots,j$ are present then:

1. the service capacity is $f((r_1,s_1),\dots,(r_j,s_j))$ and positive if $j \geq 1$;
2. the fraction of this total capacity given to the job with number-pair (r_i,s_i) is $\varphi((r_1,s_1),\dots,(r_j,s_j)|(r_i,s_i))$. It is also possible to define $\varphi((r_1,s_1),\dots,(r_j,s_j)|(r_i,s_i))$ as the probability that the full capacity is given to this job;
3. the probability that an incoming job with job/class-number r_{j+1} obtains service-number s_{j+1} is $\varphi((r_1,s_1),\dots,(r_j,s_j)|(r_{j+1},s_{j+1}))$.

Assumption 1.

$$\sum_{i=1,\dots,j} \varphi((r_1,s_1),\dots,(r_j,s_j)|(r_i,s_i)) = 1,$$

$$\sum_{s_{j+1}} \delta((r_1,s_1),\dots,(r_j,s_j)|(r_{j+1},s_{j+1})) = 1.$$

In the sequel also the following realistic assumptions are made. The characteristic (f,φ,δ) is invariant for permutations of the number-pairs (r_i,s_i); $i=1,\dots,j$; $j=1,2,\dots$. i.e.: For each permutation $\pi((r_1,s_1),\dots,(r_j,s_j))$ of $(r_1,s_1),\dots,(r_j,s_j)$:

$$f((r_1,s_1),\dots,(r_j,s_j)) = f(\pi((r_1,s_1),\dots,(r_j,s_j)));$$

$$\varphi((r_1,s_1),\dots,(r_j,s_j)|(r_i,s_i)) = \varphi(\pi((r_1,s_1),\dots,(r_j,s_j))|(r_i,s_i))$$

and

$$\delta((r_1,s_1),\dots,(r_j,s_j)|(r_{j+1},s_{j+1}))$$

$$= \delta(\pi((r_1,s_1),\dots,(r_j,s_j))|(r_{j+1},s_{j+1})).$$

These assumptions imply that the characteristic functions only depend on the number-pairs present at a node.

Assumption 2. Each job has an irreducible routing-matrix and the stationary probabilities on the states are unique.

Below we introduce a condition with which the stationary distribution can be obtained. It generalizes the symmetric condition of section 1. We call this condition the "invariance-property". Actually it turns out that this "invariance-property" is equivalent to the "job-local-balance property". This in turn implies that the stationary distribution is valid for general service requirements. These results will be discussed in a subsequent paper. Here we will only show that the "invariance-property" implies "local-balance". "Invariance-property": For each ℓ; $\ell = 0,1,\dots,N$, any $n \in \mathbb{N}$ and n-tuple $(r_1,s_1),\dots,(r_n,s_n)$:

$$\prod_{j=1}^{n} \frac{\delta_\ell((r_1,s_1),\dots,(r_{j-1},s_{j-1})|(r_j,s_j))}{f_\ell((r_1,s_1),\dots,(r_j,s_j)) \cdot \varphi_\ell((r_1,s_1),\dots,(r_j,s_j)|(r_j,s_j))} \tag{2.1}$$

is invariant under permuting the number-pairs $(r_1,s_1),\dots,(r_n,s_n)$.

Note that we allow the product given in (2.1) to have a factor for which the numerator as well as the denominator are zero. Since, in that case such a product can be defined equal to a product for a permutation for which all denominators are positive. Note that such a product exists by the definition of service-capacity and assumption 1.

Under the assumption of the "invariance-property" the stationary probabilities are given by:

$$P\{(r_{\ell i},s_{\ell i});\ \ell = 0,1,\dots,N;\ i = 1,\dots,x_\ell\} =$$

$$c \prod_{\ell=0}^{N} \prod_{j=1}^{x_\ell} \frac{\rho_{\ell r_{\ell j}}\, \delta_\ell((r_{\ell 1},s_{\ell 1}),\dots,(r_{\ell j-1},s_{\ell j-1})|(r_{\ell j},s_{\ell j}))}{f_\ell((r_{\ell 1},s_{\ell 1}),\dots,(r_{\ell j},s_{\ell j}))\cdot \varphi_\ell((r_{\ell 1},s_{\ell 1}),\dots,(r_{\ell j},s_{\ell j})|(r_{\ell j},s_{\ell j}))} \tag{2.2}$$

with

$$\lambda_j(k) = \sum_{i=0}^{N} \lambda_i(k)\, p_{ij}(k) \text{ and } \rho_{\ell k} = \lambda_\ell(k)\, \mu_{\ell k}^{-1}$$

for the closed model and

$$P\{(r_{\ell i}, s_{\ell i});\ \ell = 1,\dots,N;\ i = 1,\dots,x_\ell\} =$$

$$c \prod_{\ell=1}^{N} \prod_{j=1}^{x_\ell} \frac{\rho_{\ell r_{\ell j}} \quad \delta_\ell((r_{\ell 1}, s_{\ell 1}),\dots,(r_{\ell x_\ell - 1}, s_{\ell x_\ell - 1}) \mid (r_{\ell x_\ell}, s_{\ell x_\ell}))}{f_\ell((x_{\ell 1}, s_{\ell 1}),\dots,(r_{\ell x_\ell}, s_{\ell x_\ell}))}$$

$$\cdot \frac{1}{\varphi_\ell((r_{\ell 1}, s_{\ell 1}),\dots,(r_{\ell x_\ell}, s_{\ell x_\ell}) \mid (r_{\ell x_\ell}, s_{\ell x_\ell}))} \qquad (2.3)$$

with

$$\lambda_j(k) = \gamma_\ell(k) + \sum_{i=1}^{N} \lambda_i(k)\, p_{ij}(k), \quad \rho_{\ell k} = \lambda_\ell(k)\, \mu_{\ell k}^{-1}$$

for the open model. In both models c is a normalizing constant.

As in section 1 the verification that these probabilities are stationary is done by showing "job-balance" for each job. In fact we will prove that the stream out of a state due to a certain job equals the stream into that state due to that job. From the "invariance-property" it follows that the probabilities as defined by (2.2) and (2.3) are also invariant under permuting the number-pairs. Together with assumption 1 this implies that in notating the states we may use an arbitrary permutation of the number-pairs at a node. In the derivations below we shall frequently use convenient permutations of number-pairs.

We first consider closed models. The number of jobs is M. We will prove "local-balance" for the job with job-number k, $k = 1,\dots,M$. In the following derivation it is convenient to permute the number-pairs such that job k has always the highest serial-number at a node. Say,

$r_{\ell x_\ell} = k$. Let $[r_{jt}, s_{jt}]$ denote the state in which job k has the highest serial-number at node j, i.e., $t = x_j + 1$ for $j \neq \ell$ and $t = x_\ell$ for $j = \ell$, and with service-number s_{jt}, and the other jobs have number-pairs, if necessary after shifts of service-numbers under protocol 1, as given in: $\{(r_{\ell i}, s_{\ell i});\ \ell = 0,1,\ldots,N;\ i = 1,\ldots,x_\ell\}$.

Let $P\,[r_{jt}, s_{jt}]$ denote the probability on state $[r_{jt}, s_{jt}]$ as given by (2.2) and (2.3). Note that $P[r_{jt}, s_{jt}]$ can be zero.

The stream out of state $[r_{\ell x_\ell}, s_{\ell x_\ell}]$ due to job k is equal to:

$$P\,[r_{\ell x_\ell}, s_{\ell x_\ell}] \tag{2.4}$$

$$\cdot\ \{f_\ell((r_{\ell 1}, s_{\ell 1}),\ldots,(k, s_{\ell x_\ell}))\ \varphi_\ell((r_{\ell 1}, s_{\ell 1}),\ldots,(k, s_{\ell x_\ell})\,|\,(k, s_{\ell x_\ell}))\ \mu_{\ell k}\}.$$

The stream out of state $[r_{jt}, s_{jt}]$ into state $[r_{\ell x_\ell}, s_{\ell x_\ell}]$ is given by:

$$P[r_{j\ x_j+1}, s_{j\ x_j+1}]\{\mu_{jk}\ f_j((r_{j1}, s_{j1}),\ldots,(r_{jx_j+1}, s_{jx_j+1}))$$

$$\cdot\ \varphi_j((r_{j1}, s_{j1}),\ldots,(r_{jx_j+1}, s_{jx_j+1})\,|\,(r_{jx_j+1}, s_{jx_j+1}))p_{j\ell}(k)$$

$$\cdot\ \delta_\ell((r_{\ell 1}, s_{\ell 1}),\ldots,(r_{\ell x_\ell -1}, s_{\ell x_\ell -1})\,|\,(k, s_{\ell x_\ell}))\}\ ,$$

for $j \neq \ell$ and any $s_{j\ s_j+1}$.

$$P[r_{\ell x_\ell}, \tilde{s}_{\ell x_\ell}]\ \{f_\ell((r_{\ell 1}, s_{\ell 1}),\ldots,(r_{\ell x_\ell}, \tilde{s}_{\ell x_\ell}))$$

$$\cdot\ \varphi_\ell((r_{\ell 1}, s_{\ell 1}),\ldots,(r_{\ell x_\ell}, \tilde{s}_{\ell x_\ell})\,|\,(r_{\ell x_\ell}, \tilde{s}_{\ell x_\ell}))\mu_{\ell k}\ p_{\ell\ell}(k)$$

$$\cdot\ \delta_\ell((r_{\ell 1}, s_{\ell 1}),\ldots,(r_{\ell x_\ell -1}, s_{\ell x_\ell -1})\,|\,(k, s_{\ell x_\ell}))\}$$

for $j = \ell$ and any $\tilde{s}_{\ell x_\ell}$ and 0 elsewhere.

By summation over j; $j = 0,1,\ldots,N$; s_{jx_j+1} for $j \neq \ell$ and $\tilde{s}_{\ell x_\ell}$ we find that the total stream into state $[r_{\ell x_\ell}, s_{\ell x_\ell}]$ is given by:

$$\sum_{\ell \neq j, s_{jx_j+1}} P[r_{jx_j+1}, s_{jx_j+1}] \tag{2.5}$$

$$\cdot \{\mu_{jk}\ f_j((r_{j1}, s_{j1}), \ldots, (r_{jx_j+1}, s_{jx_j+1}))$$

$$\cdot \varphi_j((r_{j1}, s_{j1}), \ldots, (r_{jx_j+1}, s_{jx_j+1}) | (r_{jx_j+1}, s_{jx_j+1})) p_{j\ell}(k)$$

$$\cdot \delta_\ell((r_{\ell 1}, s_{\ell 1}), \ldots, (r_{\ell x_\ell - 1}, s_{\ell x_\ell - 1}) | (k, s_{\ell x_\ell}))\} +$$

$$\sum_{\tilde{s}_{\ell x_\ell}} P[r_{\ell x_\ell}, s_{\ell x_\ell}]\{\mu_{\ell k}\ f_\ell(r_{\ell 1}, s_{\ell 1}), \ldots, (r_{\ell x_\ell}, \tilde{s}_{\ell x_\ell}))$$

$$\cdot \varphi_\ell((r_{\ell 1}, s_{\ell 1}), \ldots, (r_{\ell x_\ell}, \tilde{s}_{\ell x_\ell}) | (r_{\ell x_\ell}, \tilde{s}_{\ell x_\ell})) p_{\ell\ell}(k)$$

$$\cdot \delta_\ell((r_{\ell 1}, s_{\ell 1}), \ldots, (r_{\ell x_\ell - 1}, s_{\ell x_\ell - 1}) | (k, s_{\ell x_\ell}))\}.$$

In the following derivation we first consider all denominators to be positive. For $\ell \neq j$, s_{jx_j+1} arbitrary:

$$\frac{P[r_{jx_j+1}, s_{jx_j+1}]}{P[r_{\ell x_j}, s_{\ell x_\ell}]} = \frac{\mu_{\ell k}\ f_\ell((r_{\ell 1}, s_{\ell 1}), \ldots, (r_{\ell x_j}, s_{\ell x_j}))}{\mu_{jk}\ f_j((r_{j1}, s_{j1}), \ldots, (r_{jx_j+1}, s_{jx_j+1}))}$$

$$\cdot \frac{\varphi_\ell((r_{\ell 1}, s_{\ell 1}), \ldots, (r_{\ell x_\ell}, s_{\ell x_\ell}) | (r_{\ell x_\ell}, s_{\ell x_\ell})) \lambda_j(k)}{\varphi_j((r_{j1}, s_{j1}), \ldots, (r_{jx_j+1}, s_{jx_j+1}) | (r_{jx_j+1}, s_{jx_j+1})) \lambda_\ell(k)}$$

$$\cdot \frac{\delta_j((r_{j1}, s_{j1}), \ldots, (r_{jx_j}, s_{jx_j}) | (k, s_{jx_j+1}))}{\delta_\ell((r_{\ell 1}, s_{\ell 1}), \ldots, (r_{\ell x_\ell - 1}, s_{\ell x_\ell - 1}) | (k, s_{\ell x_\ell}))} .$$

For any $\tilde{s}_{\ell x_\ell}$:

$$\frac{P[r_{\ell x_\ell}, \tilde{s}_{\ell x_\ell}]}{P[r_{\ell x_\ell}, s_{\ell x_\ell}]} = \frac{f_\ell((r_{\ell 1}, s_{\ell 1}), \ldots, (r_{\ell x_\ell}, s_{\ell x_\ell}))}{f_\ell((r_{\ell 1}, s_{\ell 1}), \ldots, (r_{\ell x_\ell}, \tilde{s}_{\ell x_\ell}))}$$

$$\cdot \frac{\varphi_\ell((r_{\ell 1}, s_{\ell 1}), \ldots, (r_{\ell x_\ell}, s_{\ell x_\ell}) \mid (r_{\ell x_\ell}, s_{\ell x_\ell}))}{\varphi_\ell((r_{\ell 1}, s_{\ell 1}), \ldots, (r_{\ell x_\ell}, \tilde{s}_{\ell x_\ell}) \mid (r_{\ell x_\ell}, \tilde{s}_{\ell x_\ell}))}$$

$$\cdot \frac{\delta_\ell((r_{\ell 1}, s_{\ell 1}), \ldots, (r_{\ell x_\ell - 1}, s_{\ell x_\ell - 1}) \mid (k, \tilde{s}_{\ell x_\ell}))}{\delta_\ell((r_{\ell 1}, s_{\ell 1}), \ldots, (r_{\ell x_\ell - 1}, s_{\ell x_\ell - 1}) \mid (k, s_{\ell x_\ell}))}$$

Now we remark that if $P[r_{\ell x_\ell}, s_{\ell x_\ell}] = 0$ then (2.4) and (2.5) are both equal to zero and if a denominator in one of the right hand sides is zero then also, according to (2.1), the corresponding numerator is zero. Using these arguments as well as the derivations given above we find that (2.5) is equal to:

$$P[r_{\ell x_\ell}, s_{\ell x_\ell}]\ f_\ell((r_{\ell 1}, s_{\ell 1}), \ldots, (r_{\ell x_\ell}, s_{\ell x_\ell}))$$

$$\cdot\ \varphi_\ell((r_{\ell 1}, s_{\ell 1}), \ldots, (r_{\ell x_\ell}, s_{\ell x_\ell}) \mid (r_{\ell x_\ell}, s_{\ell x_\ell}))\mu_{\ell k} \left\{ \sum_{j \neq \ell} p_{j\ell}(k) \frac{\lambda_j(k)}{\lambda_\ell(k)} \right.$$

$$\cdot \sum_{s_{jx_j+1}} \delta_j((r_{j1}, s_{j1}), \ldots, (r_{jx_j}, s_{jx_j}) \mid (k, s_{jx_j+1})) + p_{\ell\ell}(k) \frac{\lambda_\ell(k)}{\lambda_\ell(k)}$$

$$\left. \cdot \sum_{\tilde{s}_{\ell x_\ell}} \delta_\ell((r_{\ell 1}, s_{\ell 1}), \ldots, (r_{\ell x_\ell - 1}, s_{\ell x_\ell - 1}) \mid (k, \tilde{s}_{\ell x_\ell})) \right\} .$$

Now suppose that $\lambda_j(k)$, $j = 0, 1, \ldots, N$ is the unique solution of:

$$\lambda_j(k) = \sum_{i=0}^{N} p_{ij}(k)\ \lambda_i(k), \quad j = 0, 1, \ldots, N$$

then by using the fact that $\delta \cdot (\cdot \mid (\cdot\ s))$ is a non-defective probability i.e., $\sum_s \delta \cdot (\cdot \mid (\cdot\ x)) = 1$, we find that (2.4) and (2.5) are equal. Hence, the probabilities given by (2.2) satisfy the "local-balance" equations and consequently are stationary probabilities.

As in Section 1 the results for an open model can be obtained from a closed network with the number of jobs tending to infinity. Here, we prefer to give a direct proof. For the open model nodes $1,\ldots,N$ and Q classes of customers are considered.

Let $\gamma_\ell(k)$ be the rate of the Poisson-input at node ℓ of jobs of class k; $\ell = 1,2,\ldots,N$; $k = 1,\ldots,Q$. The verification of the "job-local-balance" is almost the same as for the closed model. The state only specifies the number-pairs of the jobs at nodes $1,2,\ldots,N$, and the term in expression (2.5), which corresponds to the stream out of state $[r_{0x_0+1}, s_{0x_0+1}]$ into state $[r_{\ell i}, s_{\ell i}]$ has to be replaced by:

$$P\{(r_{ji}, s_{ji});\ i=1,\ldots,x_j \text{ for } j \neq \ell;\ i=1,\ldots,x_\ell-1 \text{ for } j=\ell\}\ [\gamma_\ell(k)$$

$$\cdot\ \delta_\ell((r_{\ell 1}, s_{\ell 1}),\ldots,(r_{\ell x_\ell -1}, s_{\ell x_\ell -1}) \mid (k, s_{\ell x_\ell}))]$$

Moreover, in the open model $\lambda_j(k)$ is the solution of:

$$\lambda(k) = \gamma_\ell(k) + \sum_{j=1}^{N} \lambda_j(k)\, p_{j\ell}(k); \quad \ell = 1,\ldots,N \text{ and } k = 1,\ldots,Q.$$

Remark. Under the conditions of this section again the "job-local-balance" property is satisfied. Therefore simultaneous distributions of number-pairs and residual service-requirements can be derived as in Section 1. These distributions are given by the probabilities of (2.2) and (2.3) multiplied with the appropriate product of factors as

$$\frac{1}{\tau_{\ell k}} \int_0^{t_{\ell i}(r_{\ell i})} (1-F_{\ell r_{\ell i}}(u))\,du,$$

for the job with number-pair $(r_{\ell i}, s_{\ell i})$ and with $t_{\ell i}(r_{\ell i})$ denoting the residual service-requirement of this job. (See for instance (1.10).)

Section 3

Examples.

1. Generalized-symmetric-disciplines. In this model it is assumed that the characteristics of a node satisfies:

$$\delta((r_1,s_1),\dots,(r_{j-1},s_{j-1})\,|\,(r_j,s_j)) = \varphi((r_1,s_1),\dots,(r_j,s_j)\,|\,(r_j,s_j))$$

for each $(r_1,s_1),\dots,(r_j,s_j)$; $j = 1,2,\dots$. Then the "invariance-property" (2.1) will also be satisfied if:

$$\prod_{j=1}^{n} f((r_1,s_1),\dots,(r_j,s_j)) \quad \text{is invariant}$$

under permuting $(r_1,s_1),\dots,(r_n,s_n)$.

Remark. The symmetric model of Section 1 is included, since the place-numbers there can be seen as service-numbers here under protocol 1, and the capacity only depends on the number of jobs present.

Here we allow the capacity function to be a product of functions $\Phi_i(n_i)$ where i is used for indexing different types of number-pairs and n_i is the number of jobs with number-pairs of type i, present. For instance, jobs of a special type can cause an acceleration of the total capacity. For example: possibilities for considering number-pairs to be of the same type are:

1. Whenever the service-numbers are the same regardless their job/class-numbers. So here, the type of a job is only determined by its chosen service-number at that node, for instance, representing a chosen type of server.

2. Whenever the job/class-numbers belong to a same defined set of numbers, regardless their job/class-numbers.

Now the type of a job only depends on its fixed job/class-number.

Example. A number-pair at a node is called to be of type 1 whenever its job/class-number is odd and of type 2 whenever its job/

class-number is even. Assume

$$\delta_\ell((r_1,s_1),..,(r_{j-1},s_{j-1})\,|\,(r_j,s_j)) = \varphi_\ell((r_1,s_1),..,(r_j,s_j)\,|\,(r_j,s_j))$$

for $j = 1,2,\ldots$; $s_j = 1,\ldots,j$; and

$$f_\ell((r_1,s_1),\ldots,(r_j,s_j)) = \Phi_{\ell 1}(n_{\ell 1}) \quad \Phi_{\ell 2}(n_{\ell 2})$$

with n_ℓ the number of number-pairs of type i in $(r_1,s_1),\ldots,(r_j,s_j)$ at node ℓ, $i = 1,2,\ldots$. The stationary probabilities as given in (2.2) are:

$$P\{(r_{\ell i},s_{\ell i})\ \ell = 0,\ldots,N;\ i = 1,\ldots,x_\ell\}$$

$$= c \prod_{\ell=0}^{N} \prod_{k=1}^{n_{\ell 1}} \frac{1}{\Phi_{\ell 1}(k)} \prod_{k=1}^{n_{\ell 2}} \frac{1}{\Phi_{\ell 2}(k)} \prod_{i=1}^{x_\ell} \rho_{\ell r_{\ell i}}$$

with c a normalizing constant.

Remark. In the model of this section each time a job enters a node the capacity is determined by its job/class-number together with its realized service-number. In other words, the capacity function depends in a random way on the arriving jobs.

2. Priority models. Here we introduce some kind of 'priority' by allowing the capacity given to a job explicitly to depend on its service-number. For example jobs with higher service-number may receive more capacity.

a. Assume:

$$\prod_{j=1}^{n} \delta_\ell((r_1,s_1),\ldots,(r_{j-1},s_{j-1})\,|\,(r_j,s_j))$$

is invariant under permuting the number-pairs $(r_1,s_1),\ldots,(r_n,s_n)$. A job which arrives at a node is assigned a service-number which equals the job/class-number, i.e.,

$$\delta((r_1,s_1),\ldots,(r_{j-1},s_{j-1})\,|\,(r_j,r_j)) = 1.$$

A number-pair is said to be of type 1 whenever its service-number is odd and of type 2 whenever its service-number is even. Let $n_{\ell i}$ denote the number of jobs at node ℓ with number-pair of type i, $i = 1,2$. Also assume:

$$f_\ell((r_1,r_1),\dots,(r_j,r_j)) = \Phi_{\ell 1}(n_{\ell 1}) + \Phi_{\ell 2}(n_{\ell 2})$$

and $$\varphi_\ell((r_1,r_1),\dots,(r_j,r_j)\,|\,(r_j,r_j)) = \frac{n_{\ell i}^{-1}\ \Phi_{\ell 1}(n_{\ell i})}{\Phi_{\ell 1}(n_{\ell 1}) + \Phi_{\ell 2}(n_{\ell 2})}$$

when the number-pair (r_j,r_j) is of type i, $i = 1,2$. So,

$$f_\ell((r_1,r_1),\dots,(r_j,r_j))\ \varphi_\ell((r_1,r_1),\dots,(r_j,r_j)\,|\,(r_j,r_j))$$

$$= n_{\ell i}^{-1}\ \Phi_{\ell 1}(n_{\ell i})\ .$$

It is easy to verify that (2.1) is satisfied. The stationary probabilities as given in (2.2) are:

$$P\{(r_{\ell i},r_{\ell i});\ \ell = 0,1,\dots,N;\ i = 1,\dots,x_\ell\}$$

$$= c\prod_{\ell=0}^{N}\frac{1}{n_{\ell 1}!}\,\frac{1}{n_{\ell 2}!}\,\Big[\prod_{k=1}^{n_{\ell 1}}\frac{1}{\Phi_{\ell 1}(k)}\Big]\Big[\prod_{k=1}^{n_{\ell 2}}\frac{1}{\Phi_{\ell 2}(k)}\Big]\cdot\prod_{i=1}^{x_\ell}\rho_{\ell r_{\ell i}}$$

where c is a normalizing constant.

Remark. In this model the capacity given to a job at a node is not the same for all jobs present at the node but depends on its job/class-number. In other words, jobs with certain job/class-numbers receive priority. This priority may even depend on the number of such jobs present.

b. In this example will be shown that for satisfying the "invariance-property" (2.1) it is even allowed that the behaviour of a job of a certain type may depend on the number of jobs of other types present.

Suppose that to each job which arrives at a node is assigned a service number 1 or 2, representing two different types of servers. Let $n_{\ell i}$ denote the number of service-numbers i at node ℓ then the probability $\delta_\ell(\cdot|(\cdot\ i))$ that an incoming job is assigned service-number i can be denoted as $p_\ell(n_{\ell 1}, n_{\ell 2}|i)$. The fraction of the total capacity given to a job with server-number i, $\varphi_\ell(\cdot|(\cdot\ i))$, can be denoted as $\Psi_\ell(n_{\ell 1}, n_{\ell 2}|i)$. Assume:

$$p_\ell(n_{\ell 1},n_{\ell 2}|1) = \begin{cases} \frac{1}{4} & \text{if } n_1 = n_2; \\ 0 & n_1 > n_2; \\ 1 & n_1 < n_2; \end{cases}$$

$$p_\ell(n_{\ell 1},n_{\ell 2}|2) = \begin{cases} \frac{3}{4} & \text{if } n_1 = n_2; \\ 1 & n_1 > n_2; \\ 0 & n_1 < n_2; \end{cases}$$

$$\Psi_\ell(n_{\ell 1},n_{\ell 2}|1) = \begin{cases} \frac{3}{4} n_1^{-1} & n_1 = n_2; \\ 0 & n_1 < n_2; \\ n_1^{-1} & n_1 > n_2; \end{cases}$$

$$\Psi_\ell(n_{\ell 1},n_{\ell 2}|2) = \begin{cases} \frac{1}{4} n_2^{-1} & n_1 = n_2; \\ n_2^{-1} & n_1 < n_2; \\ 0 & n_1 > n_2; \end{cases}$$

and $f_\ell((r_1,s_1),\dots,(r_j,s_j)) = j$. With these assumptions the "invariance property" is satisfied. There is an obvious dependence in the behaviour of jobs of different type. Also note that the discipline is not symmetric.

With these characteristics the stationary probabilities as given in (2.2) are:

$$P\{(r_{\ell i},s_{\ell i});\ \ell=0,1,\ldots,N;\ i=1,\ldots,x_\ell\}$$

$$= c\prod_{\ell=0}^{N}\left[\frac{1}{x_\ell!}\,3^{[n_{\ell 2}-n_{\ell 1}]}\prod_{i=1}^{x_\ell}\rho_{\ell r_{\ell i}}\right]$$

with c a normalizing constant. Note that in this example is alternatively given priority to jobs with service-numbers 1 resp. 2.

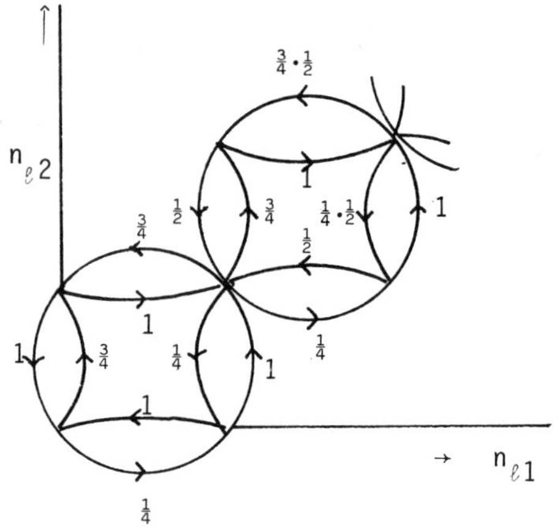

c. If at a node ℓ, jobs with service-numbers $s^1_{\ell 1}$, $s^2_{\ell i_2}$, $s^3_{\ell i_3},\ldots,$ $s^{j-1}_{\ell i_{j-1}}$ are present, then to an incoming job is assigned service-number $s^j_{\ell i_j}$ with probability

$$\delta_\ell((s^1_{\ell 1}),(s^2_{\ell i_2}),\ldots,(s^{j-1}_{\ell i_{j-1}})\,|\,(s^j_{\ell i_j}));\ i_j=1,\ldots,j.$$

Further it is assumed that the sets of numbers $\{s^1_{\ell 1}\}$, $\{s^2_{\ell i_2};\ i_2=1,2\},\ldots$ $\ldots,\{s^j_{\ell i_j};\ i_j=1,\ldots,j\}$ are disjoint. The total capacity is given to the last entered job present, i.e.,

$$\varphi_\ell((r_{\ell 1},s^1_{\ell 1}),\ldots,(r_{\ell j},s^j_{\ell i_j})\,|\,(r_{\ell j},s^j_{\ell i_j}))=1.$$

The capacity $f_\ell((r_{\ell 1},s^1_{\ell 1}),\ldots,(r_{\ell j},s^j_{\ell i_j}))$ may be arbitrary.

Now for each ℓ the product as given in the invariance property (2.1) is only positive for one permutation. For the other permutations at least one denominator as well as numerator are zero. So with the appropriate convention (2.1) is satisfied. The stationary probabilities as given in (2.2) become:

$$P\{(r_{\ell j},s^j_{\ell i_j});\ \ell = 0,1,\ldots,N;\ j = 1,\ldots,x_\ell\}$$

$$= c \prod_{\ell=0}^{N} \prod_{j=1}^{x_\ell} \rho_{r_{\ell j}} \frac{\delta_\ell((s^1_{\ell 1}),\ldots,(s^{j-1}_{\ell i_{j-1}}) \mid (s^j_{\ell i_j}))}{f_\ell((r_{\ell 1},s^1_{\ell 1}),\ldots,(r_{\ell j},s^j_{\ell i_j}))} ,$$

with c a normalizing constant. Note that in this example the capacity function is randomly determined even depending on all the service-numbers present.

Remark. In all examples only the stationary probabilities corresponding to closed models are given. Analogous stationary probabilities for the open models can be derived from (2.3).

Section 4

Conclusions and remarks:

1. The examples show that there exist queueing models in which job-local-balance is satisfied and which consequently posses the insensitivity property, although the discipline is not symmetric. (cf. Kelly [7]; Chandy, Howard and Towsley [3]).
2. From the invariance property we learn that whenever there is assigned a service-number to a certain job, i.e., $d(\cdot \mid \cdot) > 0$, then that job must also receive a positive service-capacity, i.e., $\varphi(\cdot\ \cdot \mid \cdot) > 0$. Consequently, there can be no job-local-balance,

and thereby no insensitivity in models without immediate service.

3. If the invariance-property is satisfied then the stationary distribution factorizes to the different nodes. On the other hand, under some extra conditions this property can also be derived as a necessary condition for job-local-balance. So in many models job-local-balance and thereby insensitivity also requires a factorizing form of the stationary distribution. (see "Networks of queues and insensitivity" by Hordijk and Van Dijk, to appear).

4. In the reference given in 3 it is shown that if the routing is reversible, then the invariance-property (2.1) will be sufficient for job-local-balance also if $d(\cdot\ \cdot \mid \cdot)$ and $\varphi(\cdot\ \cdot \mid \cdot)$ are defective. In such models blocking is allowed. For instance, an open model with one service-facility has a reversible routing (seen as a closed model as described in p. 10) and so jobs may be blocked for entering or leaving the facility. In particular, the blocking may depend on the number of jobs of all types present.

5. Models with a general routing as well as blocking are analysed in "Networks of queues with blocking" (Hordijk and Van Dijk [10]), and the reference given in 3. Conditions for job-local-balance are derived, and models are given in which these conditions are satisfied.

Section 5. References

[1] Barbour, A. D. (1976), "Networks and the method of stages," Adv. in Appl. Probab., 8, 584-591.

[2] Baskett, Chandy, Muntz, Palacois (1975), "Open, Closed and Mixed Networks of Queues with Different Classes of Customers," J. Assoc. Comput. Mach., 22, 248-250.

[3] Chandy, Howard, Towsley (1977), "Product Form and Local Balance in Queueing Networks," J. Assoc. Comput. Mach., 24, 250-263.

[4] Cohen, J. W. (1979), "The Multiple Phase Service Network with Generalized Process Sharing," Acta Inform., 12, 245-284.

[5] Hordijk, A. (1978), "Weak Convergence for queueing processes." Seminar on queueing theory (in Dutch).

[6] Hordijk, A. and Schassberger, R. (1977), "Weak convergence of generalized semi-Markov processes," Report No. 77-8 (Institute of Applied Mathematics and Computer Science, University of Leiden). To appear in Stochastic Process. Appl.

[7] Kelly, F. P. (1978), Reversibility and stochastic networks, Wiley. New York.

[8] Kingman, J. P. C. (1969), "Markov population processes," J. Appl. Probab., 6, 1-18.

[9] Whitt, W. (1980), "Continuity of general semi-Markov process," Math. Oper. Res., 4, 494-501.

[10] Hordijk, A. and Van Dijk, N. "Networks of queues with blocking," Proceedings of Performance '81, 51-65.

Institute of Applied Mathematics and Computer Science, University of Leiden, Wassenaarseweg 80, Leiden, The Netherlands.

A CLASS OF CLOSED MARKOVIAN QUEUEING NETWORKS: INTEGRAL REPRESENTATIONS, ASYMPTOTIC EXPANSIONS, GENERALIZATIONS

J. McKenna†
D. Mitra†
and
K. G. Ramakrishnan††

Abstract

This is an abstract of work reported on in detail in [1]. Additional references may be found in [2].

The theoretical results on the product form of the stationary distributions of large classes of Markovian queueing networks continue to have a profound influence on computer communications, computer systems analysis and traffic theory. The subclass of closed networks of queues, with their inherent feedback, is more difficult to analyze than the open networks. However, the incentive for investigating the closed networks does exist since they have been used to model multiple-resource computer systems, multiprogrammed computer systems, time-sharing, and window flow control in computer communication networks. Networks with external inputs subject to blocking require the analysis of a large number of closed networks. The closed network model that is used in [1] for illustrative purposes arose in the modeling of a central processor in a node of a computer network with many classes of users. [1] considers the problem of efficiently computing the partition function (the normalizing constant), the only element of the product form solution requiring significant computation.

Existing recursive techniques when applied to the large problems which are of particular interest in the Bell System, wherein the constituents of the closed chains are many and the number of chains are many, are observed to have severe shortcomings in the amount of computing time and memory required and the accuracy that can be attained. Even in our specific illustrative problem of rather modest dimensions, the existing recursive techniques are largely ineffectual.

We present a new way to view the problem which surmounts many of the difficulties associated with large networks. The approach is broadly applicable, as Section 10 of [1] on 'Generalizations' will indicate, even though that paper is a detailed account of applications to a specific class of closed networks. The new approach consists, first, of recognizing that the partition function may be written as an integral with a large parameter N present in the integrand to reflect the large size of the network. Next, the classical techniques of asymptotic analysis are applied to derive asymptotic power series, typically in descending powers of N. The integrand will have fundamentally different properties in different ranges of the system parameters and this will require correspondingly different expansions. Thus, in [1] we develop three separate series expansions, in Proposition 3 (Section 4), Proposition 12 (Section 7), Proposition 17 (Section 8), each corresponding to a specific range of values of the usage parameter α. It is worth emphasizing that, commensurate with an objective of providing solutions with any desired accuracy, we give procedures for generating multiple terms in the asymptotic expansions, not just the dominant term. In Section 8, we unify the preceding results by giving a common expansion which holds uniformly in the system parameters. The uniform expansions introduce in a natural way the parabolic cylinder (or Weber) functions, a classical family of special functions with many antecedents and ties with other special functions.

Besides duplicating the specialized expansions derived earlier, the uniform expansion makes available for use the many well-documented and tested expansions that are known in connection with parabolic cylinder functions.

Section 9 of [1] ("Computational Notes") describes a user-oriented, software package that has been written in C-language to implement the approach developed here. We supply results obtained by the package in applications on four test problems which arose in analyzing performance of a Bell System project; also reported are the results of a comparison with a well known, commercially marketed package which obtains solutions recursively. The new package is able to solve the large problems, which are well beyond the range of the other package, and surprisingly, the small problems as well with errors that have small upper bounds.

Section 10 of [1] ("Generalizations") provides the basis for extending the approach developed here to quite general, multi-processor, multi-discipline queueing networks. It is shown that for most networks which have been shown to have the product form in their stationary probability distribution, the partition function has a representation as a multiple integral in which the multiplicity equals the number of non-infinite server nodes in the network. The expansions appropriate for its computation are not considered in [1]. However, the complete theory of the asymptotic expansion of these multiple integrals for the normal usage case is given in [2].

Not surprisingly the new representation of the partition function as an integral, the starting point of our computational procedures, may be exploited anew to derive analytical estimates and bounds of the quantities of interest, such as throughput, mean response time, etc. As is shown in [1], Section 5.3 in particular, these formulas explicitly exhibit the system parameters and as such are rather useful as design and synthesis aids. (The bounds are also useful as checks on the computational procedure.) Purely computational procedures by themselves do not yield this particular form of insight into system behavior.

The asymptotic sequences used typically are power series in N^{-1}, where recall N is the generic large parameter. Thus, the number of terms required to achieve the desired accuracy decreases with increasing N. In contrast, with recursive solutions the computational complexity grows with the network size. Thus, the contrasting techniques are not replacements for each other but complementary: loosely speaking, the recursions are most effective for smaller networks while the asymptotic expansions are most effective for large networks.

1. References

[1] J. McKenna, D. Mitra and K. G. Ramakrishnan, "A Class of Closed Markovian Queueing Networks: Integral Representations, Asymptotic Expansions, and Generalizations," Bell Systems Tech. J., May-June, 1981, pp. 599-641.

[2] J. McKenna, D. Mitra, "Integral Representations and Asymptotic Expansions for Closed Markovian Queueing Networks: Normal Usage", Proceedings of the 8th International Symposium on Computer Performance Modelling, Measurement and Evaluation, November 1981.

†Bell Laboratories, Murray Hill, New Jersey 07974

††Bell Laboratories, Holmdel, New Jersey 07733

QUEUEING MODELS IN PERFORMANCE ANALYSIS, II

Benjamin Melamed, Chairman

L. W. Dowdy & R. M. Budd

T. Rolski & R. Szekli

C. L. Samelson & W. G. Bulgren

FILE PLACEMENT USING PREDICTIVE QUEUING MODELS

Lawrence W. Dowdy*
and
Rosemary M. Budd**

Abstract

Queuing network models have been used extensively in recent years to describe the behavior of computer systems. The validation of a computer system model lies in the predictive ability of the model. The predictive ability is also the model's primary asset.

In this paper, we test the predictive ability of a computer system model when user files are relocated. A queuing network model is constructed. An "optimal" assignment of user files to system I/O devices is found by perturbing this system model. The file assignment is then implemented. Actual and predicted performances are compared.

1. Motivation

Computer system performance prediction is a difficult problem. System tuning, capacity planning, and upgrade analysis all require performance predictions. The natural approach is to abstract the most critical system features into a system model. Several good simulation and analytical models have been suggested [Schw 78] [Buze 78] [Bard 78] [KKT 80] [Bard 80]. Once a model is constructed, it serves as a tool to upgrade or tune the system. For example, the model can be used to compare the performance impact of adding main memory versus the performance impact of adding a new drum. The best cost/benefit upgrade can be implemented.

The primary difficulty with system modeling is that of validation. It is one issue to describe a system by a model, whose parameters are based upon a measurement period, where the model reproduces the performance observed in that measurement period. It is quite another issue to independently alter the model and the actual system and still have consistent performance measures between the model and the system. That is, a model's usefulness, and its validity, is determined by its ability to predict performance under a system change.

In this paper, a simple queuing network model of a Univac 1100/42 system is constructed. Known load balancing techniques are applied to the model to determine the optimal placement of user files across the I/O subsystem. The new file assignment is then implemented. The performance improvement which is predicted by the model closely matches the actual performance measures. This is in spite of the usual problems that are encountered when modeling dual channeled devices.

Section 2 outlines the details of the approach used to construct the model and optimal file assignment. Section 3 gives the methodology for predicting the preformance under alternative file assignments. Section 4 compares actual versus predicted performance. Section 5 summarizes our findings and their implications.

2. Approach

2.1 Model

The particular system we study is a Univac 1100/42 running under the EXEC 8, level 33, operating system. A central server queuing model describing the system is shown in Figure 1.

Standard queuing theoretical assumptions are made [CS 78]. All I/O device queues are serviced first-come-first-served. The CPU queue is processor shared. All queues have exponential holding times. The CPU server, the 8424 disc server, and the 8433 disc server are load dependent servers. The CPU queue is serviced by two processors, while the disc servers are each dual channeled.

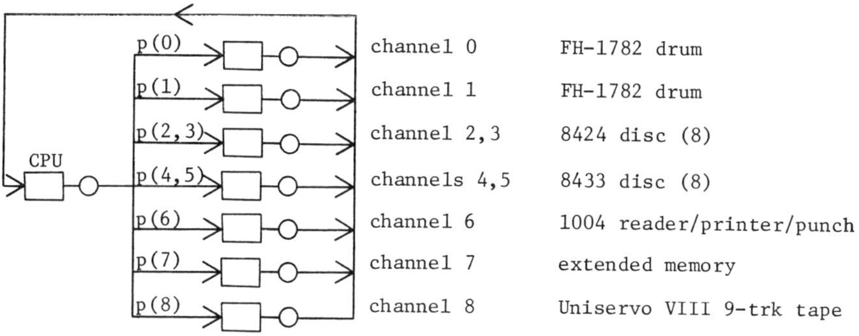

Figure 1. Univac 1100/42 Queuing Model

The model derives its parametric values from the Univac Software Instrumentation Package, SIP, release 5.R1A and Univac's IOTRACE package release 33.R1A:11. SIP and IOTRACE are assumed to be running constantly and are considered to be a fixed part of the operating system. These packages report such measures as the number of transactions and busy time of each channel per file. The ratio of transactions to busy time gives an estimate of the mean service rate at each channel. The load dependency factors for the dual channeled discs are found from the respective service rates of the individual channels. For example, suppose the mean service rates for channels 2 and 3 are X and Y, respectively. Channel 2 is the primary 8424 disc channel (i.e., whenever a transaction arrives in the 8424 queue, it is served by channel 3 only when channel 2 is busy and channel 3 is free). When one customer is in the 8424 queue, the mean service rate is taken to be X. When two or more customers are in the 8424 queue, the queue is serviced with mean rate X + Y. Two obvious inaccuracies exist with the load dependent rates. First, when two customers are in queue, they may require service from the same disc pack. The mean service rate in this case should still be X, not X + Y. Second, when more than two customers are in queue, overlapping seek activity by several requests may speed the mean service rate to more than X + Y. These two inaccuracies have opposing effects upon the mean service rate. The 8433 dual channeled discs, channels 4 and 5, are similarly modeled.

The load dependency factors for the CPU server are found in the same way as the dual channeled devices. However, the inaccuracies described above do not apply to the CPU queue, and the load dependent server appears to be an accurate representation. The branching probability, which reflects the stationary probability of requiring channel service from channel i immediately following CPU activity, is found by taking the ratio of the number of channel i transactions to the number of total I/O transactions. The only remaining parameter of the model is the degree of multiprogramming (DMP). The proper DMP is not reported by SIP and is found by calibrating the DMP to the value where all device utilizations match the SIP reported utilization values. Linear interpolation around integral DMP values is used when matching the utilization values.

2.2 Workload

For the following file assignment, performance prediction, and model validation, the workload of the actual system must be held as constant as possible. To assure this, a benchmark is used.

The benchmark should be an accurate representation of the actual system workload. A typical time period of the Univac 1100/42 operation was selected and SIP and IOTRACE information was collected. A synthetic benchmark was constructed which accurately reproduced the actual observed SIP and IOTRACE statistics (e.g., CPU burst times, device utilizations, swapping activity, interactive and batch usages, memory requirements, etc.). The benchmark driver is a Univac script driven simulator, CS1100 [CS11]. The benchmark construction technique is reported elsewhere [Gord 81]. The relevant point here is that a representative benchmark is used.

2.3 File Assignment

The CS1100 benchmark was run and SIP and IOTRACE statistics were collected. The model described in Section 2.1 was constructed and parameterized as shown in Figure 2. The service rates, branching probabilities, and the degree of multiprogramming are indicated.

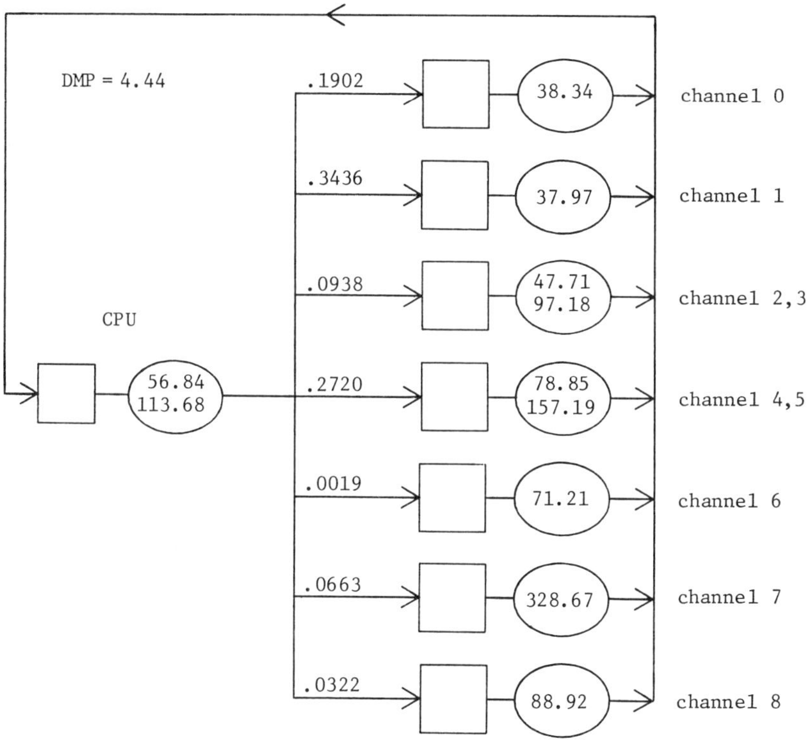

Figure 2. Benchmark Model

We now seek to tune the Univac 1100/42 system. The only variable under our control is the assignment of user files to the I/0 channels. The placement of system files (e.g., directories, swap files, etc.) is not a control variable and is assumed to be fixed. This is not a minor limitation since 70% of all I/0 transactions are to system files. However, given the placement of the system files, the placement of the user files can still significantly influence system performance.

Table 1 gives the breakdown of the system versus user load placed on each channel during the CS1100 benchmark.

Table 1. Channel Load Breakdown

channel	branching probability	system portion	user portion
0	.1902	.1250	.0652
1	.3436	.2878	.0558
2,3	.0938	.0746	.0192
4,5	.2720	.1170	.1550
6	.0019	.0019	.0000
7	.0663	.0663	.0000
8	.0332	.0332	.0000

We note that only system files reside on channels 6, 7, and 8.

The statement of our restricted file assignment problem is: assign the user files to channels 0, 1, 2, 3, 4, and 5 such that throughput (i.e., the number of transactions processed by all devices per unit time) is maximized. The approach [FDA 81] finds the optimal file assignment to the benchmark model (Figure 2). The solution methodology is:

1) Using the given mean service rates, find the optimal branching probabilities. Throughput is viewed as a non-linear function of the branching probabilities. Any of several non-linear programming techniques (e.g., conjugate direction methods, quasi-Newton methods) are applicable [Luen 73]. [Note: The probabilities found by these search methods may only be

locally optimal. Price [Pric 74] shows that for load independent closed central server queuing networks, throughput is unimodal, implying that any local optimum is also the global optimum. This result has not as yet been extended to load dependent networks.]

2) Reassign the user files to realize the branching probabilities found in step 1 as closely as possible. This reassignment must not violate any storage capacity constraint on any device. In general, several file assignments may be feasible, and selecting the best alternative is an unsolved problem. However, in the present experiment, only a single file assignment is feasible which matches the branching probabilities found in step 1.

3) Adjust the mean service rates at the devices to reflect the new file assignment. The technique for this adjustment is given in section 3.

4) Iterate through steps 1, 2, and 3 until convergence.

When this methodology is applied to the benchmark model (Figure 2), under the constraints of the required system loads (Table 1), it is determined that all user files should be moved to the 8433 discs. This configuration is shown in Table 2.

Table 2. Optimal Feasible Loading Probabilities

channel	optimal feasible loading probability
0	.1250
1	.2878
2,3	.0746
4,5	.4122
6	.0019
7	.0633
8	.0332

The technique for predicting the performance of such a file assignment is now detailed.

3. Prediction

Two prediction experiments are conducted for contrast. One experiment is to predict the performance under the best user file assignment as suggested in the previous section -- assign all user files to channels 4 and 5, the 8433 discs. The second experiment is to predict the performance under the worst user file assignment. The motivation for this experiment is not for implementation purposes, but for validating the model at a second point. As described in Section 4, this experiment gives insight for constructing future, even more accurate, models. This file assignement is to assign all user files to the slowest, most utilized channel -- channel 1, the FH-1782 drum.

The techniques for the predictions of both experiments are the same. The approach is to predict the branching probabilities to a channel to be the sum of the relative access frequencies to the files which are resident on each of the channel's devices. The predicted mean service rate of a channel is predicted using, 1) the mean access time of the channel's devices, 2) the mean number of words transferred per transaction on the channel, which depends upon the particular files accessed by the channel, and 3) the transfer speed of the channel's devices. The actual parametric predictions for the two experiments follow.

The parameters for the queuing model (Figure 1) must be predicted. Since only channels 0 - 5 are affected by the user file placement, all parameters shown in Figure 2 except for the branching probabilities and service rates for channels 0 - 5 are predicted to remain the same.

The predicted branching probabilities for both experiments are shown in Table 3.

Table 3. Predicted Branching Probabilities

channel	predicted probabilities, user files on channel 1	predicted probabilities, user files on channels 4 and 5
0	.1957	.1957
1	.5123	.2171
2,3	.0746	.0746
4,5	.1170	.4122
6	.0019	.0019
7	.0663	.0663
8	.0332	.0332

These values are obtained from Table 1, with one exception. In the time period between the original CS1100 benchmark run and the file reassignment experiments, one system file (SYSTEMFILE) was moved from channel 1 to channel 0. This system change does not alter the optimal file assignment found in the previous section, and the performance impact of this system change is included in the predictions. From IOTRACE of the original benchmark, the branching probability to SYSTEMFILE is .0707. The predicted branching probabilities shown in Table 3 are now evident. For example, when all user files are moved to channel 1, the prediction for the branching probability to channel 0 is the sum of the system portion (.1250 from Table 1) and the SYSTEMFILE portion (.0707). For channel 1, we add the system portion, .2878, to the total user activity, .2952, and subtract SYSTEMFILE activity, .0707. For channels 2 - 8, their respective system portions alone are the predicted branching probabilities. The predictions when all user files are moved to channels 4 and 5 are similarly obtained.

The predicted service rates for both experiments are shown in Table 4. A load dependent service rate of 47.7 94.9 indicates that when a single customer is at a server, it is serviced at rate 47.7; when two or more customers are present, the mean service rate is 94.9. We describe how the predicted service rates are obtained when all user files are assigned to channel 1. The service rates for all devices except channels 0 - 5 are predicted to remain unchanged from the values observed in Figure 2.

Table 4. Predicted Mean Service Rates

channel	predicted rate, user files on channel 1	predicted rate, user files on channels 4 and 5
0	39.4	39.4
1	40.6	32.6
2,3	47.7 94.9	47.7 94.9
4,5	86.6 173.4	78.8 157.4
6	71.2	71.2
7	328.7	328.7
8	88.9	88.9
CPU	56.8 113.7	56.8 113.7

For channel 0, the average number of words transferred per transaction in the original benchmark from SIP is 1587. The reported hardware transfer rate for a FH-1782 drum is 240 words per millisecond (ms), implying an average of 6.6 ms is spent in transfer time per transaction. The mean service time per transaction reported by SIP is 26.1 ms. The average access time (i.e., rotational latency) is 19.5 ms -- the difference between mean service time and mean transfer time. When the user files are removed, and SYSTEMFILE is added, the average number of words transferred per transaction is predicted to be 1417. This value is obtained from weighting the number of words in each system file resident on channel 0 (from IOTRACE) by the branching probability to each file (also from IOTRACE). The predicted transfer time is 5.9 ms. Adding in the average access time and inverting, the predicted mean service rate for channel 0 is 39.4 transactions per second. The predicted service rate for channel 1 follows analogously.

For channel 4, the average number of words transferred per transaction in the original benchmark is 430. The reported hardware transfer rate is 179 words per millisecond, implying that an average of

2.4 ms is spent in transfer time per transaction. SIP reports a mean service time of 12.7 ms. Thus, the average access time (e.g., seeking, overlap seeking, rotational delay, dual channel effects, etc.) is 10.3 ms. From IOTRACE, when the user files are removed, leaving only the resident system files, the predicted number of words per transaction is 222. The predicted transfer time is 1.2 ms. Adding in the 10.3 ms average access time and inverting gives a predicted mean service rate of 86.6 transactions per second. Similar calculations for channel 5 yield a mean service rate of 86.8 transactions per second. The load dependent service rate for the 8433 disc server is, therefore, 86.6 and 173.4 for one customer and more than one customer, respectively.

Service rates for the other dual channeled discs and for when all user files are moved to channels 4 and 5 are predicted analogously.

4. Comparison

The queuing network models with the predicted parameters for the file reassignments were solved to obtain predicted performance measures. The files were then reassigned and the CS1100 benchmark was rerun to measure actual performance. The results are given in Table 5. Several observations are made.

1) The model slightly, but consistently, overestimates performance (e.g., TPUT). In the case of the worst file assignment, the model conservatively estimates the degree of system degradation, while the model predicts a greater than actual improvement when user files are optimally placed. (Point 6 provides an explanation for this overestimation.)

Table 6. Predicted versus Actual Comparison

	originally	worst file assignment			optimal file assignment		
		predicted	actual	% error	predicted	actual	% error
channel 0							
words	1587	1417	1589	10.8	1417	961	47.5
rate	38.3	39.4	38.9	1.3	39.4	43.4	9.2
prob	.1902	.1957	.1947	0.5	.1957	.1784	9.7
UTIL	39.4	33.8	29.8	13.4	43.1	34.8	23.9
TPUT	15.1	13.3	11.6	14.7	17.0	15.1	12.3
channel 1							
words	1912	1522	2036	25.2	2894	2413	19.9
rate	38.0	40.6	33.1	22.7	32.6	35.4	7.9
prob	.3436	.5123	.5297	3.3	.2171	.2569	15.5
UTIL	71.8	85.7	95.5	10.3	57.7	61.3	5.9
TPUT	27.3	34.8	31.6	10.1	18.8	21.7	13.2
channel 2,3							
words	312	313	304	3.0	313	321	2.5
rate	47.7,97.2	47.7,94.9	46.6,92.5	2.4	47.7,94.9	45.7,93.4	4.4
prob	.0938	.0746	.0464	60.8	.0746	.0730	2.2
UTIL	14.6	10.1	5.8	74.1	12.8	12.7	0.8
TPUT	7.5	5.1	2.8	83.7	6.7	6.2	4.6
channel 4,5							
words	433	222	217	2.3	430	433	0.7
rate	78.9,157.2	86.6,173.4	89.9,176.4	3.7	78.8,157.4	80.2,160.5	1.7
prob	.2720	.1170	.1253	6.6	.4122	.3944	4.5
UTIL	24.4	8.8	8.0	10.0	37.6	35.0	7.4
TPUT	21.6	8.0	7.5	6.6	35.7	33.3	7.3

(Continued)

Table 6. (Continued)

	originally	worst file assignment			optimal file assignment		
		predicted	actual	% error	predicted	actual	% error
channel 6							
words	1	1	1	0.0	1	1	0.0
rate	71.2	71.2	71.4	0.3	71.2	71.4	0.3
prob	.0019	.0019	.0026	26.9	.0019	.0018	5.6
TPUT	0.2	0.1	0.2	50.0	0.2	0.2	0.0
channel 7							
words	920	920	829	11.0	920	899	2.3
rate	328.7	328.7	358.9	8.4	328.7	322.7	1.9
prob	.0663	.0663	.0692	4.2	.0663	.0634	4.6
TPUT	5.3	4.5	4.1	9.8	5.7	5.4	5.6
channel 8							
words	248	248	248	0.0	248	248	0.0
rate	88.9	88.9	89.0	0.1	88.9	89.0	0.1
prob	.0322	.0322	.0321	0.3	.0322	.0321	0.3
TPUT	2.6	2.2	1.9	10.5	2.8	2.7	3.7
CPU							
rate	56.8,113.7	56.8,113.7	66.4,122.7	14.5	56.8,113.7	56.2,112.4	2.8
UTIL	84.3	76.2	64.0	19.1	88.8	88.1	0.8
TPUT	79.5	68.0	59.6	14.1	86.7	84.5	2.6
DMP	4.44	4.44	4.97	10.6	4.44	4.23	5.1

legend:
- words - average number of words transferred per transaction
- rate - mean service rate (transactions/second)
- prob - branching probability
- UTIL - utilization
- TPUT - throughput (transactions/second)

2) The sensitivity of the system to user file placement is illustrated in Table 6.

Table 6. Performance Change Comparisons

original versus optimal file assignment	
% load moved	-- 14.6%
% CPU TPUT improvement	-- 6.3%
original versus worst file assignment	
% load moved	-- 17.4%
% CPU TPUT degradation	-- 25.0%
worst versus optimal file assignment	
% load moved	-- 29.5%
% CPU TPUT improvement	-- 41.8%

It is interesting to note that moving less than 30% of the I/0 transactions from the drum channel to the disc channel results in an improvement of system throughput of over 40%.

3) The predictions for the dual channel discs, channels 2-5, are fairly accurate.

4) The drum units, channels 0 and 1, have the highest prediction errors. The reason for this is the dynamics of the computer system environment. The time lag between the original CS1100 benchmark and the implementations of the file reassignments was six weeks. During this time the file space changes and system file residency changes between the drums frequently occured, SYSTEMFILE being the major contributor. To be more realistic, the drum channels should be viewed together. This is done in Table 7 which indicates more accurate predictions.

Table 7. Comparisons with Channels 0 and 1 Combined

	worst file assignment			optimal file assignment		
	predicted	actual	% error	predicted	actual	% error
prob	.7080	.7244	2.3	.4128	.4353	5.2
TPUT	48.1	43.2	11.3	35.8	36.1	0.8

5) Consistent with the assumptions made, parameters on the unaffected channels, channels 6 - 8, remained fixed. This assumption has been noticed to be invalid in previous prediction studies when an extra CPU was configured [DAGT 79]. However, under file assignment changes, no such problem has been observed.

6) Owing to the fact that it has the higher percentage prediction errors, the worst file assignment experiment illustrates a failing in the basic model. The relevant facts are (compared to the prediction): a) The CPU became faster, b) the degree of multiprogramming increased, c) the channel 1 drum slowed and saturated, and d) the word size per transfer increased on channel 1. The scenario behind these facts is simple but not modeled. Channel 1 is the primary swapping channel. The average word size per swap transaction is larger than an average transaction, around 4000 words. As user files are moved to channel 1, which is a slower channel than the channels where the user files originally resided, the CPU becomes more idle. The operating system loads more jobs to "take advantage" of the CPU's idleness. More swapping activity results. The average word size per transfer to channel 1 increases, reducing its mean service rate, and saturation occurs. The CPU looks faster because of an increase in the number of short overhead requests -- a short CPU burst required to initiate a swap. The cycle repeats. The system thrashes and performance suffers more than expected.

This scenario is substantiated by SIP (e.g., more swapping occurs, more jobs are resident in core). The scenario, in reverse fashion, also explains the small inaccuracies in the prediction of the optimal file assignment. That is, user I/O wait time decreases; the CPU is busier; fewer jobs are allowed in core; and swapping decreases.

5. Summary

In this paper, a simple queuing model of an actual system was built. The model was then theoretically optimized with respect to the user file placement. The file placement was implemented and comparisons between actual and predicted performance were made.

The degree to which actual and predicted performance match is the validation of the model. In addition to being the model's validation, performance prediction is the single most important application of the model. The model constructed in section 2 has good predictive capability for file assignment problems. The behavior of the dual channeled devices, in particular, is accurately modeled.

A primary benefit of conducting the described experiments is model improvement. The experiments expose those features in the actual system that influence performance and are not modeled. For example, the interaction between the swapping rate, the drum service rate, the CPU service rate, and the degree of multiprogramming are all interrelated and cannot be assumed to be independent of each other (i.e., observation 6 in the previous section). These interdependencies suggest the need for: 1) a better understanding of parametric relationships, and 2) queuing network models which model state dependent service rates, branching probabilities, and multiprogramming levels.

6. Acknowledgements

We are grateful for the suggestions and contributions made by the Systems Design and Analysis Group and the Systems Staff of the Computer

Science Center at the University of Maryland.

7. References

[Bard 78] Y. Bard, "The VM/370 Performance Predictor," Comput. Surveys, 10, (September 1978), 333-342.

[Bard 80] Y. Bard, "A Model of Shared DASD and Multipathing," Comm ACM, 23, (October 1980), 564-583.

[Buze 78] J. P. Buzen, "A Queueing Network Model of VMS," Comput. Surveys, 10, (September 1978), 319-331.

[CS 78] Comput. Surveys, Special Issue: Queueing Network Models of Computer System Performance, Vol. 10, (September 1978).

[CS11] CS1100 Communications Simulator Reference Manual, Sperry Rand Corporation.

[DAGT 79] L. W. Dowdy, A. K. Agrawala, K. D. Gordon, and S. K. Tripathi, "Computer Performance Prediction via Analytical Modeling -- An Experiment," Proc. of the Conf. on Simu., Meas., and Modeling of Comp. Sys., (August 1979), 13-18.

[FDA 81] D. V. Foster, L. W. Dowdy, and J. E. Ames, "File Assignment in a Computer Network," Comput. Networks, 1981, 5, (September 1981), 341-349.

[Gord 81] K. D. Gordon, "On the Construction of Representative Test Workloads," Ph. D. Dissertation, Dept. of Comp. Sci., Univ. of Maryland, College Park, Maryland, 1981.

[KKT 80] P. S. Kritzinger, A. E. Krzesinski, and P. Teunissen, "Incorporating System Overhead in Queueing Network Models," IEEE Trans. Software Engrg., Vol. SE-6, (July 1980), 381-390.

[Luen 73] D. G. Luenberger, Introduction to Linear and Nonlinear Programming, Addison-Wesley, Reading, Massachusetts, 1973.

[Pric 74] T. G. Price, "Probability Models of Multiprogrammed Computer Systems," Ph. D. Dissertation, Dept. of Elec. Eng., Stanford Univ., Palo Alto, California, (December 1974).

[Schw 78] H. D. Schwetmann, "Hybrid Simulation Models of Computer Systems," Comm. ACM, 21, (September 1978), 718-723.

This research was supported in part by the Computer Science Center of the University of Maryland.

*Department of Computer Science, Vanderbilt University, Nashville, Tennessee 37235.

**GTE Subscriber Network Products, McLean, Virginia 22102.

NETWORKS OF WORK-CONSERVING NORMAL QUEUES

Tomasz Rolski
and
Ryszard Szekli

1. Introduction

In this paper we deal with a network of parallel queues. It consists of a switch and s server queues. Such a network we call a module; see Figure 1. Each queue of the module has unlimited waiting room. Queueing disciplines at queues of the network need not be the same and they belong to a broad class of so called work conserving normal (WCN) disciplines. The class of WCN disciplines was introduced by Rolski (1981a). In this paper we follow a definition given by Szekli (1981) which is recalled in Section 2. The sample history of a module is completely determined by a so-called generic sequence $\{T,S,X\} = \{(T_i,S_i,X_i),\ i = 0,1,\ldots\}$ where T_i denotes the inter-arrival time between the i-th and (i+1)-st customer in the module, S_i denotes the service time of the i-th customer, and X_i denotes the queue to which the i-th customer is routed. We assume that the module is initially empty. The triple (T_i,S_i,X_i) is called the basic datum associated to the i-th customer to arrive at the module.

The notion of an ergodically stable sequence of random elements was introduced by Rolski (1981b). We recall the definition and some basic properties of this notion in Section 3. For now, we merely remark that $Z = \{Z_i,\ i = 0,1,\ldots\}$ is ergodically stable if and only if there exists a stationary ergodic sequence $\hat{Z} = \{\hat{Z}_i,\ i = 0,1,\ldots\}$ such that for any

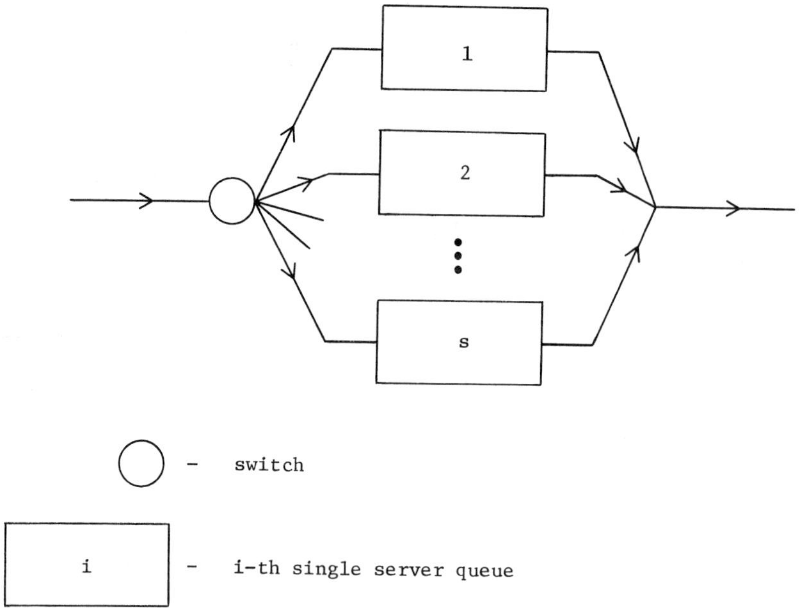

Figure 1

measurable non-negative function g with probability 1

$$\lim_{j\to\infty} \frac{1}{j} \sum_{i=0}^{j-1} g(Z_i, Z_{i+1}, \ldots) = Eg(\hat{Z}_0, \hat{Z}_1, \ldots).$$

In this paper we shall consistently append a ^ to any stationary sequence corresponding to an ergodically stable sequence. It was shown by Rolski (1981b) that if the generic sequence of a single server FIFO queue is ergodically stable and $\rho = E\hat{S}_0/E\hat{T}_0 < 1$, then the sequence of waiting times $W = \{W_i,\ i = 0,1,\ldots\}$ or the sequence $\{(W_i, T_i, S_i),\ i = 0,1,\ldots\}$ is ergodically stable. Waiting times will refer here to the time spent in the waiting line only. Thus with probability 1

$$\lim_{j\to\infty} \frac{1}{j} \sum_{i=0}^{j-1} T_i = \lambda^{-1}, \tag{1.1}$$

$$\lim_{j\to\infty} \frac{1}{j} \sum_{i=0}^{j-1} (W_i + S_i) = E(\hat{W}_0 + \hat{S}_0), \tag{1.2}$$

$$\lim_{j\to\infty} \frac{W_j}{j} = 0. \tag{1.3}$$

Let $L = \{L(t),\ t \geq 0\}$ be the queue size process in the queue. Stidham (1974) proved that in any queueing system for which (1.1)-(1.3) hold we have with probability 1 that

$$\lim_{T\to\infty} \frac{1}{T} \int_0^T L(s)ds = E\hat{L}(0)$$

and

$$E\hat{L}(0) = \lambda E(\hat{W}_0 + \hat{S}_0). \tag{1.4}$$

Relation (1.4) is called Little's formula.

In Section 2 we state a theorem about the ergodic stability of the waiting time sequence in a module and at each queue of the module. As a corollary we show that Little's formula holds at each queue of the module and in the module as a whole. Moreover, we also give sufficient conditions for the sequence of inter-departure times to be ergodically stable, and for the output rate from a module to equal the input rate. Having established the above results we may analyse queueing networks consisting of independent modules which form a tree.

All proofs are given in outline form in Appendix A.

2. WCN modules

Consider a single server queue with unlimited waiting room with a generic sequence $\{T,S\}$. We assume that the queue is initially empty.

Definition 1. The service discipline is work-conserving (WC) if the work-load process decreases with velocity 1 (provided the work-load is positive) at any instant not being an arrival one; at any arrival instant the work-load jumps up and the size of jump equals the service time of the arrived customer.

Note that in single server WC queues busy cycles, busy periods, and the number of customers served in a busy cycle are independent of WC disciplines and depend only on the generic sequence $\{T,S\}$.

Definition 2. A WC discipline is normal (WCN) if the waiting times of all customers in the same busy cycle are functions of all basic data associated to these customers. These functions are measurable and are identical on cycles having the same length.

Specifically let $W^* = \{W_i^*, i = 0,1,\dots\}$ be the sequence of waiting times in a FIFO queue with the same generic sequence $\{T,S\}$. We must assume that with probability 1, $W_i^* = 0$ infinitely often. Let

$$\zeta_1 = 0,$$

$$\zeta_i = \min\{j > \zeta_{i-1}, W_j^* = 0\} \quad (j = 1,2,\dots)$$

be the index sequence of customers which inaugurate a busy cycle and let $\varphi = \{\varphi_k, k = 1,2,\dots\}$ be the sequence of measurable functions from Definition 2 where $\varphi_k : \mathbb{R}_+^k \to \mathbb{R}_+^k$. Then a WC discipline is normal if the waiting time process $W = \{W_i, i = 0,1,\dots\}$ in the queue is given, with probability 1, by

$$(W_{\zeta_i},\dots,W_{\zeta_{i+1}-1}) = \varphi_{\zeta_{i+1}-\zeta_i}((T_{\zeta_i},S_{\zeta_i}),\dots,(T_{\zeta_{i+1}-1},S_{\zeta_{i+1}-1})).$$

Any WCN discipline is characterized by a sequence of functions φ. The class of WCN disciplines includes: First In First Out (FIFO), Last In First Out (LIFO), Shortest Residual Processing Time, Round Robin, Instantaneous Feedback.

Definition 3. A WCN module is a network consisting of a finite number of single server WCN queues connected in parallel; the queueing discipline need not be the same. If at each queue of the module the queueing discipline is FIFO, then the module is called a FIFO module.

Consider a WCN module with a generic sequence $\{T,S,X\}$. Recall that the module is initially empty. Let $T^{(j)} = \{T_i^{(j)}, i = 0,1,\dots\}$ be the sequence of inter-arrival times at the j-th queue ($j = 1,\dots,s$), and let $S^{(j)} = \{S_i^{(j)}, i = 0,1,\dots\}$ be the sequence of service times of customers routed to the j-th queue ($j = 1,\dots,s$). Let

$W^{(j)} = \{W_i^{(j)}, i = 0,1,\ldots\}$ be the sequence of waiting times of customers routed to the j-th queue ($j = 1,\ldots,s$), and let $W = \{W_i, i = 0,1,\ldots\}$ be the sequence of waiting times in the module.

Theorem 1. Let $\{T,S,X\}$ be ergodically stable and

$$\Pr(\hat{X}_0 = i) > 0, \quad (i = 1,\ldots,s), \tag{2.1}$$

$$\rho_i = \frac{E(\hat{S}_0|\hat{X}_0 = i)\Pr(\hat{X}_0 = i)}{E\hat{T}_0} < 1. \tag{2.2}$$

Then

(i) $\{W^{(j)}, T^{(j)}, S^{(j)}\}$ is ergodically stable ($j = 1,\ldots,s$).

(ii) $\{W,T,S,X\}$ is ergodically stable.

Let $\{L^{(j)}(t), t \geq 0\}$ be the queue size process at the j-th queue ($j = 1,\ldots,s$) and $\{L(t), t \geq 0\}$ be the process of number of customers in the module at time t.

Corollary 1. The following limits exists and are constant with probability 1:

$$\lim_{n\to\infty} \frac{1}{n} \sum_{i=0}^{n-1} T_i^{(j)} = (\lambda^{(j)})^{-1} \quad (\lambda = E\hat{T}_0^{(j)}) \ (j = 1,\ldots,s), \tag{2.3}$$

$$\lim_{n\to\infty} \frac{1}{n} \sum_{i=0}^{n-1} T_i = \lambda^{-1}, \tag{2.4}$$

$$\lim_{n\to\infty} \frac{1}{n} \sum_{i=0}^{n-1} W_i^{(j)} = w^{(j)}, \quad (j = 1,\ldots,s), \tag{2.5}$$

$$\lim_{n\to\infty} \frac{1}{n} \sum_{i=0}^{n-1} W_i = w, \tag{2.6}$$

$$\lim_{T\to\infty} \frac{1}{T} \int_0^T L^{(j)}(t)dt = \ell^{(j)}, \quad (j = 1,\ldots,s), \tag{2.7}$$

$$\lim_{T\to\infty} \frac{1}{T} \int_0^T L(t)\,dt = \ell \tag{2.8}$$

and

$$\ell^{(j)} = \lambda^{(j)} w^{(j)}, \quad (j = 1,\ldots,s), \tag{2.9}$$

$$\ell = \lambda w. \tag{2.10}$$

Let $U^{(j)} = \{U_i^{(j)}, i = 0,1,\ldots\}$ be the sequence of inter-departure times from the j-th queue in the module ($j = 1,\dots,s$) and $U = \{U_i,\ i = 0,1,\dots\}$ be the sequence of inter-departure times from the module.

Corollary 2.

(i) The sequence $\{W^{(j)}, T^{(j)}, S^{(j)}, U^{(j)}\}$ is ergodically stable and

$$E\hat{U}_0^{(j)} = E\hat{T}_0^{(j)} \qquad (j = 1,\dots,s).$$

(ii) The sequence $\{W,T,S,U\}$ is ergodically stable and

$$E\hat{U}_0 = E\hat{T}_0.$$

Example 1. (s-server queue with the cyclic discipline). A FIFO module with a generic sequence $\{T,S,X\}$ is an s-server queue with the cyclic discipline if we put $X_i = i$ modulo$(s) + 1$. In such a case $\{T,S,X\}$ is ergodically stable if for example $\{(T_i,S_i,\ i = 0,1,\dots\}$ is a sequence of independent identically distributed (i.i.d.) random vectors, independent of X (it is not sufficient to assume that $\{T,S\}$ is stationary ergodic; see Brown (1976) Proposition 1.6). Clearly $\Pr(\hat{X}_0 = i) = \frac{1}{s}$ ($i = 1,\dots,s$). Hence the stability condition (2.2) reduces to

$$\frac{ES_0}{s\ ET_0} < 1.$$

The cyclic s-server queue was considered by Franken, et al. (1981) under some assumption on $\{T,S,X\}$.

Example 2. Define a generic sequence of a WCN module as follows. The sequence of inter-arrival times T is ergodically stable while $\{S,X\} = \{(S_i,X_i),\ i = 0,1,\dots\}$ is the sequence of i.i.d. random vectors, also independent of T. Then $\{T,S,X\}$ is ergodically stable; see Brown (1976), Proposition 1.6. The condition (2.2) is equivalent to

$$\rho_i = \frac{E(S_0|X_0 = i)\Pr(X_0 = i)}{E\hat{T}_0} < 1 \qquad (i = 1,\dots,s).$$

However in queueing theory it is more natural to specify the generic sequence of each queue in the module rather than given the generic sequence of the module as a whole. For example, we may assume that the sequence of service times in the j-th queue $S^{(j)} = \{S_i^{(j)}, i = 0,1,\ldots\}$ consists of i.i.d. random variables $(j = 1,\ldots,s)$ and that the sequences $\{T,X,S^{(j)}\}$ $(j = 1,\ldots,s)$ are independent. Moreover the sequence of inter-arrival times T at the module and the switching sequence X each consist of i.i.d. random variables. Then

$$S_j = S_{\nu_k(j)} \text{ if } X_j = k,$$

where

$$\nu_k(j) = \sum_{i=0}^{j} 1_{\{k\}}(X_i) \quad (j = 0,1,\ldots,k = 1,\ldots,s). \tag{2.11}$$

In this case the generic sequence $\{T,S,X\}$ is ergodically stable and the condition (2.2) becomes

$$\rho_j = \frac{ES_0^{(j)}\Pr(X_0 = i)}{ET_0} < 1 \quad (j = 1,\ldots,s).$$

Theorem 1 and the ensuing corollaries are useful in the analysis of more complicated networks of queues, such as, networks of branching WCN modules (see Figure 2). We may assume that all service times are independent and that at each module they are identically distributed. Consider two kinds of switches. The first one switches customers to queues within a module (as described earlier) and the second one among modules. We assume that all switching decisions are i.i.d., independent of the input and service processes. If the input rate at the switch $X = \{X_i, i = 0,1,\ldots\}$ is λ, then the output rate from the switch along the j-th stream is $\lambda\Pr(\hat{X}_0 = j)$. This fact will follow from the next proposition. In this proposition we deal with a stream of customers which is split into s streams. The input data are: the inter-customer interval between the i-th and (i+1)-st customer, and the index X_i of the stream the i-th customer is routed to. In the k-th stream the inter-customer interval sequence is denoted by $T^{(k)}$ $(k = 1,\dots,s)$.

Proposition 1. If $\{T,X\}$ is ergodically stable and $\Pr(\hat{X}_0 = j) > 0$ then $T^{(j)}$ is ergodically stable and

$$E\hat{T}_0^{(j)} = \frac{E\hat{T}_0}{\Pr(\hat{X}_0 = j)} .$$

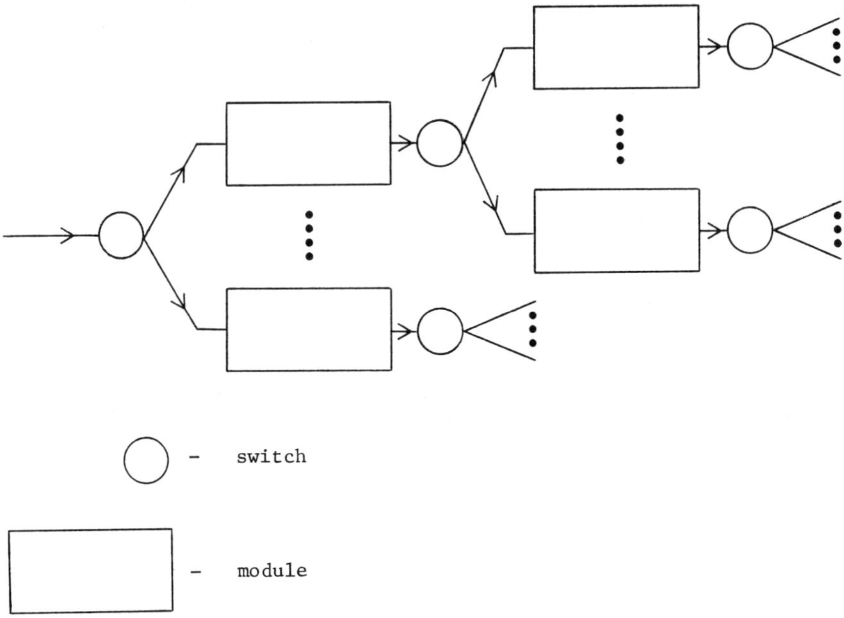

Figure 2

3. Stable Sequences of Random Elements

In this section we recall some facts concerning ergodically stable sequences of random elements (SRE). Further information may be found in Rolski (1981a), (1981b). A similar notion was independently introduced in information theory by Gray and Kiefer (1980).

We start with some notation. Let $\mathbb{R}_+$ be the non-negative reals, $\mathbb{Z}$ the integers, $\mathbb{Z}_+$ the non-negative integers, $\mathbb{N}$ the positive integers and $\mathscr{E}$ a metric space. The Borel σ-field of subsets of a space $\mathscr{E}$ is denoted $\mathscr{B}(\mathscr{E})$. The set of all possible mappings of A into B is denoted by B^A. A sequence Z indexed by $\mathscr{J}$ is denoted by $z = \{z_i, i \in \mathscr{J}\}$. In this paper $\mathscr{J} = \mathbb{Z}, \mathbb{Z}_+$ or $\mathbb{N}$. If $x^{(1)} \in \mathscr{E}_1^{\mathscr{J}}$ and $x^{(2)} \in \mathscr{E}_2^{\mathscr{J}}$ then

$$\{x^{(1)},x^{(2)}\} = \{(x_i^{(1)},x_i^{(2)}), i \in \mathscr{J}\}.$$

The shift transformation

$$\alpha: \mathscr{E}^{\mathscr{J}} \to \mathscr{E}^{\mathscr{J}}$$

is defined by

$$\alpha\{z_i, i \in \mathscr{J}\} = \{z_{i+1}, i \in \mathscr{J}\}.$$

To introduce the second transformation β_K set $\mathscr{J} = \mathbb{Z}$ or $\mathbb{Z}_+$,

$$\mathscr{F} = (\mathscr{E} \times \{1,\dots,s\})^{\mathscr{J}}.$$

and

$$\mathscr{F}_K = \{\{z,x\} \in \mathscr{F}, x_0 \in K\} \subset \mathscr{F}$$

where $K \subset \{1,\dots,s\}$. The transformation β_K is defined by

$$\beta_K(\{z,x\}) = \alpha^{v_K(\{z,x\})}(\{z,x\}),$$

where $\alpha^n = \alpha \circ \alpha^{n-1}$ is the n-fold shift and

$$v_K(\{z,x\}) = \begin{cases} \min\{i > 0,\ x_i \in K\} & \text{if } x_i \in K \text{ for some } x_i \\ \infty & \text{if } x_i \notin K \text{ for all } x_i \end{cases}$$

Singleton sets $\{i\}$ will be denoted by i e.g., $\mathcal{F}_{\{2\}} = \mathcal{F}_2$, $\beta_{\{1\}} = \beta_1$.

All random elements are defined on a common probability space $(\Omega,\mathcal{A},\mathrm{Pr})$. The distribution of a random element Z is denoted by P_Z. We refer to Breiman (1968) or Brown (1976) for definitions and theorems from ergodic theory. Recall that $(\Omega,\mathcal{A},\mathrm{Pr},\gamma)$ is an ergodic dynamical system if $\gamma\colon \Omega \to \Omega$ is an ergodic measure-preserving transformation on $(\Omega,\mathcal{A},\mathrm{Pr})$. A subset $B \in \mathcal{A}$ is said to be γ-invariant, where $\gamma\colon \Omega \to \Omega$ is a measurable mapping, if $\gamma^{-1}(B) = B$.

Definition 4.

(i) An SRE $Z = \{Z_i,\ i \in \mathcal{J}\}$ is stationary if $\alpha(Z)$ and Z are identically distributed.

(ii) Stationary SRE Z is ergodic if for any α-invariant set $B \in \mathcal{B}(\mathcal{E}^{\mathcal{J}})$

$$\mathrm{Pr}(Z \in B) = 0 \text{ or } 1.$$

(iii) An SRE Z is ergodically stable if

$$\lim_{j\to\infty} \frac{1}{j} \sum_{i=0}^{j-1} \mathrm{Pr}(\alpha^i Z \in B) = P_{\hat{Z}}(B) \qquad (B \in \mathcal{B}(\mathcal{E}^{\mathcal{J}})),$$

$P_{\hat{Z}}$ is a probability measure on $\mathcal{E}^{\mathcal{J}}$ and $(\mathcal{E}^{\mathcal{J}},\mathcal{B}(\mathcal{E}^{\mathcal{J}}),P_{\hat{Z}},\alpha)$ is an ergodic dynamical system. We call $P_{\hat{Z}}$ the stationary distribution of Z; any SRE $\hat{Z}$ with distribution $P_{\hat{Z}}$ is called a stationary representation of Z.

Definition 5. Two SRE $Z^{(1)}$ and $Z^{(2)}$ are stably equivalent (denoted $Z^{(1)} \overset{s}{\equiv} Z^{(2)}$) if they have a common stationary distribution.

For example if $Z_i^{(1)} = i$ modulo$(s) + 1$ $(i = 0,1,\ldots)$ and $Z^{(2)}$ is such that

$$\mathrm{Pr}(Z^{(2)} = z_j) = \frac{1}{s} \qquad (j = 1,\ldots,s),$$

where

$$z_i = \{(i+j) \text{ modulo}(s) + 1,\ j = 0,1,\ldots\} \quad (i = 1,\ldots,s)$$

then $Z^{(1)} \overset{s}{\equiv} Z^{(2)}$. Another example of stably equivalent SRE's are a stationary SRE $Z^{(1)}$ and an SRE $Z^{(2)}$ such that for a random variable $N \in \mathbb{Z}_+$

$$Z^{(1)}_{N+i} = Z^{(2)}_{N+i} \quad (i = 0,1,\ldots);$$

see Lemma 1.2 of Rolski (1981a). The SRE's $Z^{(1)}$ and $Z^{(2)}$ are single-ended although in Rolski (1981a) the Lemma was also stated for double-ended SRE's. However, the latter case is false.

In the next definition $Z_i \in \mathscr{E}$, $X_i \in \{1,\ldots s\}$ and $K \subset \{1,\ldots,s\}$.

Definition 6.

(i) An SRE $\{Z,X\}$ is β_K-stationary if $\beta_K(\ Z,X\ )$ and $\{Z,X\}$ are identically distributed.

(ii) A β_K-stationary SRE $\{Z,X\}$ is β_K-ergodic if the dynamical system $(\mathscr{F}_K, \mathscr{B}(\mathscr{F}_K), P_{\{Z,X\}}, \beta_K)$ is ergodic.

(iii) An SRE $\{Z,X\}$ is β_K-ergodically stable if

$$\lim_{j\to\infty} \frac{1}{j} \sum_{i=0}^{j-1} \Pr(\beta_K^i\{Z,X\} \in B) = P_{\{Z^o,X^o\}}(B) \quad (B \in \mathscr{B}(\mathscr{F}))$$

exists, $P_{\{Z^o,X^o\}}$ is a probability measure on $\mathscr{F}_K$ and the dynamical system $(\mathscr{F}_K, \mathscr{B}(\mathscr{F}_K), P_{\{Z^o,X^o\}}, \beta_K)$ is ergodic. We call $P_{\{Z^o,X^o\}}$ a β_K-stationary distribution of $\{Z,X\}$ and $\{Z^o,X^o\}$ a β_K-stationary representation of $\{Z,X\}$.

We now state three propositions which explore the relationship between ergodic and β_K-ergodic stability. Part "if" of the next proposition is proved in Rolski (1981a). Results related to both parts "if" and "only if" can be found in Zähle (1980).

<u>Proposition 2</u>. An SRE $\{Z,X\}$ is ergodically stable and $\Pr(\hat{X}_0 \in K) > 0$ if and only if $\{Z,X\}$ is β_K-ergodically stable and $Ev_K(\{Z^o,X^o\}) < \infty$.

Moreover

$$P_{\{Z^o,X^o\}}(\cdot) = \Pr(\{\hat{Z},\hat{X}\} \in \cdot \ |\hat{X}_0 \in K).$$

Thus for two SRE's $\{Z^{(1)},X^{(1)}\} \overset{s}{\equiv} \{Z^{(2)},X^{(2)}\}$ we have

$P_{\{Z^{(1)},X^{(1)}\}}(B) = P_{\{Z^{(2)},X^{(2)}\}}(B)$ for any β_K-invariant set $B \in \mathscr{B}(\mathscr{F})$

provided $\Pr(\hat{X}_0^{(1)} \in K) > 0$.

Proposition 3. (Rolski (1981b)): An SRE $Z^{(1)}$ is ergodically stable and $Z^{(1)} \overset{s}{\equiv} Z^{(2)}$ if and only if there exists a stationary sequence $\hat{Z}$ such that for any measurable function g: $\mathscr{E}^{\mathscr{J}} \to \mathbb{R}_+$

$$\lim_{j\to\infty} \frac{1}{j} \sum_{i=0}^{j-1} g(\alpha^i Z^{(1)}) = \lim_{j\to\infty} \frac{1}{j} \sum_{i=0}^{j-1} g(\alpha^i Z^{(2)}) = Eg(\hat{Z}).$$

with probability 1.

The next proposition is a special case of a general one from Rolski (1981a).

Proposition 4.

(i) Let μ: $\mathscr{E}_1^{\mathscr{J}} \to \mathscr{E}_2^{\mathscr{J}}$ be measurable such that

$\mu\alpha = \alpha\mu$.

If $Z^{(1)}$ is stationary ergodic (ergodically stable) then $Z^{(2)} = \mu(Z^{(1)})$ is stationary ergodic (ergodically stable).

(ii) Let μ: $(\mathscr{E}_1 \times \{1,\ldots,s\})^{\mathscr{J}} \to \mathscr{E}_2^{\mathscr{J}}$ be measurable such that

$\mu\beta_K = \alpha\mu$.

If $\{Z^{(1)},X^{(1)}\}$ is β_K-stationary ergodic (β_K-ergodically stable) then $Z^{(2)} = (\{Z^{(1)},X^{(1)}\})$ is stationary ergodic (ergodically stable).

4. Appendix A

Proof of Theorem 1. The proof consists in two parts. In part A we prove the theorem for FIFO modules. In part B we extend the proof of general WCN modules. Without loss of generality we may assume $s = 2$.

Part A: Consider a FIFO module with a generic sequence $\{T,S,X\}$ such that conditions (2.1) and (2.2) are fulfilled. The module is initially empty.

1°. Customer switching to queues. Divide all customers into groups, the i-th group consisting of the i-th arrival in the 1-st queue and all subsequent arrivals in the 2-nd queue up to but not including the (i+1)-st arrival in the 1-st queue. The collection of basic data of all customers in the i-th group is denoted by Φ_i $(i = 0,1,\ldots)$. These data may be arranged in a sequence

$$\Phi_i = \{(T_i^{(1)},S_i^{(1)}),\ N_i,\ (T_{i1}^{(2)},S_{i1}^{(2)}),\ldots,(T_{i,N_i}^{(2)},S_{i,N_i}^{(2)}),0,0,\ldots\}$$

where N_i denotes the number of customers from the 2-nd queue in the i-th group, $T_{ij}^{(2)}$ denotes the inter-arrival time between the j-th and (j+1)-st customer in the i-th group $(j = 1,2,\ldots)$ arriving at the 2-nd queue, $S_{ij}^{(2)}$ denotes the service times of the j-th customer in the i-th group $(j = 1,2,\ldots)$ at the 2-nd queue. Denote $\Phi = \{\Phi_i,\ i = 0,1,\ldots\}$. There exists a measurable transformation ψ such that

$$\Phi = \psi(\{T,S,X\}).$$

Notice that the SRE $\{T,S,X\}$ may be considered as a random element assuming values in $\mathscr{F} = (\mathbb{R}_+ \times \mathbb{R}_+ \times \{1,2\})^{\mathbb{Z}_+}$. Recalling the definition of β_1 and α given in Section 3 we can demonstrate that

$$\alpha\psi = \psi\beta_1.$$

If $\{T^o,S^o,X^o\}$ is a β_1-stationary representation of $\{T,S,X\}$ (namely the distribution of $\{T^o,S^o,X^o\}$ is $\Pr(\{\hat{T},\hat{S},\hat{X}\} \in \cdot \ |\hat{X}_0 = 1))$ then by Proposition 4 the SRE $\Phi^o = \psi(\{T^o,S^o,X^o\})$ is stationary ergodic. We shall consistently append a circle to all elements of Φ^o, for example $T^{o(2)}$. Any single-ended stationary SRE may be extended to a double-ended stationary one; see e.g., Breiman (1968), Proposition 6.5. We may therefore assume that Φ^o is a stationary double-ended SRE, whenever necessary.

2°. Palm representation of a module. The sequence of waiting times of customers in the i-th group is denoted by

$$\mathcal{W}_i = (W_i^{(1)}, W_{i1}^{(2)}, \ldots, W_{iN_i}^{(2)}, 0, 0, \ldots), \tag{4.1}$$

where $W_{ij}^{(2)}$ denotes the waiting time of the j-th customer (j = 1,2,...) in the i-th group at the 2-nd queue. We have

$$W_j^{(2)} = \begin{cases} W_{0,j+1}^{(2)} & \text{if } 0 \le j < N_0 \\ W_{i+1,j-\sum_{k=0}^{i} N_k+1}^{(2)} & \text{if } \sum_{k=0}^{i} N_i \le j < \sum_{k=0}^{i+1} N_k, \ i = 0,1,\ldots \end{cases}$$

Notice that the waiting times of customers from the (i+1)-st group depend on Φ_{i+1}, the waiting times of the i-th customer from the 1-st queue and the last customer served at the 2-nd queue before the (i+1)-st group. Thus there exists a function f such that $\mathcal{W}_{i+1} = f(\mathcal{W}_i, \Phi_{i+1})$ (i = 0,1,...). This function fulfills the conditions of Loynes' lemma (see Loynes (1962)), as the queueing discipline in both queues is FIFO. The original lemma was proved in a less general setting; however, in his paper Loynes points out the possibility of generalizations. Thus by Loynes' lemma there exists an SRE $\mathcal{W}^o = \{\mathcal{W}_i^o, \ i = \ldots,-1,0,\ldots\}$ satisfying

$$\mathcal{W}_{i+1}^o = f(\mathcal{W}_i^o, \Phi_{i+1}^o) \quad (i = \ldots,-1,0,1,\ldots) \tag{4.2}$$

$$\{\mathcal{W}^o, \Phi^o\} \text{ is stationary ergodic.} \tag{4.3}$$

The assumption (2.2) of Theorem 1 ensures that elements of $\mathcal{W}_i^o = (W_i^{o(1)}, W_{i1}^{o(2)}, W_{i2}^{o(2)}, \ldots, W_{iN_i^o}^{o(2)}, 0, 0, \ldots)$ are finite with probability 1. The random element $\{\mathcal{W}^o, \Phi^o\}$ is called a Palm representation of the module. The first element $W_i^{o(1)}$ of $\mathcal{W}_i^o$ is the waiting time of the i-th customer at the 1-st queue. The sequence $W_i^{o(1)} = \{W_i^{o(1)}, \ i = 0,1,\ldots\}$ is stationary ergodic. The sequence of waiting times of customers at the second queue is given by

$$W_j^{o(2)} = \begin{cases} W_{0,j}^{o(2)} & \text{if } 0 \le j < N_0^o \\ W_{i+1,\,j-\sum_{k=0}^{i} N_k^o+1}^{o(2)} & \text{if } \sum_{k=0}^{i} N_k^o \le j < \sum_{k=0}^{i+1} N_k^o, \quad i = 0,1,\ldots \,. \end{cases}$$

Notice that the sequence $W^{o(2)} = \{W_j^{o(2)},\ j = 0,1,\ldots\}$ is not stationary; however, from Theorem 1 of Rolski (1981b) it is ergodically stable. Similarly we have

$$T_j^{o(2)} = \begin{cases} T_{0,j+1}^{o(2)} & \text{if } 0 \le j < N_0^o \\ T_{i+1,\,j-\sum_{k=0}^{i} N_k^o+1}^{o(2)} & \text{if } \sum_{k=0}^{i} N_k^o \le j < \sum_{k=0}^{i+1} N_k^o, \quad i = 0,1,\ldots \,. \end{cases}$$

$$S_j^{o(2)} = \begin{cases} S_{0,j+1}^{o(2)} & \text{if } 0 \le j < N_0^o \\ S_{i+1,\,j-\sum_{k=0}^{i} N_k^o+1}^{o(2)} & \text{if } \sum_{k=0}^{i} N_k^o \le j < \sum_{k=0}^{i+1} N_k^o, \quad i = 0,1,\ldots \,. \end{cases}$$

3°. $(\mathscr{W},\Phi) \overset{s}{\equiv} (\mathscr{W}^o,\Phi^o)$. The random element $(\mathscr{W},\Phi)$ assumes values in $(\mathbb{R}_+ \times \mathbb{R}_+ \times \mathbb{R}_+ \times [\mathbb{N} \times (\mathbb{R}_+ \times \mathbb{R}_+ \times \mathbb{R}_+)^{\mathbb{N}}]^{\mathbb{Z}_+}$. There exists a measurable mapping χ such that

$$\{\mathscr{W},\Phi\} = \chi(\{T,S,X\}).$$

For $a \in \mathbb{R}_+$ and $g\colon\ (\mathbb{R}_+ \times \mathbb{R}_+ \times \mathbb{R}_+)^{\mathbb{Z}_+} \to \mathbb{R}_+$ set

$$\mathscr{I}_{g,a} = \{\{t,s,x\}\in \mathscr{F}_1,\ \lim_{j\to\infty} \frac{1}{j} \sum_{i=0}^{j-1} g(\alpha^i\chi(\{t,s,x\}) = a$$

$$\lim_{j\to\infty} \sum_{i=0}^{j} (s_i - t_i) = -\infty\,.$$

Similar to the proof of Theorem 1 of Rolski (1981b) we can demonstrate that $\mathscr{I}_{g,a}$ is β_1-invariant. Since $\{T^o,S^o,X^o\} \overset{s}{\equiv} \{T,S,X\}$ and bearing in mind condition (2.1), we have by Proposition 2

$$P_{\{T^o,S^o,X^o\}}(\mathscr{I}_{g,Eg(\{\mathscr{W}^o,\Phi^o\})}) = P_{\{T,S,X\}}(\mathscr{I}_{g,Eg(\{\mathscr{W}^o,\Phi^o\})}). \tag{4.4}$$

By the ergodic theorem

$$1 = \Pr(\lim_{j\to\infty} \frac{1}{j} \sum_{i=0}^{j-1} g(\alpha^i\{\mathcal{W}^o,\Phi^o\}) = Eg(\{\mathcal{W}^o,\Phi^o\})).$$

This yields, by (4.4)

$$\Pr(\lim_{j\to\infty} \sum_{i=0}^{j-1} g(\alpha^i\{\mathcal{W},\Phi\}) = Eg(\{\mathcal{W}^o,\Phi^o\})) = 1$$

which by Proposition 3 demonstrates that $\{\mathcal{W},\Phi\} \overset{s}{\equiv} \{\mathcal{W}^o,\Phi^o\}$.

4°. Proof of (i). By (4.3) the sequence $\{W^{o(1)},T^{o(1)},S^{o(1)}\}$ is stationary ergodic. Since $\{\mathcal{W}^o,\Phi^o\} \overset{s}{\equiv} \{\mathcal{W},\Phi\}$ we have

$$\{W^{(1)},T^{(1)},S^{(1)}\} \overset{s}{\equiv} \{W^{o(1)},T^{o(1)},S^{o(1)}\}$$

which proves that $\{W^{(1)},T^{(1)},S^{(1)}\}$ is ergodically stable. We prove now that $\{W^{(2)},T^{(2)},S^{(2)}\}$ is ergodically stable. In view of Theorem 1 in Rolski (1981b) it suffices to prove that $\{T^{(2)},S^{(2)}\}$ is ergodically stable. Instead of $\{T^{(2)},S^{(2)}\}$ consider a sequence $\{T^{(2)},S^{(2)},Z^{(2)}\}$ where $Z_i^{(2)} = 1$ if the i-th customer in the 2-nd queue is first in the group and $Z_i^{(2)} = 2$ otherwise. Such a sequence is β_1-ergodically stable and by Proposition 2 and in view of condition (2.1) it is ergodically stable.

5°. Proof of (ii). For a generic sequence $\{T^o,S^o,X^o\}$, let $W^o = \{W_i^o,\ i = 0,1,\dots\}$ be the stationary sequence of waiting times in the module. There exists a measurable mapping ν such that

$$\nu(\{\mathcal{W}^o,\Phi^o\}) = \{W^o,T^o,S^o,X^o\}$$

and

$$\nu\alpha = \beta_1\nu.$$

Thus by Proposition 4 the sequence $\{W^o,T^o,S^o,X^o\}$ is stationary ergodic. Using again Proposition 4 we can prove that $\{W,T,S,X\} \overset{s}{\equiv} \{W^o,T^o,S^o,X^o\}$ because $\{\mathcal{W}^o,\Phi^o\} \overset{s}{\equiv} \{\mathcal{W},\Phi\}$.

Part B. Consider simultaneously a WCN module and a FIFO module with the same generic sequence $\{T,S,X\}$. All notation from Part A carries over to Part B.

1°. Palm representation of WCN modules. Let $\{\mathcal{V}_i,\ i = 0,1,\ldots\}$ be the sequence of waiting times in the WCN module of all customers in the i-th group (cf. the sequence $\mathcal{W}$ defined in (4.1)); here

$$\mathcal{V}_i = (V_i^{(1)}, V_{i1}^{(2)}, \ldots, V_{iN_i}^{(2)}, 0, 0, \ldots)$$

where $V_i^{(1)}$ denotes the waiting time of the i-th customer server at the 1-st queue, $V_{ij}^{(2)}$ denotes the waiting time of the j-th customer in the i-th group $(1 \le j \le N_i)$ at the 2-nd queue. Notice that

$$V_j^{(2)} = \begin{cases} V_{0,j+1}^{(2)} & \text{if } 0 \le j < N_0 \\ V^{(2)}_{i+1,\,j-\sum_{k=0}^{i} N_k+1} & \text{if } \sum_{k=0}^{i} N_k \le j < \sum_{k=0}^{i+1} N_k, \quad i = 0,1,\ldots\ . \end{cases}$$

Consider the sequences $\{W_i^{o(k)}, T_i^{o(k)}, S_i^{o(k)},\ i = \ldots,-1,0,1,\ldots\}$ $(k = 1,2)$ defined in the proof of Part A, Section 2°. Since $\rho_k < 1$ $(k = 1,2)$ we have that $W_i^{o(k)} = 0$ infinitely often for $i \to \pm\infty$ with probability 1. Let $\{\varphi^{(k)}\}$ be a sequence of functions defining a WCN discipline at the k-th queue. Having $\{W_i^{o(k)},\ i = \ldots,-1,0,1,\ldots\}$ we may define $\{V_i^{o(k)},\ i = \ldots,-1,0,1,\ldots\}$, according to the procedure described at the beginning of Section 2. Let

$$V_{ij}^{o(2)} = \begin{cases} V_{j-1}^{o(2)} & \text{if } 1 \le j \le N_0^o, \quad i = 0 \\ V^{o(2)}_{\sum_{k=0}^{i-1} N_k^o+j-1} & \text{if } 1 \le j \le N_i^o, \quad i = 1,2,\ldots \\ 0 & \text{otherwise.} \end{cases}$$

Set

$$\mathcal{V}_i^o = (V_i^{o(1)}, V_{i1}^{o(2)}, V_{i2}^{o(2)}, \ldots, V_{iN_i^o}^{o(2)}, 0, 0, \ldots).$$

The SRE $\{\mathcal{V}^o, \Phi^o\}$ is stationary ergodic and is called a Palm representation of the module.

2°. $\{\mathcal{V}, \Phi\} \overset{s}{\equiv} \{\mathcal{V}^o, \Phi^o\}$. The proof of this fact proceeds similarly to the proof of Part A, Section 3°.

3°. The proofs of (i) and (ii) are similar to the proof of Part A, Sections 4° and 5°.

Proof of Corollary 1. The already proven ergodical stability of $\{W,T,S,X\}$ and $\{W^{(j)},T^{(j)},S^{(j)}\}$ implies that (2.3)-(2.6) hold. To prove (2.7) and (2.8) we use the result of Stidham (1974). From his proof it follows that instead of assuming that $w < \infty$ and $w^{(j)} < \infty$ $(j = 1,\dots,s)$, it suffices to suppose that with probability 1

$$\lim_{i\to\infty} \frac{W_i(j)}{i} = 0, \tag{4.5}$$

$$\lim_{i\to\infty} \frac{W_i}{i} = 0. \tag{4.6}$$

Szekli (1981) proved that in any WCN queue with a stationary ergodic generic sequence $\{\hat{T},\hat{S}\}$ and a waiting time sequence $W' = \{W_i', i = 0,1,\dots\}$ we have with probability 1

$$\lim_{i\to\infty} \frac{W_i'}{i} = 0.$$

However $\{W,T,S\}$ and $\{W',\hat{T},\hat{S}\}$ are stably equivalent (see Rolski (1981b)), which yields (4.5). To prove (4.6) denote

$$\nu_k(j) = \sum_{i=0}^{j} 1_{\{k\}}(X_i) \quad (j = 0,1,\dots\ , k = 1,\dots,s). \tag{4.7}$$

Since

$$W_j \le \max(W^{(1)}_{\nu_1(j)},\dots,W^{(s)}_{\nu_s(j)})$$

and with probability 1.

$$\lim_{j\to\infty} \frac{W^{(i)}_{\nu_i(j)}}{j} = \lim_{j\to\infty} \frac{\nu_i(j)}{j} \frac{W^{(i)}_{\nu_i(j)}}{\nu_i(j)} = 0 \quad (i = 1,\ldots,s)$$

we have that (4.6) is fulfilled. This completes the proof of Corollary 1.

Proof of Corollary 2. To prove (i) notice that for $j = 1,\ldots,s$

$$U_i^{(j)} = W_{i+1}^{(j)} - W_i^{(j)} + S_{i+1}^{(j)} - S_i^{(j)} + T_i^{(j)}, \quad (i = 0,1,\ldots).$$

Hence by Lemma 3.3 of Rolski (1981a) $\{W^{(j)},T^{(j)},S^{(j)},U^{(j)}\}$ is ergodically stable. Now

$$\lim_{n\to\infty} \frac{1}{n} \sum_{i=0}^{n-1} U_i^{(j)} = \lim_{n\to\infty} \frac{W_n^{(j)} - W_0^{(j)} + S_n^{(j)} - S_0^{(j)}}{n} + \lim_{n\to\infty} \frac{1}{n} \sum_{i=0}^{n-1} T_i^{(j)}.$$

Since $E\hat{S}_0^{(j)} < \infty$ we have by the Borel-Cantelli lemma that with probability 1

$$\lim_{i\to\infty} \frac{S_i^{(j)}}{i} = 0.$$

Thus in view of the stable equivalence of S to $\hat{S}$, we have with probability 1

$$\lim_{i\to\infty} \frac{S_i^{(j)} - S_0^{(j)} - W_0^{(j)}}{i} = 0.$$

Finally by (2.3) and (4.5) we obtain

$$E\hat{U}_0^{(j)} = \lim_{n\to\infty} \frac{1}{n} \sum_{i=0}^{n-1} U_i{}^{j} = E\hat{T}_0^{(j)} \quad (j = 1,\ldots,s)$$

which completes the proof of (i). The proof of (ii) is similar.

Proof of Proposition 1. Let $j = 1,\ldots,s$ be fixed and

$$Y_{j0} = 0,$$

$$Y_{ji} = \min\{k > Y_{i-1}, X_k = j\}.$$

Then the sequence $T^{(j)}$ is defined by

$$T_i^{(j)} = \sum_{k=Y_{j,i}}^{Y_{j,i+1}} T_k \quad (i = 0,1,\ldots).$$

The mapping $T^{(j)} = \mu(\{T,X\})$ fulfills the assumption of Proposition 4. By Proposition 2, the SRE $\{T,X\}$ is β_j-ergodically stable and hence by Proposition 2 the SRE $T^{(j)}$ is ergodically stable. Denote by $\{T^o,X^o\}$ a β_j-stationary representation of $\{T,X\}$. Then $T^{o(j)} = \mu(\{T^o,X^o\})$ is a stationary representation of $T^{(j)}$. By a result in Example 2.2 of Rolski (1981a) (see (2.12)) we obtain that

$$E\hat{T}_0^{(j)} = \frac{E\hat{T}_0}{Pr(\hat{X}_0=j)}$$

5. Acknowledgement

We would like to thank the referee for suggesting several improvements in the exposition of the paper.

6. References

[1] Breiman, L. (1968), Probability, Addison-Wesley, New York.

[2] Brown, J. R. (1976), Ergodic Theory and Topological Dynamics, Academic Press, New York.

[3] Gray, R. M. and Kieffer, J. C. (1980), "Asymptotically Mean Stationary Measures, Ann. Probab., 8, 962-973.

[4] Franken, P., König, D., Arndt, U. and Schmidt, V. (1981), Queues and Point Processes, Akademie-Verlag/Wiley, Berlin/New York.

[5] Loynes, R. M. (1962), "The Stability of a Queue with Non-independent Inter-arrival and Service Time, Proc. Cambridge Phil. Soc., 58, 497-520.

[6] Rolski, T. (1981a), Stationary Random Processes Associated with Point Processes, Lecture Notes in Statistics 5, Springer-Verlag, New York.

[7] Rolski, T. (1981b), "Queues with Non-stationary Input Stream: Ross's Conjecture, Adv. in Appl. Probab., 13, 603-618.

[8] Stidham, Sh. (1974), "A Last Word on L = λW," Oper. Res., 22, 417-422.

[9] Szekli, R. (1981), "Little's Formula for Single Server Work-conserving Queues," (in preparation).

[10] Zähle, M. (1980), "Ergodic Properties of Random Fields and Random Geometric Figures in the n-dimensional Space with Embedded Point Processes," Friedrich-Shiller-Universität Jena. Forschungsergebnisse, N/80/13.

Mathematics Institute, Wrocław University, Plac Grunwaldski 2/4, 50-384 WROCŁAW, POLAND.

PRODUCT FORM SOLUTION FOR QUEUEING NETWORKS WITH POISSON ARRIVALS AND GENERAL SERVICE TIME DISTRIBUTIONS WITH FINITE MEANS

Christopher L. Samelson

and

William G. Bulgren

Abstract

Product form was first introduced by Jackson [Operations Research, 1957; Management Science, 1963] who considered only negative exponential service distributions with queue length dependent service rates. Posner and Bernholtz [Operations Research, 1968] showed that one has product form in the case where there are many classes of customers, service times are exponential with class queue length dependent service rates, and lag times (the time for a customer to travel between two servers is a random variable with continuously differentiable distribution function). Baskett, Chandy, Muntz and Palacios [JACM, 1975] extended these results to networks with multiple classes of customers and service time distributions which have rational Laplace transforms vanishing at ∞. Chandy, Howard and Towsley [JACM, 1977] extended this further to the case where the service time distribution functions are continuously differentiable. They also introduced the concept of station balance and established some important relationships between station balance, local balance and product form. In particular, for the closed queueing network they showed that if each queue in the network satisfies local balance in isolation with Poisson arrivals then one has product form solution for the network. This paper extends previous results to arbitrary service with finite mean (the nondifferentiable case included). This is done by introducing supplementary variables for the remaining service requirements at the various queues in the system. By using generalized function (in the sense of Schwartz) a new technique is presented for handling queueing networks in which service distributions may be general.

The University of Kansas, Lawrence, Kansas 66045.